MATERIALS CONCEPTS FOR SOLAR CELLS

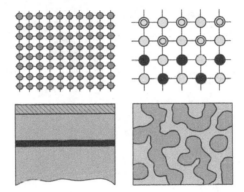

ENERGY FUTURES – Vol. 1

MATERIALS CONCEPTS
FOR SOLAR CELLS

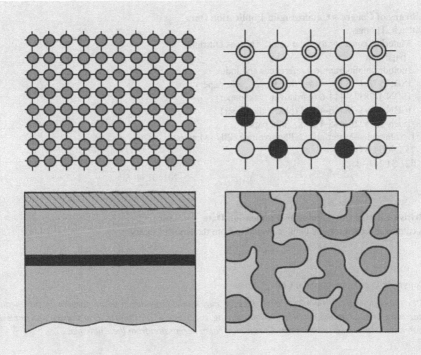

Thomas Dittrich

Helmholtz Center Berlin for Materials and Energy, Germany

Imperial College Press

ICP

Published by

Imperial College Press
57 Shelton Street
Covent Garden
London WC2H 9HE

Distributed by

World Scientific Publishing Co. Pte. Ltd.
5 Toh Tuck Link, Singapore 596224
USA office: 27 Warren Street, Suite 401-402, Hackensack, NJ 07601
UK office: 57 Shelton Street, Covent Garden, London WC2H 9HE

Library of Congress Cataloging-in-Publication Data
Dittrich, Thomas.
　　Materials concepts for solar cells / Thomas Dittrich.
　　　pages cm
　　Includes bibliographical references and index.
　　ISBN 978-1-78326-444-5 (hardcover : alk. paper)
　　ISBN 1-78326-444-6 (hardcover : alk. paper)
　　ISBN 978-1-78326-445-2 (pbk. : alk. paper)
　　ISBN 1-78326-445-4 (pbk. : alk. paper)
　　1. Solar cells--Materials. 2. Photovoltaic cells--Materials. I. Title.
　TK2960.D57 2014
　621.31'244--dc23

2014013406

British Library Cataloguing-in-Publication Data
A catalogue record for this book is available from the British Library.

Typeset by Stallion Press
Email: enquiries@stallionpress.com

Printed in Singapore

To my wife

Preface

This textbook results from lecture courses I have been giving to students who are interested in principles of solar energy conversion and in solar cell materials. Students listening to my courses usually study in the fields of engineering, materials science or natural science at universities around the world. I have had, and I continue to have, the great pleasure to teach not only in Europe but also in Asia, Africa and Latin America. A broad understanding of materials and combinations of materials for photovoltaic solar energy conversion helps to implement renewable energy worldwide. The textbook gives a comprehensive introduction to materials concepts for solar cells including basic principles and materials specific concepts.

The success in the development and application of solar cells is closely related to countless improvements of materials and to the development of new materials for solar cells. Quality criteria of solar cell materials are consequences of basic principles of solar cells. Materials and technological concepts follow from the ways in which different photovoltaic absorbers can be realized. Materials concepts of solar cells are based (i) on the growth of large crystals (wafer-based technology), (ii) on the growth of sophisticated layer systems on substrate crystals (epitaxy-based technology), (iii) on the deposition of absorber layers on foreign substrates (thin-film photovoltaics) and (iv) on the combination of very different materials on a nanometer scale (nano-composite solar cells). Interfaces between materials for charge separation, electric contacts to external leads and passivation are considered for all kinds of solar cells.

I would like to express my appreciation to all of my students, but especially those from the Free University Berlin for their interest and questions, to my former teacher Fred Koch for inspiration, to Günther

Seliger for the opportunity to teach at GPE (Global Production Engineering for Solar Technology) at the Technical University Berlin, to friends at the Kasetsart University in Bangkok and other universities around the world for supporting my teaching projects, to Martha Lux-Steiner at the Helmholtz Centre Berlin for giving me the opportunity to teach alongside my scientific work, to Bernhard Reinhold for discussing aspects of this textbook, to Brian Edlefsen Lasch for his critical reading of the manuscript, and to Catharina Weijman from Imperial College Press for excellent collaboration.

Thomas Dittrich
Berlin, 2013

Contents

Symbols and Abbrevations

Those Beginning with Latin Letters

A, A$^-$	acceptor, ionized acceptor
A	surface area
A	ampere, unit of the current
a	distance between two grid fingers
A/cm^2	unit of the current density
A$_{cr}$	area of a cross section
a$_0$	Bohr radius of the hydrogen atom (0.051 nm)
a$_0$	interatomic distance
A$_{cell}$	area of the solar cell
a$_{elh1,b}$	radius of an exciton with the lowest energy in bulk semiconductor
a$_{lc}$	lattice constant
A$_{fc}$	area of the front contact
ALD	atomic layer deposition
AM	air mass
AM0	air mass zero
AM1.5d	air mass 1.5 for direct irradiation
AM1.5G	air mass 1.5 for global irradiation (P_{sun}(AM1.5G) $\equiv$ 1000 W/cm^2)
As	unit of the charge (ampere-second or Coulomb)
a-Si:H	hydrogenated amorphous silicon
a-SiC:H	hydrogenated amorphous silicon carbide
AU	astronomical unit

B	width
B	radiative recombination rate constant
B_{ideal}	radiative recombination rate constant of an ideal absorber
C	heat capacitance
C_A	Auger recombination rate constant
$C_{A,e}$	Auger recombination rate constant of electrons
$C_{A,h}$	Auger recombination rate constant of holes
c	velocity of light ($2.99 \cdot 10^8$ m/s)
C_{i0}	initial concentration of impurity atoms
C_{is}	concentration of impurity atoms in the growing crystal
C_m	concentration of impurity atoms in the melt
c-Si	crystalline silicon
CSS	close space sublimation
Cu_{In}^{2-}	copper replacing an In atom at a lattice side of a chalcopyrite
$CuInS_2$	copper indium disulfide
$Cu(In,Ga)Se_2$	copper indium gallium diselenide
CVD	chemical vapor deposition
D	diffusion coefficient
D, D^+	donor, ionized donor
d	thickness
d_{abs}	thickness of the absorber
D_e	density of states for free electrons in the conduction band
dE	interval of energy
d_{em}	thickness of the emitter layer
$d(h\nu)$	interval of photon energies
D_h	density of states for free holes in the valence band
D_i	diffusion constant of an impurity
d_k	thickness of kuvette
d_{local}	local thickness
D_n	diffusion coefficient of free electrons
D_p	diffusion coefficient of free holes
dbs	dangling bonds

$DOS_{st,a}$	density of donor-like surface states
$DOS_{st,d}$	density of acceptor-like surface states
dR	resistance of a very thin slice
D_{th}	thermal diffusion constant
dx	interval of distances
DSSC	dye-sensitized solar cell
E	energy (in units of eV)
e, e^-	free electron
E_A	energy of an acceptor
E_A	activation energy
E_{Ai}	activation energy of a diffusing impurity
$E_A^{SRH,pn}$	activation energy for Shockley–Read–Hall recombination in a pn-junction
E_{ab}	energy of an anti-bonding state
$E_{abs}(r_{QD})$	absorption edge of a semiconductor quantum dot
E_b	energy of a bonding state
E_C	energy of the conduction band edge, conduction band edge
$E_{C(n)}$	conduction band edge in an n-type doped region
$E_{C(p)}$	conduction band edge in a p-type doped region
$E_{C,abs}$	conduction band edge of the photovoltaic absorber
$E_{C,pass}$	conduction band edge of the passivation layer
E_{Cs}	conduction band edge at the surface of a photovoltaic absorber
E_D	energy of a donor
E_{e1h1}	lowest energy of an exciton
$E_{e1h1,b}$	lowest energy of an exciton in a bulk semiconductor
E_F	Fermi-energy
E_{F0}	Fermi-energy in thermal equilibrium
E_{Fm}	Fermi-energy of a metal
E_{Fn}	Fermi-energy of free electrons
$E_{Fn(n)}$	Fermi-energy of free electrons in an n-type doped region
$E_{Fn(p)}$	Fermi-energy of free electrons in a p-type doped region
$E_{F(n)}^0$	Fermi-energy in thermal equilibrium in an n-type doped region

E_{Fp}	Fermi-energy of free holes
$E_{Fp(n)}$	Fermi-energy of free holes in an n-type doped region
$E_{Fp(p)}$	Fermi-energy of free holes in a p-type doped region
$E_{F(p)}^{0}$	Fermi-energy in thermal equilibrium in a p-type doped region
E_{Fn}-E_{Fp}	Fermi-level splitting
E_{Fs}	Fermi-energy at the surface of a photovoltaic absorber
E_{g}	energy of the forbidden band gap
E_{g0}	energy of the forbidden band gap at the temperature of 0 K
E_{HL}	energy of the HOMO–LUMO gap of an organic semiconductor
$E_{HL}^{absorption}$	energy of the HOMO–LUMO gap for the absorption of an organic semiconductor
$E_{HL}^{transport}$	energy of the HOMO–LUMO gap for the transport in an organic semiconductor
E_{HOMO}	lowest energy of an unoccupied state in the HOMO band of an organic semiconductor
E_{i}	intrinsic Fermi-level
$E_{ion}^{D,A}$	ionization energies of donor and acceptor states
E_{ion}^{H}	ionization energy of the hydrogen atom (13.56 eV)
$E_{kin,e}$	kinetic energy of a free electron
$E_{kin,h}$	kinetic energy of a free hole
E_{LUMO}	lowest energy of an occupied state in the LUMO band of an organic semiconductor
E_{og}	energy of the optical band gap
E_{ph}	photon energy
E_{redox}	potential of a redox reaction
$E_{S/S+}$	energy of the ground state and of the ionized state of a dye molecule
E_{S}^{*}	energy of the excited state of a dye molecule
E_{t}	energy characterizing exponential absorption tails
E_{t}	energy of a trap state in the forbidden band gap
E_{V}	energy of the valence band edge, valence band edge

$E_{V(n)}$	valence band edge in an n-type doped region
$E_{V(p)}$	valence band edge in a p-type doped region
$E_{V,abs}$	valence band edge of the photovoltaic absorber
$E_{V,pass}$	valence band edge of the passivation layer
E_{Vs}	valence band edge at the surface of a photovoltaic absorber
E_{vac}	energy of the vacuum
eV	electron Volt, unit of energy
f	occupation probability of an electron
FF	fill factor
FTO	fluorine-doped tin oxide
FZ	float zone
G	generation rate of free charge carriers
G_0	thermal generation rate of an ideal absorber
G_e	emission rate of free electrons
G_{fn}	geometry factor for the diode saturation current density of electron diffusion
G_{fp}	geometry factor for the diode saturation current density of hole diffusion
G_h	emission rate of free holes
G_{sun}	photo-generation rate of an ideal absorber
GaAs	gallium arsenide
GaP	gallium phosphide
Ge	germanium
H	height
h	Planck constant ($6.626 \cdot 10^{-34}$ Js)
$\hbar$	Planck constant in terms of $h/(2\pi)$
h, h^+	free hole
H_n	thickness of the n-type doped region
H_p	thickness of the p-type doped region
HF	hydrofluoric acid
HIT	hetero-junction with intrinsic thin layer
HOMO	highest occupied molecular orbital

I	current, current density
I_0	diode saturation current, diode saturation current density
I_{00}	pre-factor of the diode saturation current density
$I_{0,SRH}$	diode saturation current density related to Shockley–Read–Hall recombination
I_D	current across a diode
I_{mp}	current, current density in the maximum power point
I_{ph}	photo current, photo current density
I_S	ionization energy of a semiconductor (in units of eV)
I_{SC}	short-circuit current (in units of A), short circuit current density (in units of A/cm^2)
I_{SC}	short-circuit current density in the diode equation
I_{SC}^*	short-circuit current density following from the I–V characteristics
I_{SC}^{abs}	short-circuit current density under consideration of transmission losses
I_{SC}^{max}	maximum short-circuit current density
I_{SRH}	diode current related to Shockley–Read–Hall recombination
$I_{tunn,e}$	tunneling current density for electrons
$I_{tunn,h}$	tunneling current density for holes
ILGAR®	ion layer gas reaction
In_{Cu}^{2+}	indium replacing a copper atom at a lattice side of a chalcopyrite
InP	indium phosphide
ITO	indium tin oxide
I–V	current–voltage (characteristics)
J_S	solar constant (1356 W/m^2)
k	wave vector
k	heat conductivity
k_0	segregation coefficient in equilibrium
k_B	Boltzmann constant ($1.38 \cdot 10^{-23}$ J/K)
k_e	momentum of a free electron
k_h	momentum of a free hole

k_s	segregation coefficient
KCN	potassium cyanide
kWh	kilowatt hour, unit of the electrical energy
L	length
L	diffusion length
L_{drift}	drift length
L_f	length of a grid finger
L_n	diffusion length of electrons
L_p	diffusion length of holes
L_{th}	thermal diffusion length
LPE	liquid phase epitaxy
LUMO	lowest unoccupied molecular orbital
m_e	electron mass in free space ($9.1 \cdot 10^{-31}$ kg)
m_e^*	effective mass of free electrons in the conduction band of a semiconductor
m_h^*	effective mass of free holes in the valence band of a semiconductor
MBE	molecular beam epitaxy
mc-Si	multi-crystalline silicon
MOCVD	metal organic chemical vapor deposition
MOVPE	metal organic vapor phase deposition
MOSFET	metal oxide semiconductor field-effect transistor
N	number
N	density of particles (units of cm^{-3})
N_C	effective density of states in the conduction band
N_{db}	density of dangling bonds
N_e	number of photo-generated electrons
N_{i0}	initial number of impurity atoms
N_{im}	number of impurity atoms
N_{ph}	number of incident photons
N_{st}	density of surface defects
$N_{st,a}$	density of acceptor-like surface defects
$N_{st,d}$	density of donor-like surface defects
N_t	density of defects in the forbidden band gap

N_V	effective density of states in the valence band
n	ideality factor in the diode equation
n	density of free electrons (unit in cm^{-3})
n_i	intrinsic density of free charge carriers
n^+ (n^{++})	density of free electrons in a highly (very highly) n-type doped semiconductor
n_0	density of free electrons in thermal equilibrium
n_n	density of free electrons in an n-type doped region
n_n^0	density of free electrons in an n-type doped region in thermal equilibrium
n_p	density of free electron in a p-type doped region
n_p^0	density of free electron in a p-type doped region in thermal equilibrium
n_s	density of free electrons at the surface of a photovoltaic absorber
n_t	density of occupied traps
n_{air}	refractive index of air
n_{AR}	refractive index of an antireflection coating layer
n_S	refractive index of a photovoltaic absorber
n_{TCO}	refractive index of a TCO layer
nwSC	nanowire solar cell
OVC	ordered vacancy compound
OSC	organic solar cell
P	electric power (in units of W)
P_e	sun's power received on earth
P_i	phosphorus atom at an interstitial lattice side
P_{inst}	installed power (in units of W_p)
P_{light}	light intensity (in units of W/cm^2)
P_{loss}	electric power losses due to resistive heating
P_{mp}	electric power in the maximum power point
P_{Si}	phosphorus atom at a lattice side in a silicon crystal
P_{Si}^+	ionized phosphorus atom at a lattice side in a silicon crystal
P_{sun}	areal density of the power of the sun received on earth (in units of W/m^2)

$\langle P_{sun} \rangle$	average areal density of the power of the sun received on earth
P_{sun}^{total}	total power emitted by the sun
PECVD	plasma-enhanced chemical vapor deposition
PERL	passivated emitter rear locally diffused
PV	photovoltaic
PVD	physical vapor deposition
p	density of free holes (in units of cm^{-3})
p^+ (p^{++})	density of free holes in a highly (very highly) p-type doped semiconductor
p_0	density of free holes in thermal equilibrium
p_n	density of free holes in an n-type doped region
p_n^0	density of free holes in an n-type doped region in thermal equilibrium
p_p	density of free holes in a p-type doped region
p_p^0	density of free holes in a p-type doped region in thermal equilibrium
p_s	density of holes at the surface of a photovoltaic absorber
Q_{is}	total amount of a diffusing species
q	elementary charge ($1.602 \cdot 10^{-19}$ As)
Q_{SC}	charge in the space charge region
$Q_{st,a}$	charge in acceptor-like surface states
$Q_{st,d}$	charge in donor-like surface states
qc-Si	quasi-mono-crystalline silicon
QD	quantum dot
QD-SC	quantum dot-based solar cell
QE	quantum efficiency
R	recombination rate of free charge carriers (in units $s^{-1}cm^{-3}$)
R	reflection coefficient
R	resistance (in units of Ω)
R_{Auger}	Auger recombination rate (in units of $s^{-1}cm^3$)
R_{base}^{c-Si}	resistance of the base of a c-Si solar cell
R_C	contact resistance

$R_{C,t}$	tunneling resistance
$R_{C,tunn,e}$	tunneling resistance of free electrons
$R_{C,tunn,h}$	tunneling resistance of free holes
R_e	radius of the earth (in units of km)
R_e	capture rate of free electrons (in units of $s^{-1}cm^3$)
R_{em}^{c-Si}	emitter resistance of a c-Si solar cell
R_{fc}^{c-Si}	resistance of the front contact of a c-Si solar cell
R_h	capture rate of free holes
R_{if}	distance between initial and finite states
R_L	load resistance
R_{min}	minimum reflection coefficient
R_p	parallel resistance, shunt resistance
$R_{p,tol}$	tolerable parallel resistance, tolerable shunt resistance
r_{QD}	radius of a semiconductor quantum dot (in units of nm)
R_{rad}	radiative recombination rate
R_s	series resistance
$R_s^{contact}$	series resistance of contacts
$R_s^{contact-pn}$	series resistance of contacts between two pn-junctions
R_{se}	capture rate for electrons at surface states
R_{sh}	capture rate for holes at surface states
R_{SRH}	Shockley–Read–Hall recombination rate
$R_{s,tol}$	tolerable series resistance
R_{sun}	radius of the sun
R_{TCO}	resistance of a TCO layer
$R_\square$	sheet resistance
RD	reflection difference
RF	radio frequency
RHEED	reflection high-energy electron diffraction
S	empirical parameter for describing the dependence of the barrier height
Si_i	silicon atom at an interstitial lattice side
Si_{Si}	silicon atoms on the lattice side of a silicon crystal
SR	spectral response (in units of A/W)
SRH	Shockley–Read–Hall recombination
STC	standard test conditions

T	absolute temperature (in units of K)
t	time (in units of s)
T_0	ambient temperature for the operation of a solar cell
$t_{tr,e}$	transit time of electrons
$t_{tr,h}$	transit time of holes
T_{tr}	transmission probability of a barrier
$T_{tr,e}$	transmission probability of a barrier for electrons
$T_{tr,h}$	transmission probability of a barrier for holes
T_{otr}	optical transmission coefficient
T_S	temperature at the surface of the sun
T_{SC}	temperature of a solar cell under operation
TCO	transparent conducting oxide
TMG	trimethyl gallium
U	voltage, potential (in units of V)
U_D	diffusion potential
U_n	potential drop across the depletion region at an n-type doped side
U_p	potential drop across the depletion region at a p-type doped side
U_{pot}	potential energy (in units of eV)
UHV	ultra-high vacuum
V	volume (in units of cm^3)
V	volt, unit of the electrical potential
V_0	initial volume
v_{cg}	velocity of crystal growth
v_e	electron drift velocity
V_m	volume of a melt
V_{mp}	potential in the maximum power point
V_{OC}	open-circuit voltage
V_S	volume of a growing crystal
V_{Si}	vacancy at a silicon lattice side
v_{th}	thermal velocity
V_{Cu}	copper vacancy (in chalcopyrites, $CuInSe_2$)
V_{Cu}^-	charged copper vacancy (in chalcopyrites, $CuInSe_2$)

VLS   vapor–liquid–solid
vtaSC   very thin absorber solar cell

W_{el} electric energy (in units of kWh)
$W_{ideal\ gas}$ kinetic energy of an ideal gas (in units of eV)
W_p watt peak, unit of the installed power

X   concentration factor
X_{max} maximum concentration factor
X_{path} factor by which an optical path can be increased
x   parameter for describing the stoichiometry in
                         compounds and alloys
x   depth
x_n width of the space charge region in an n-type doped
                         semiconductor
x_p width of the space charge region in a p-type doped
                         semiconductor
x_{scr} width of the space charge region
x_{scr}^{n+} width of the space charge region in a highly n-type
                         doped semiconductor
x_{scr}^{p+} width of the space charge region in a highly p-type
                         doped semiconductor

ZnO   zinc oxide

Those Beginning with Greek Letters

α absorption coefficient
α coefficient in the temperature dependence of the band
                         gap
α^{-1} absorption length
α_{tunn} reciprocal tunneling length

β specific temperature in the temperature dependence of
                         the band gap
β_e emission rate constants for free electrons
β_h emission rate constants for free holes

Γ	phonon, quantum of lattice and atom vibrations
γ	photon, quantum of light
Δ	drop of the potential energy across a dipole layer at a metal–semiconductor junction
Δa_{lc}	difference of lattice constants
ΔG	photo-generation rate
ΔE_C	offset of the conduction band edges
ΔE_{dipole}	drop of the potential energy across a dipole layer at a hetero-junction
ΔE_{HOMO}	width of the HOMO band
ΔE_{LUMO}	width of the LUMO band
ΔE_V	offset of the valence band edges
ΔI	current difference
Δp	density of photo-generated holes
Δk	uncertainty of momentum
Δn_p	density of photo-generated electrons in a p-type doped region
Δp	density of photo-generated holes
Δp_n	density of photo-generated holes in an n-type doped region
ΔQ_{pol}	polarization charge
ΔR	recombination rate of photo-generated free charge carriers
ΔU	voltage difference
$\Delta \Phi_B$	barrier lowering due to image charge at a metal–semiconductor contact
Δx	uncertainty of localization
ε	electric field (in unit of V/cm)
ε_0	dielectric constant of vacuum ($8.854 \cdot 10^{-12}$ As/(Vm))
ε_{mol}	molar extinction coefficient
ε_{in}	relative dielectric constant in a quantum dot
ε_{out}	relative dielectric constant in the surrounding of a quantum dot
ε_r	relative dielectric constant

Φ_0	photon flux of blackbody radiation
Φ_B	barrier height at a metal–semiconductor contact (in unit of eV)
Φ_m	metal work function (in unit of eV)
Φ_s	work function of a semiconductor (in unit of eV)
Φ_{ph}	photon flux
Φ_{sun}	photon flux from the sun (in unit of photon/(cm^2 s))
Φ_{sun}^{abs}	part of the photon flux from the sun absorbed by the absorber of the solar cell
Φ_{sun}^{refl}	part of the photon flux from the sun reflected from the absorber of the solar cell
Φ_{sun}^{trans}	part of the photon flux transmitted by the absorber of the solar cell
φ	electric potential
φ_0	surface band bending
χ_s	electron affinity of a semiconductor
Ψ	wave function
η	solar energy conversion efficiency
η_C	efficiency of the Carnot process
η_{multi}	solar energy conversion efficiency of multi-junction solar cells
η_{record}	solar energy conversion efficiency of record solar cells
η_{SQ}	solar energy conversion efficiency in the Shockley–Queisser limit
η_{th}	efficiency at the thermodynamic limit
$\eta_{ultimate}$	ultimate solar energy conversion efficiency
λ	wavelength (in units of nm)
λ_B	de Broglie wavelength
μ	mobility of free charge carriers (in units of cm^2/Vs)
μ_e	electron drift mobility
μ_{eh}	effective mass of an exciton (in units of m$_e$)
μ_{n+}	mobility of free electrons in the highly doped emitter
μ_p	mobility of free holes

ν	photon frequency
Ω	ohm (unit of the electrical resistance)
Ω_S	space angle
θ	angle of incidence
θ_C	critical angle
ρ	density (in units of g/cm^3)
ρ	density of charge (in units of $1/cm^3$)
ρ	specific electrical resistance (in units of Ωcm)
ρ_e	specific emitter resistance
σ	electrical conductivity (in units of S/cm)
σ_e	capture cross section for free electrons
σ_h	capture cross section for free holes
σ_p	conductivity of a p-type semiconductor
σ_S	Stefan–Boltzmann constant ($5.67 \cdot 10^{-8}\,Wm^{-2}K^{-4}$)
σ_{se}	capture cross section for free electrons at surface states
σ_{sh}	capture cross section for free holes at surface states
τ	time constant (in units of s)
τ	life time
τ_0	minimum Shockley–Read–Hall recombination lifetime
τ_{Auger}	Auger recombination lifetime
$\tau_{e,SRH}^{min}$	minimum Shockley–Read–Hall recombination lifetime of free electrons
$\tau_{e,SRH}^{min}$	minimum Shockley–Read–Hall recombination lifetime of free holes
τ_{min}	minimum lifetime
τ_{rad}	radiative recombination lifetime
τ_{se}	effective surface recombination lifetime of free electrons
τ_{sh}	effective surface recombination lifetime of free holes
τ_{total}	total lifetime
ω_{tunn}	tunneling rate for Miller–Abrahams tunneling (in units of s^{-1})
$\omega_{0,tunn}$	maximum tunneling rate

Part I

Basics of Solar Cells
and Materials Demands

1

Basic Characteristics
and Characterization
of Solar Cells

Solar cells convert the power of sunlight into electric power. As an introduction, therefore, Chapter 1 is devoted to a brief characterization of sunlight and the basic electrical parameters of solar cells. The power of the sun is given in terms of the solar constant, the power spectrum and power losses in the atmosphere. The basic characteristics of a solar cell are the short-circuit current (I_{SC}), the open-circuit voltage (V_{OC}), the fill factor (FF) and the solar energy conversion efficiency (η). The influence of both the diode saturation current density and of I_{SC} on V_{OC}, FF and η is analyzed for ideal solar cells. The importance of concentrated sunlight for increasing η is shown. Tolerable series and parallel resistances are introduced as an evaluation criterion for resistive losses in real solar cells. The influence of the series resistance (R_s) and parallel resistance (R_p) on I_{SC}, V_{OC}, FF and η is investigated. The specific role of R_s and R_p is discussed in detail for the dependence of η on I_{SC}. Concepts are described for measuring the basic characteristics of solar cells and their dependencies on light intensity, temperature and light spectra. Attention is paid to principle work with various kinds of load resistances, to the function of a pyranometer, of a sun simulator and to the measurement of the quantum efficiency of solar cells.

1.1 Solar Radiation and Two Fundamental Functions
of a Solar Cell

The sun is a hot sphere radiating energy in the form of light or photons into space. The temperature of the outer photo sphere of the sun (temperature

3

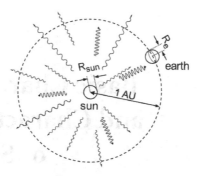

Figure 1.1. Sun emitting photons and earth receiving a proportion of photons emitted by the sun. The radius of the sun, the radius of earth and the distance between sun and earth are denoted by R_{sun}, R_e and AU, respectively.

of the sun, T_S) is about 5778 K. The radius of the sun (R_{sun}) is $6.96 \cdot 10^5$ km. The power of thermal radiation and therefore the total power emitted by the sun (P_{sun}^{total}) can be calculated by using the Stefan–Boltzmann equation and the surface area of the sun:

$$P_{sun}^{total} = \sigma_S \cdot T_S^4 \cdot 4\pi \cdot R_{sun}^2 \qquad (1.1)$$

The Stefan–Boltzmann constant (σ_S) is $5.67 \cdot 10^{-8}\,\text{Wm}^{-2}\text{K}^{-4}$.

The sun's power received on earth (P_e) is proportional to the cross section of the earth and to the reciprocal area of a sphere with the radius equal to one astronomical unit (AU), the distance between the sun and the earth. This is shown schematically in Figure 1.1. The shortest and longest distances between the sun and the earth, the so-called perihelion and aphelion, are $1.47 \cdot 10^8$ km and $1.52 \cdot 10^8$ km, respectively. The radius of the earth (R_e) is about 6400 km.

$$P_e = P_{sun}^{total} \cdot \frac{\pi \cdot R_e^2}{4\pi \cdot (AU)^2} \qquad (1.2)$$

The solar constant (J_s) is defined as the power of the sun (P_{sun}) received on earth over $1\,\text{m}^2$.

$$J_s \equiv \frac{P_{sun}^{total}}{4\pi \cdot (AU)^2} \qquad (1.3)$$

In reality the solar constant is not a constant since the distance between the earth and the sun and the temperature distribution at the surface of the sun are not constant. A solar constant of 1356 W/m^2 will be taken into account in the following.

A photon is the smallest portion or quantum of light, the energy of which is proportional to the frequency of the light (υ). The proportionality factor between the energy of a photon (E_{ph}) and the frequency is called the Planck constant (h = 6.626 $\cdot$ 10^{-34} Js).

$$E_{ph} = h \cdot \upsilon \qquad (1.4)$$

The power spectrum of the sun, i.e. the dependence of power emitted within an interval of photon energies (E_{ph}, $E_{ph} + dE_{ph}$), can be approximated by the radiation of a blackbody with the temperature T_S given by the Planck equation.

$$\frac{dJ_S(h\upsilon)}{d(h\upsilon)} = \frac{8\pi}{h^3 c^2} \cdot \frac{(h\upsilon)^3}{exp\left(\frac{h\upsilon}{k_B \cdot T_S}\right) - 1} \qquad (1.5)$$

The velocity of light (c) is equal to 2.99 $\cdot$ 10^8 m/s and the Boltzmann constant (k_B) is given as 1.38 $\cdot$ 10^{-23} J/K. The spectrum of blackbody radiation for 5800 K is shown in Figure 1.2.

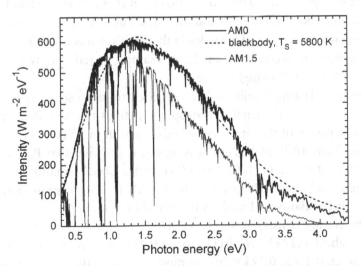

Figure 1.2. Spectrum of sunlight (sun spectrum) outside the earth's atmosphere (AM (air mass) 0, thick solid line), on earth for a zenith angle of 48.2° (AM1.5, thin solid line) and of a blackbody with a temperature of 5800 K (dashed line).

Spectra are usually measured in units of wavelength (λ). The wavelength of light is proportional to the reciprocal frequency while the proportionality factor is the velocity of light.

$$\lambda = \frac{c}{\upsilon} \tag{1.6}$$

The photon energy is given in units of eV, which means that the energy is divided by the elementary charge ($q = 1.602 \cdot 10^{-19}$ As). Therefore the product of the photon energy and the wavelength is given by:

$$E_{ph} \cdot \lambda = \frac{h \cdot c}{q} = 1240 \,\text{eV} \cdot \text{nm} \tag{1.7}$$

The wavelength of light can easily be transformed into the photon energy by using Equation (1.7). The photon energy and the wavelength in the maximum of the corresponding spectra of blackbody radiation are obtained in tasks T1.3 and T1.4 (see end of this chapter), respectively, for a blackbody temperature of 5800 K.

The power received over a certain area on earth depends on geographical location, on the rotation of the earth (day–night cycle), on the inclination angle of the earth's axis (sun in summer and winter) and on the given distance between the sun and the earth. Further, light is scattered and absorbed by molecules and particles in the earth's atmosphere. The average power of sunlight received on earth, for example, is reduced by a factor of π as a result of the day–night cycle.

The optical path of sunlight through the earth's atmosphere normalized to the thickness of the earth's atmosphere defines the so-called air mass (AM). The power of the sun corresponds at zenith to AM1 and at a zenith angle of about 48.2° at AM1.5. It is agreed worldwide that P_{sun} is equal to $1 \,\text{kW/m}^2$ at AM1.5G (IEC, 2008) (G denotes global radiation including direct light and scattered light, see also the AM1.5 spectrum in Figure 1.2; where direct radiation is less than the global radiation by a factor of about 1.1). The average density of the power of the sun on the earth ($\langle P_{sun} \rangle$) is, for example, about $0.12 \,\text{kW/m}^2$ in Germany, $0.2 \,\text{kW/m}^2$ in the south of Spain and between 0.2 and $0.27 \,\text{kW/m}^2$ in most regions of the so-called sun belt (i.e. those between about −30° and +30° of latitude).

A solar cell converts P_{sun} into electric power (P), i.e. the product of electric current (I) and electric potential or voltage (U).

$$P = I \cdot U \tag{1.8}$$

With respect to Equation (1.8), the two fundamental functions of a solar cell are

(i) photocurrent generation and
(ii) the generation of a photovoltage.

Photocurrent generation means the creation of mobile photo-generated charge carriers by absorbing light and their collection at external contacts. A potential difference between external contacts is necessary for the collection of photo-generated charge carriers. It is not important for the measurement of a photocurrent whether the potential difference between external contacts originates from an external or internal voltage source. Charge separation at a charge-selective contact inside the solar cell is the origin of an internal voltage source and leads to the generation of a photovoltage.

The current or photocurrent is often normalized to the area of the solar cell under consideration. Current and current density will not be distinguished in the following since their meaning follows directly from the unit used (A for current and A/cm^2 for current density) or from the context.

The electric power of solar cells and photovoltaic (PV) modules is on the order of $1 \ldots 300$ W. PV power plants can be installed for the kW–MW range, and even higher. The scalability of solar cells and PV power plants over many orders of magnitude makes the application of PV solar energy conversion very flexible. The installed power of a PV power plant is defined as the maximum power which can be given by the power plant at AM1.5 during the first year after installation (P_{inst}). The power of a PV power plant is proportional to the area covered by PV modules (A), the solar energy conversion efficiency (η, Equation (1.15)) and the power of sunlight (P_{sun}). P_{inst} is given in W_p (Watt peak).

$$P_{inst} = A \cdot \eta \cdot P_{sun}(AM1.5) \tag{1.9}$$

The electric energy (W_{el}) which can be produced with a PV power plant is the product of A, η, $\langle P_{sun} \rangle$ and the time period of operation (t).

$$W_{el} = A \cdot \eta \cdot \langle P_{sun} \rangle \cdot t \qquad (1.10)$$

PV power plants can be described by the ratio between the electric power produced and P_{inst}.

$$\frac{W_{el}}{P_{inst}} = \frac{\langle P_{sun} \rangle}{P_{sun}(AM1.5)} \cdot t = \left[\frac{kWh}{kW_p} \right] \qquad (1.11)$$

The solar energy conversion efficiency of PV modules decreases during operation time (known as degradation) at a rate of between 0.5 and 1.5% per year (Jordan and Kurtz, 2013). Low degradation rates of PV modules are important for high energy payback. Changing environmental parameters, such as temperature and humidity, cause stress in materials and combinations of materials. Therefore, degradation rates depend on the stability of materials and combinations of materials under stress induced, for example, by different thermal expansion coefficients of materials.

1.2 Basic Characteristics of a Solar Cell

An illuminated solar cell can provide a certain photovoltage at a given photocurrent. A combination of values of photocurrent and photovoltage at which a solar cell can be operated is called a working point. A particular working point of a solar cell is fixed with a load resistance (R_L) due to Ohm's law.

$$R_L = \frac{U}{I} \qquad (1.12)$$

In terms of Ohm's law, the photovoltage is very low at very low R_L, and the photocurrent is very low at very high R_L. The short- and open-circuit operation conditions of a solar cell are defined as a R_L which is equal to zero or which is infinitely high, respectively. The values of the photocurrent and of the photovoltage at short- and open-circuit conditions are called short-circuit current (I_{SC}) and open-circuit voltage (V_{OC}), respectively. The electric power is equal to zero at short- and open-circuit operation of a solar cell.

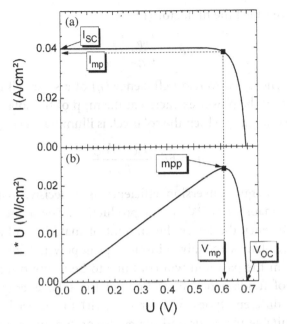

Figure 1.3. Example of a current-voltage characteristic (a) and of the corresponding power-voltage characteristic (b) of a solar cell under illumination. The voltage, current density and maximum power point are denoted by V_{mp}, I_{mp} and mpp, respectively.

A current–voltage characteristic (I–V characteristic) of a solar cell is a plot of all possible working points in a considered range. Figure 1.3 shows schematically the I–V characteristic of a solar cell under illumination.

There is one combination of current and voltage at which the power of the solar cell has its maximum (I_{mp} and V_{mp}, respectively). This point on the I–V characteristic of an illuminated solar cell is called the maximum power point (mpp).

$$P_{mpp} = I_{mp} \cdot U_{mp} \qquad (1.13)$$

The values of I_{SC} and V_{OC} can be measured easily. Therefore, it is convenient to characterize the maximum power of a solar cell with I_{SC}, V_{OC} and an additional parameter instead of characterizing the solar cell with I_{mp} and U_{mp}. The additional parameter sets the product of I_{mp} and U_{mp} in relation to the product of I_{SC} and V_{OC}. This parameter describes the amount by which the I_{SC}–V_{OC} rectangle is filled by the I_{mp}–U_{mp} rectangle and is therefore called the fill factor (FF).

$$FF = \frac{I_{mp} \cdot U_{mp}}{I_{SC} \cdot V_{OC}} \tag{1.14}$$

The solar energy conversion efficiency (η) of a solar cell is defined as the ratio between the power extracted at the mpp of the solar cell and the power of the sunlight at which the solar cell is illuminated (P_{sun}).

$$\eta = FF \cdot \frac{I_{SC} \cdot V_{OC}}{P_{sun}} \tag{1.15}$$

The solar energy conversion efficiency is a decisive parameter for costs and sustainability of PV energy production. The higher the energy conversion efficiency the lower the amount of material and area needed for a PV power plant with a given installed peak power. A lot of effort has been directed in the past and will continue to be in the future (i) to the investigation of fundamental limitations of the solar energy conversion efficiency for different types of solar cells, (ii) to materials science for finding out suitable materials and materials combinations for high solar energy conversion efficiencies and (iii) to the development of technologies allowing the realization of maximum solar energy conversion efficiencies in mass production and with a minimum of resources.

1.3 The Ideal Solar Cell under Illumination

1.3.1 *Diode equation of the ideal solar cell*

The two fundamental functions of a solar cell are expressed by two different elements in the equivalent circuit of the ideal solar cell (Figure 1.4). The first element is a current source driven by illumination, i.e. the photocurrent generator. The second element has to fulfill the condition of charge

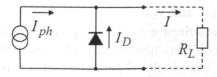

Figure 1.4.   Equivalent circuit of an ideal solar cell containing a photocurrent generator, a diode for charge separation and connected to a R_L.

separation, i.e. the current can pass through the element in one direction but not in the other. This is the characteristic property of a diode, the second element in the equivalent circuit of an ideal solar cell.

A current across a diode (I_D) is described by the diode equation:

$$I_D = I_0 \cdot \left[exp\left(\frac{q \cdot U}{k_B \cdot T} \right) - 1 \right] \qquad (1.16)$$

The diode equation contains the proportionality factor which is called the diode saturation current or diode saturation current density (I_0). The diode saturation current is a specific characteristic of each solar cell depending on the absorber and contact materials as well on their geometry. The diode saturation current is multiplied with an exponential of the voltage while the voltage is multiplied with the elementary charge and divided by the product of the Boltzmann constant and the absolute temperature (T). The diode current increases exponentially with increasing positive voltage (forward direction or forward bias). At zero voltage, the diode current should be equal to zero. It is for this reason that 1 has to be subtracted from the exponential term in the diode equation. At negative voltage, the diode current decreases to the very low negative diode saturation current (reverse direction or reverse bias).

Any charge-selective contact can be described by a diode equation. The diode equation will be derived for the pn-junction in Chapter 4 and for the ideal solar cell in the so-called Shockley–Queisser limit in Chapter 6.

The resistance of an ideal diode is extremely low in the forward direction and very large in the reverse direction of the diode. The photocurrent flows through the R_L, i.e. the photocurrent is not shunted by the diode. Therefore the photocurrent is a reverse current and has to be subtracted from the diode current.

$$I = I_0 \cdot \left[exp\left(\frac{q \cdot U}{k_B \cdot T} \right) - 1 \right] - I_{ph} \qquad (1.17)$$

Figure 1.5 shows I–V characteristics of an ideal solar cell with $I_0 = 10^{-13}$ A/cm^2 in the dark and under illumination ($I_{ph} = 0.04$ A/cm^2).

Equation (1.17) is used for the analysis of I–V characteristics of illuminated solar cells by setting I_{ph} equal to I_{SC}. Typical values of the I_{SC} density are of the order of tens of mA/cm^2 at AM1.5.

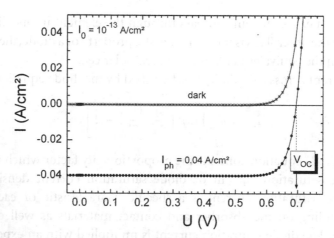

Figure 1.5. Current–voltage characteristics of an ideal solar cell with $I_0 = 10^{-13}$ A/cm^2 in the dark (stars) and under illumination at $I_{ph} = 0.04$ A/cm^2 (circles).

The I–V characteristics of solar cells under illumination are often plotted in the first quadrant for convenience. For this purpose, the negative current is plotted in the I–V characteristic of a solar cell. As a convention, $-I_{ph}$ corresponds to I_{SC}.

$$-I = I_0 \cdot \left[exp\left(\frac{q \cdot U}{k_B \cdot T} \right) - 1 \right] - I_{SC} \qquad (1.18)$$

1.3.2 *Open-circuit voltage and diode saturation current*

The V_{OC} of an ideal solar cell can be obtained from Equation (1.18) by setting the current to zero.

$$V_{OC} = \frac{k_B \cdot T}{q} \cdot ln\left(\frac{I_{SC}}{I_0} + 1 \right) \qquad (1.19)$$

The value of I_0 can change over many orders of magnitude depending on the materials and on interfaces between the materials forming the diode. The importance of I_0 for the performance of a solar cell follows from the limitation of V_{OC} by I_0. The minimization of I_0 is crucial for the optimization of the performance of solar cells in general.

The value of V_{OC} increases with increasing I_{SC}. The dependence of V_{OC} on I_{SC} is shown in Figure 1.6(a) for three different values of I_0.

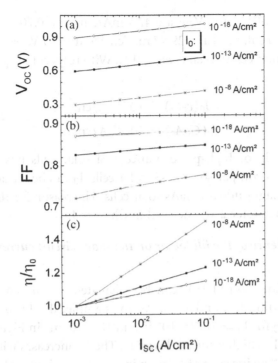

Figure 1.6. Dependence of V_{OC} (a), FF (b) and of the normalized solar energy conversion efficiency (c) on I_{SC} for an ideal solar cell with $I_0 = 10^{-18}, 10^{-13}$ and 10^{-8} A/cm² (squares, circles and stars, respectively).

It is useful to remember that $k_B \cdot T/q$ amounts to 0.026 V at room temperature and that the natural logarithm of 10 is equal to 2.3. Therefore, for an ideal solar cell, V_{OC} increases at room temperature by 60 mV if I_{SC} increases by one order of magnitude.

$$\left[\frac{\Delta V_{OC}}{\text{decade of } I_{SC}} \right]_{300\,K} = 60\,\text{mV per decade} \tag{1.20}$$

The order of magnitude of I_0 can be estimated for crystalline silicon (c-Si) and gallium arsenide (GaAs) record solar cells if assuming them to be ideal or almost ideal.

$$I_0 = \frac{I_{SC}}{exp\left(\frac{q \cdot V_{OC}}{k_B \cdot T}\right)} \tag{1.21}$$

The values of I_{SC} and V_{OC} are 42.7 mA/cm² and 0.705 V, respectively, for a c-Si record solar cell and 29.47 mA/cm² and 1.107 V, respectively, for a GaAs record solar cell (Green *et al.*, 2013). With respect to Equation (1.18), for I_0 one gets:

$$I_0(c\text{-}Si) \approx 10^{-13} \, \text{A/cm}^2 \qquad (1.22')$$

$$I_0(GaAs) \approx 10^{-19} \, \text{A/cm}^2 \qquad (1.22'')$$

Therefore, I_0 of high-performance c-Si solar cells has to be about 10^{-13} A/cm². In comparison to c-Si solar cells, I_0 can be reduced by about six orders of magnitude for GaAs solar cells. The reason for this difference is explored in Chapter 2 and derived in Chapter 4.

1.3.3 *Dependence of the fill factor on the short-circuit current density*

The FF (see Equation (1.14)) can be calculated by simulating Equation (1.18) and finding I_{mp} and U_{mp}. As an example, the FF is plotted as a function of I_{SC} for $I_0 = 10^{-18}, 10^{-13}$ and 10^{-9} A/cm² in Figure 1.6(b). At fixed I_{SC} the FF is higher for the lower I_0. The FF increases with increasing I_{SC} at fixed I_0. The increase of FF with increasing I_{SC} is weak for low values of I_0. For example, at $I_0 = 10^{-18}$ A/cm² the FF changes roughly between 0.87 and 0.88 for I_{SC} between 1 and 100 mA/cm². On the other hand, the increase of FF with increasing I_{SC} is stronger for lower values of I_0. For example, at $I_0 = 10^{-8}$ A/cm² the FF changes roughly between 0.72 and 0.78 for I_{SC} between 1 and 100 mA/cm².

Due to the influence of I_0, the FFs of solar cells made from similar materials, materials combinations and architecture are practically identical at a given I_{SC} if a similar fabrication technology was used for all compared solar cells. Therefore, besides the relatively weak change of the FF with I_{SC}, FF obtains specific values depending on the type of a solar cell and its production process. For example, FF is lower for thin-film solar cells (see Chapter 9) than for advanced c-Si solar cells (see Chapter 7).

1.3.4 *Dependence of the efficiency of an ideal solar cell on the light intensity*

The solar energy conversion efficiency depends on P_{sun} since FF, I_{SC} and V_{OC} depend on light intensity. The ratio of two solar energy conversion efficiencies η/η_0, where the index 0 is related to the lower value of P_{sun}, is considered in the following:

$$\frac{\eta}{\eta_0} = \frac{FF}{FF_0} \cdot \frac{I_{SC}}{I_{SC,0}} \cdot \frac{V_{OC}}{V_{OC,0}} \cdot \frac{P_{sun,0}}{P_{sun}} \tag{1.23}$$

Equation (1.23) can be simplified if it is assumed that the I_{SC} density increases linearly with increasing light intensity, i.e. I_{SC} is proportional to P_{sun}. With respect to Equation (1.19), therefore, one gets:

$$\frac{\eta}{\eta_0} = \frac{FF}{FF_0} \cdot \frac{T}{T_0} \cdot \frac{ln\frac{I_{SC}}{I_0}}{ln\frac{I_{SC,0}}{I_{0,0}}} \tag{1.24}$$

Equation (1.24) can be further simplified if heating due to increased P_{sun} is neglected and if a constant I_0 is assumed. Of course, one has to be careful with both of these since the temperature of solar cells generally increases with increasing P_{sun} and since I_0 can depend on P_{sun}. After a transformation, the following equation is obtained:

$$\frac{\eta}{\eta_0} = \frac{FF}{FF_0} \cdot \left[1 + \frac{X}{ln\frac{I_{SC,0}}{I_0}} \right] \tag{1.25}$$

X is the so-called concentration factor.

$$X = \frac{P_{sun}}{P_{sun,0}} \tag{1.26}$$

The ratio of the FFs is larger than 1, as determined in Section 1.3.3, and X and the natural logarithm of the ratio of $I_{SC,0}$ and I_0 are positive. Therefore, the solar energy conversion efficiency of an ideal solar cell is larger at the higher light intensity. This is the reason why the absolute largest energy conversion efficiencies are reached under concentrated sunlight.

The solar energy conversion efficiency of an ideal solar cell normalized to the efficiency at $I_{SC} = 1$ mA/cm^2 is plotted in Figure 1.6(c) as a function of I_{SC} for $I_0 = 10^{-18}, 10^{-13}$ and 10^{-9} A/cm^2. The relative energy conversion efficiency of an ideal solar cell increases by 13%, 20% and 50% for $I_0 = 10^{-18}, 10^{-13}$ and 10^{-9} A/cm^2 if I_{SC} increases by two orders of magnitude. However, it has to be taken into account that the assumptions of (i) linear I_{SC} response, (ii) absence of additional heating and (iii) independence of I_0

on P_{sun} may not be fulfilled for real solar cells so that Equation (1.25) can no longer be applied.

1.4 Real Solar Cells: Consideration of Series and Parallel Resistances

1.4.1 *Resistive losses and tolerable series and parallel resistances*

In reality, any solar cell has losses due to ohmic resistances. For example, contact resistances or material resistances are responsible for series resistances, and local shunts or other imperfections can cause parallel resistances in a solar cell. One common series resistance (R_s) and one common parallel resistance (R_p), which is also called shunt resistance, are considered in the equivalent circuit of a real solar cell (Figure 1.7).

The current flowing through the R_s is equal to the current flowing through R_L. Therefore, the potential across the R_L is reduced by the voltage drop across the R_s. Further, the current flowing through the R_L is reduced by the current flowing through the R_p. The voltage drop across R_s and the current shunted by R_p are taken into account in the following equation:

$$-I = I_0 \cdot \left[exp\left(\frac{q \cdot (U - I \cdot R_s)}{k_B \cdot T} \right) - 1 \right] + \frac{U - I \cdot R_s}{R_p} - I_{SC} \quad (1.27)$$

The voltage drop across R_s is assumed to be significantly smaller than U_{mp}. Then, the power losses due to resistive heating of R_s and R_p are given by:

$$P_{loss}(R_s) = R_s \cdot I_{mp}^2 \quad (1.28')$$

$$P_{loss}(R_p) \approx \frac{U_{mp}^2}{R_p} \quad (1.28'')$$

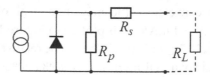

Figure 1.7. Equivalent circuit of a real solar cell containing a photocurrent generator, an ideal diode, a shunt resistance (R_p) and a series resistance (R_s), and connected with a load resistance (R_L).

It is useful at this juncture to introduce so-called tolerable series and parallel resistances ($R_{s,tol}$ and $R_{p,tol}$, respectively) in accordance with the power losses that can be tolerated in comparison to the power in the mpp. It makes sense to tolerate a power loss caused by resistive heating of about 1% of the power in the mpp of a solar cell with very high solar energy conversion efficiency. The R_L in the mpp is of the same order of magnitude as the ratio of V_{OC} and I_{SC} (R_L^*). As an approximation rule, one can write the following conditions for $R_{s,tol}$ and $R_{p,tol}$

$$R_{s,tol} < \frac{1}{100} \cdot \frac{V_{OC}}{I_{SC}} = \frac{R_L^*}{100} \tag{1.29'}$$

$$R_{p,tol} > 100 \cdot \frac{V_{OC}}{I_{SC}} = 100 \cdot R_L^* \tag{1.29''}$$

Equations (1.29) describe rather harsh conditions. For example, a solar cell based on a c-Si wafer (for details about c-Si solar cells, see Chapter 7) with an area of about $100\,cm^2$ has an I_{SC} of about 3–4 A and a V_{OC} of about 0.6–0.7 V. This means that the overall R_s of this solar cell should be less than about $2\,m\Omega$.

A major source for R_s in c-Si solar cells are the bus bars collecting the charge carriers from the whole solar cell. The bus bars are mainly based on silver. The resistance of a bus bar can be calculated if the specific conductivity ($\sigma, 6.2 \times 10^7$ S/m for silver), the length (L), the width (B) and the height (H) of the bus bar are known.

$$R = \frac{L}{\sigma \cdot B \cdot H} \tag{1.30}$$

The resistance is about $3\,m\Omega$ for a bus bar with a length of 10 cm, a width of 5 mm and a height of 0.1 mm, which is larger than the $2\,m\Omega$ demanded. Therefore, a c-Si solar cell should have several bus bars depending on the area of the solar cell.

Current and voltage can be related to area fractions of a solar cell, to solar cells connected in PV modules, to PV modules connected in strings or to strings of PV modules connected in large PV power plants. The conditions for the cross section of bus bars in c-Si solar cells or for the cross section of copper cables connecting strings of PV modules are equivalent and can be obtained by using Equations (1.29') and (1.30). For this purpose, the series and parallel connection of solar cells has to be taken into account. In a series connection, the current remains constant whereas the individual values of the voltage are added.

$$U = \sum_i U_i \qquad (1.31')$$

$$I = I_i \qquad (1.31'')$$

For a parallel connection, the voltage remains unchanged and the individual values of the current are added.

$$U = U_i \qquad (1.32')$$

$$I = \sum_i I_i \qquad (1.32'')$$

A series connection is favored if an increase of the cross section of connecting cables and therefore of weight or amount of material is unwanted. Binning of solar cells or PV modules in narrow ranges is very important for a series connection since the solar cell or the PV module with the lowest photocurrent limits the photocurrent of all other connected solar cells or PV modules. As a consequence the area fraction in an isolated solar cell, the solar cell in a PV module or the PV module in a string of PV modules with the lowest energy conversion efficiency will limit the efficiency of the whole system of which they are part, i.e. of the complete solar cell, of the PV module or of the PV power plant. This is a reason, for example, why only solar cells with an area of at least 1 cm^2 are considered for world records (AM1.5) and why the energy conversion efficiency decreases with increasing area of solar cells, PV modules or large PV power plants.

1.4.2 *Influence of series and parallel resistances on the basic characteristics of a solar cell*

The R_s and R_p can heavily influence the I–V characteristics of a solar cell. The influence of R_s and R_p on I–V characteristics will be analyzed separately in the following for a solar cell with fixed I_0 (10^{-13} A/cm^2) and fixed I_{SC} (0.04 A/cm^2), which is close to record c-Si solar cells illuminated at AM1.5.

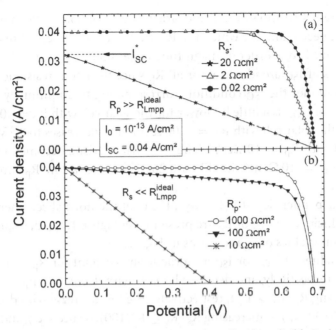

Figure 1.8. Current–voltage characteristics for a solar cell with negligible R_p (a) or R_s (b) for different values of R_s (a) and R_p (b) and $I_0 = 10^{-13}$ A/cm^2 and $I_{SC} = 0.04$ A/cm^2.

Under these conditions, V_{OC} is close to 0.7 V and R_L^* is about 17 Ωcm^2 in case of the ideal solar cell. Figure 1.8 (a) shows the I–V characteristics of solar cells for which R_p can be neglected and for which the R_s are about three orders of magnitude lower than R_L^*, one order of magnitude lower than R_L^* or very close to R_L^*. The values of I_{SC} and V_{OC} are identical for the I–V characteristics with $R_s = 0.02$ Ωcm^2 and $R_s = 2$ Ωcm^2, while FF decreases significantly for the I–V characteristic with $R_s = 2$ Ωcm^2. The value of V_{OC} remains constant with further increase of R_s since R_s has no influence if there is no current flowing. However, I_{SC} decreases to about 0.032 A/cm^2 and the I–V characteristic becomes a straight line for $R_s = 20$ Ωcm^2. The FF is 0.25 for a linear I–V characteristic. A FF of 0.25 therefore corresponds to the minimum possible FF of a solar cell which can be described with an equivalent circuit as shown in Figure 1.7. High R_s limit I_{SC} and FF.

The I_{SC} in the diode equation (1.27) and the I_{SC} obtained from the I–V characteristic (I_{SC}^*, see also Figure 1.8(a)) can be rather different depending on R_s and thus have to be distinguished.

Figure 1.8(b) shows the I–V characteristics for solar cells in which R_s can be neglected and for which the R_p are about two orders of magnitude larger than R_L^*, one order of magnitude larger than R_L^* or very close to R_L^*. The values of I_{SC} are identical for all R_p since the shunt resistance can be neglected under the I_{SC} condition corresponding to an infinitely low R_L. The value of V_{OC} is a little bit lower for the solar cell with $R_p = 100\ \Omega cm^2$ than for the solar cell with $R_p = 1000\ \Omega cm^2$ and decreases to 0.4 V for the solar cell with $R_p = 10\ \Omega cm^2$. The I–V characteristic becomes a straight line for $R_p = 10\ \Omega cm^2$, i.e. FF is 0.25. Therefore, a low R_p limits V_{OC} and FF.

The dependencies of V_{OC}, I_{SC}, FF and of the normalized energy conversion efficiency on R_s or R_p are presented in Figure 1.9(a–d, respectively) for the same values of I_0 and I_{SC} as in Figure 1.8.

The values of V_{OC} or I_{SC} are practically constant for $R_p > R_L^*$ or for $R_s < R_L^*$, respectively, and decrease linearly with decreasing R_p ($R_p < R_L^*$) or increasing R_s ($R_s > R_L^*$), respectively. The FFs decrease with decreasing R_p ($R_p \leq 100 \cdot R_L^*$) or increasing R_s ($R_s \geq R_L^*/100$), respectively, and remain practically constant at 0.25 for $R_p < R_L^*$ or for $R_s > R_L^*$, respectively. Therefore, the energy conversion efficiency is limited by FF in the ranges $R_L^* \leq R_p \leq 100 \cdot R_L^*$ and $R_L \geq R_s \geq R_L^*/100$ and by V_{OC} and I_{SC} in the ranges of $R_p < R_L^*$ and $R_s > R_L^*$, respectively.

The qualitative influence of R_p and R_s on the basic characteristics of a solar cell is generally similar to the dependencies shown in Figures 1.8 and 1.9 where R_s and R_p should be scaled with respect to I_0 and I_{SC}.

1.4.3 *Influence of series and parallel resistances on the intensity dependence of the basic characteristics of a solar cell*

The influence of R_s and R_p on the I–V characteristics of a solar cell can change strongly with changing light intensity. For this reason R_s and R_p have a tremendous influence on the intensity dependence of the basic characteristics of a solar cell. The I_{SC} in the diode equation increases linearly with increasing P_{sun}.

The influence of I_{SC} on the I–V characteristics will be analyzed separately in the following for a solar cell with fixed I_0 ($10^{-13}\ A/cm^2$) and given values of R_s ($10\ \Omega cm^2$) and R_p ($100\ \Omega cm^2$), where the respective

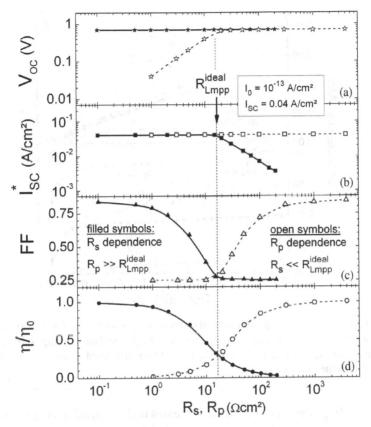

Figure 1.9. Dependence of the V_{OC} (a), the I_{SC} density (b), the FF (c) and the normalized energy conversion efficiency (d) on the R_s (filled symbols) and R_p (open symbols) for a solar cell with $I_0 = 10^{-13}$ A/cm^2, $I_{SC} = 0.04$ A/cm^2 and negligible R_s (right).

values of R_p ($R_s = 10\,\Omega$cm^2) and R_s ($R_p = 100\,\Omega$cm^2) are neglected (Figure 1.10(a) and (b), respectively).

As shown in Figure 1.10(a), the FF decreases with increasing I^*_{SC} for a given R_s. The value of I^*_{SC} is proportional to I_{SC} at values for which $I_{SC} < V_{OC}/R_s$. The FF reaches 0.25 when I^*_{SC} becomes larger than V_{OC}/R_s. In this region I^*_{SC} becomes proportional to V_{OC}. The FF decreases with decreasing I^*_{SC} for a given R_p (Figure 1.10(b)). The logarithm of the value of V_{OC} is proportional to I_{SC} at values for which $I_{SC} > V_{OC}/R_p$. The FF reaches 0.25 when I^*_{SC} becomes lower than V_{OC}/R_s. In this region V_{OC} decreases linearly with decreasing I_{SC}.

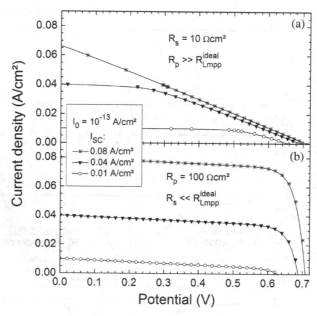

Figure 1.10. Current–voltage characteristics for solar cells with negligible R_p (a) or R_s (b) for $I_0 = 10^{-13}$ A/cm^2 and $I_{SC} = 0.08, 0.04$ and 0.01 A/cm^2 (stars, triangles and circles, respectively) for $R_s = 10$ Ωcm^2 (a) and $R_p = 100$ Ωcm^2 (b). Incidentally, I_{SC} corresponds to the value in the diode equation.

The voltage drop across the R_s increases with increasing current, which leads to a decrease of the energy conversion efficiency. This means that R_s limits the energy conversion efficiency at high values of I_{SC}. On the other hand, the influence of R_p increases with decreasing I_{SC}, R_p limits the energy conversion efficiency at low values of I_{SC}. The intensity dependencies of V_{OC}, I_{SC}^* and FF can be summarized as the dependencies of the normalized solar energy conversion efficiency on I_{SC}.

Figure 1.11 shows the dependence on I_{SC} of the normalized solar energy conversion efficiency of solar cells with $I_0 = 10^{-13}$ A/cm^2 for R_S of 0.01, 1 and 100 Ωcm^2 (a) and for R_p of 1, 100 and 10000 Ωcm^2 (b). The range of I_{SC} is chosen in such a way that the influence of concentrated sunlight or of weak illumination can be analyzed. The illumination target range of the application of a given type of solar cell has significant technological and economic consequences.

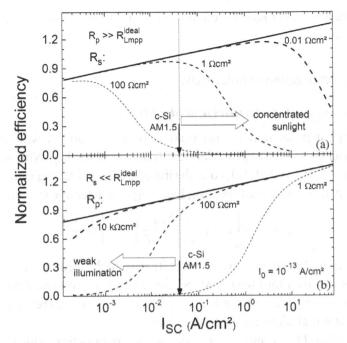

Figure 1.11. Dependencies of the normalized energy conversion efficiency on I_{SC} for a solar cell with $I_0 = 10^{-13}$ A/cm^2 at different values of R_s (a) (R_p neglected) and R_p (b) (R_s neglected). The thin and thick arrows mark I_{SC} of an ideal c-Si solar cell and the directions towards concentrated sunlight and weak illumination, respectively.

Concentration factors of up to 1000 are important for highly efficient concentrator solar cells. At very high concentration factors R_s should be significantly lower than 0.01 Ωcm^2.

The R_p of c-Si solar cells fabricated in conventional mass production is of the order of several hundred Ωcm^2, which is not favorable for highly efficient operation at reduced light intensity. More energy can be produced over the year with a PV power plant based on solar cells with higher R_p due to reduced losses in the morning and evening hours. For example, high values of R_p can be reached with amorphous silicon solar cells (see Chapter 9).

Light intensities reduced by up to about two orders of magnitude in comparison to AM1.5 can make sense for low-power indoor applications of solar cells and can be relevant for stand-by functions of electronic devices.

The highest values of R_p can be achieved with dye-sensitized solar cells (see Chapter 10).

1.5 Characterization of Solar Cells

1.5.1 I_{SC}–V_{OC} *characteristics and ideality factor*

The simplest measurements on solar cells are performed with just one multimeter by measuring the V_{OC} as a function of the I_{SC}. With regard to Equations (1.19) and (1.27), considering a low R_s and taking into account an ideality factor (n), the following I_{SC}–V_{OC} characteristic is obtained:

$$I_{SC} = I_0 \cdot \left[exp \left(\frac{q \cdot V_{OC}}{n \cdot k_B \cdot T} \right) - 1 \right] + \frac{V_{OC}}{R_p} \tag{1.33}$$

It has to be taken into account that a high accuracy for current measurement over four to six orders of magnitude is required. An intensity control of the light is not needed since I_{SC} and V_{OC} are correlated directly and measured at identical light intensities.

Equation (1.33) has I_0, n and R_p as parameters which can be determined from the I_{SC}–V_{OC} characteristic (see Figure 1.12). An I_{SC}–V_{OC} characteristic can be divided into two parts that are dominated by the

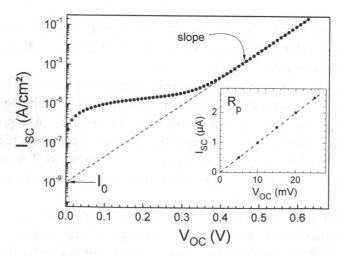

Figure 1.12. Dependence of the I_{SC} density on the V_{OC} for a given solar cell in a logarithmic scale and in a linear scale (zoom, insert).

exponential term and by the ohmic term at high and low values of V_{OC}, respectively.

The slope of the exponential term is usually significantly larger than 60 mV per decade, i.e. a higher potential is needed to increase the current by one order of magnitude. The ratio of the measured slope versus 60 mV per decade is called the ideality factor and is equal to one for an ideal solar cell. The ideality factor is usually on the order of 1.1–1.3 for c-Si solar cells. The value of I_0 is also found from the exponential part of the I_{SC}–V_{OC} characteristic. The value of R_p is found from the slope of the ohmic part of the I_{SC}–V_{OC} characteristic.

1.5.2 *Temperature-dependent diode saturation current and activation energy*

The value of I_0 can be obtained from I_{SC}–V_{OC} characteristics measured at different temperatures. The temperature dependence of I_0 is usually plotted in a so-called Arrhenius plot. Arrhenius plots generally present a temperature-dependent parameter on a logarithmic scale as a function of the absolute temperature in a reciprocal scale. Figure 1.13 shows an example of a typical Arrhenius plot for I_0.

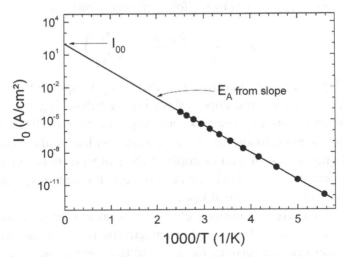

Figure 1.13. Arrhenius plot of the I_0 for a given solar cell. The logarithm of the I_0 should be multiplied with the temperature-dependent ideality factor in case of a non-ideal solar cell.

Arrhenius plots allow for the extraction of the relevant parameter controlling the temperature dependence of I_0. The point is that the temperature dependence of I_0 usually follows a straight line in the Arrhenius plot, which means that I_0 depends exponentially on temperature. The control parameter is the thermal activation energy (E_A) which can be extracted from the slope of the temperature-dependent I_0 in the Arrhenius plot. The temperature-independent pre-factor I_{00} can be obtained at the inverse temperature equal to zero. The value of I_{00} can be rather high and amounts to about 100 A/cm^2 for the example shown in Figure 1.13.

$$I_0 = I_{00} \cdot exp\left(-\frac{E_A}{k_B \cdot T}\right) \qquad (1.34)$$

The activation energy is given in the unit of eV and depends on fundamental processes in the solar cell. For c-Si solar cells, the activation energy is about 1.1 eV. For non-ideal solar cells, the ideality factor can also depend on temperature. In this case the product of the logarithm of the diode saturation current and the temperature-dependent ideality factor has to be plotted in the Arrhenius plot.

The thermal activation of the diode saturation current determines the temperature dependence of V_{OC}. For an ideal solar cell the temperature dependence of V_{OC} is given by the following equation:

$$V_{OC} = \frac{E_A}{q} - \frac{k_B \cdot T}{q} \cdot ln\left(\frac{I_{00}}{I_{SC}}\right) \qquad (1.35)$$

The V_{OC} decreases linearly with increasing temperature following Equation (1.35), while the slope scales with the natural logarithm of the ratio between I_{00} and I_{SC}. The maximum V_{OC} can theoretically be reached at 0 K and is equal to the activation energy divided by the elementary charge. Figure 1.14 shows an example of the temperature dependence of V_{OC} at different ratios of I_{00}/I_{SC}. As can be seen, the higher I_{SC}, the lower the temperature dependence of V_{OC}.

The solar energy conversion efficiency of solar cells decreases with increasing temperature due to the thermal activation of the diode saturation current. Solar cells are certified for standard test conditions, i.e. at AM1.5 and 25°C. However, the temperature of solar cells can easily exceed 25°C and increase to values on the order of 60°C, and even higher under

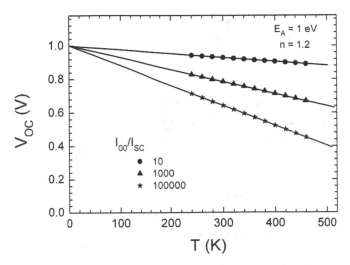

Figure 1.14. Temperature dependence of the V_{OC} for a solar cell with given activation energy and ideality factor measured at different I_{00}/I_{SC} ratios.

operation conditions (see task T1.5 at the end of this chapter). This should be taken into account when calculating the expected energy production of a projected PV power plant. Further, there are solar cells with stronger or weaker temperature dependence of V_{OC} depending on materials, materials combinations and technology. For this reason, the temperature coefficient of PV modules is usually certified as well.

1.5.3 *Measurement of I–V characteristics with loads*

The measurement of the complete I–V characteristics of illuminated solar cells is needed to calculate the FF and therefore the energy conversion efficiency. In addition, the R_s and R_p can be obtained from complete I–V characteristics. The ratio between the current and voltage is different at each point of an I–V characteristic, and therefore variable R_L for probing different working points are necessary. Resistors, electronic loads or capacitors (Figure 1.15) can be applied for I–V measurements depending on the requirements for accuracy, costs of measurement equipment, flexibility and measurement speed.

The use of a set of known resistances and a multimeter for voltage measurement is the easiest and cheapest way to measure an I–V characteristic of an illuminated solar cell. The resistance and the multimeter

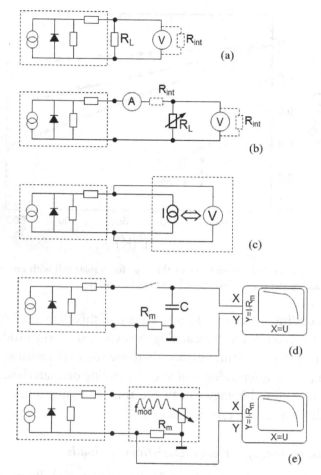

Figure 1.15. Arrangement for the measurement of I–V characteristics of solar cells for (a) a known set of resistances and multimeter, (b) a potentiometer and two multimeters, (c) an electronic load or constant current source, (d) a switch and capacitor with oscilloscope and (e) a periodically variable R_L and oscilloscope.

are connected in parallel with the illuminated solar cell (Figure 1.15(a)). The current is found by Ohm's law from the values of the resistance and the measured voltage. The values of suitable R_L can be estimated by the following procedure. First, V_{OC} and I_{SC} are measured and the ratio of V_{OC} and I_{SC}, R_L^*, is calculated. The value of R_L^* is close to the R_L in the mpp as mentioned above. Values of resistances less than or greater than R_L^* by roughly one order of magnitude or more are chosen to get points on the

I–V characteristic towards I_{SC} or towards V_{OC} where more values of R_L have to be chosen around the mpp in order to increase the accuracy for the measurement of the FF. The described method works very well for solar cells and mini-modules with relatively low power when the R_L in the mpp is about 10 Ω or larger. The point is that contact and cable resistances can become an important source of error for low R_L. A fixed correction resistance can be introduced into the analysis if the contact resistance and the resistances of cables are constant. Furthermore, the power range of a given resistor has to be considered since the resistance increases under heating to high power. The accuracy of the described method depends on the accuracy of voltage and resistance measurements.

The measurement procedures of I–V characteristic and of the R_p and R_s are depicted in Figure 1.16 ((a)–(c), respectively). The value of R_p is the negative slope of the I–V characteristic near I_{SC} where the diode current can be neglected.

$$R_p = -\left.\frac{\Delta U}{\Delta I}\right|_{U \to 0} \tag{1.36}$$

The R_s has a strong influence on the I–V characteristic towards V_{OC}. Two I–V characteristics at different intensities should be measured for obtaining R_s from a simple procedure. The two intensities have to be chosen in such a way that the respective I_{SC} (I_{SC1} and I_{SC2}) differ by a factor of roughly two. The differences between I_{SC1} and I_{SC2} and an identical current difference ΔI are determined (I_1 and I_2). It is recommended that a value of ΔI between half of I_{SC} and about 90% of I_{SC} be chosen at the lower intensity for the measurement of R_s. The potentials U_1 and U_2 corresponding to I_1 and I_2 are found on the respective I–V characteristics (Figure 1.16 (a)). The R_s is then given by:

$$R_s = \frac{U_2 - U_1}{I_{SC1} - I_{SC2}} \tag{1.37}$$

The described procedure of measuring R_s works well for conventional solar cells when the FF is not very low and when R_p and R_s can be well distinguished.

A potentiometer for faster variation of R_L and multimeters for current and voltage measurements can also be used for the measurement of I–V characteristics of illuminated solar cells (Figure 1.15(b)). However, one has

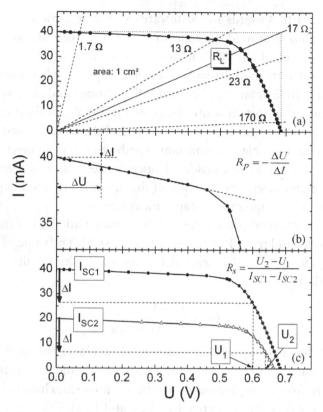

Figure 1.16. Procedure for the measurement of I–V characteristics (a) and determination of the R_p (b) and R_S (c).

to be careful with the current measurement since the internal resistance of the multimeter can become a serious source of error, especially when changing the range of sensitivity.

A load can be simulated with a constant current source connected with the solar cell (Figure 1.15(c)). The resulting voltage drop across the solar cell is then measured. Constant current sources, also called source measure units or electronic loads, allow I–V measurements to a high degree of accuracy over a wide range of power depending on the dimensioning of the electronic load. Constant current sources can be fabricated at excellent precision so that currents down to the range of pA and less can be measured. Dark I–V characteristics and I–V characteristics of solar cells illuminated over a wide range of intensities can only be investigated with source measure units. As an aside, double-shielded cables have to be used for measurements at very low currents to avoid potential drops across long cables. Four-point probes or so-called Kelvin contacts have to be applied for measurements at high currents and/or very low R_L. In this case, the potential drop is measured between the contacts of the current input to avoid an influence of voltage drops at contact resistances for the current input and to avoid an influence of an inhomogeneous current flow.

Fast and reliable measurements of I_{SC}, V_{OC} and FF are needed for in-line control and binning of solar cells. For example, the tact cycle of a production line for c-Si solar cells is of the order of 1 s or less. Time for the handling, contacting and sorting of a solar cell is needed within one tact cycle in addition to the measurement time. This means that a c-Si solar cell has to be characterized within a time even less than 0.1 s. Very fast measurements of I–V characteristics become possible with a load changing its resistance automatically and rapidly during illumination with a light flash. The principle of such a flasher is given in Figure 1.15(d). The heart of a flasher is a capacitor (or an electronically simulated capacitor). The resistance of a capacitor varies over a huge range during charging. At the beginning of charging, the resistance of a capacitor is extremely low so that the maximum current which can be provided by the solar cell can flow. Therefore, the current flowing through the capacitor is equal to I_{SC} at the beginning of charging. At the end of charging, the resistance of the capacitor is extremely high since the charged capacitor behaves like an insulator. The maximum voltage across the charged capacitor is equal to the maximum voltage which can be provided by the solar cell, i.e. V_{OC}. The resistance of the capacitor increases continuously during the charging process so that the exact and complete I–V characteristic of the solar cell can be monitored with an oscilloscope. In the simplest case, the voltage drop across the capacitor is applied to the input of the x-channel of the oscilloscope and the current is converted with a low measurement resistance to a voltage signal and applied to the y-channel of the same oscilloscope. It is worth noting that a switch connecting and disconnecting the solar cell can be applied instead of using a light pulse for flash characterization.

Rapid but not very fast measurements of I–V characteristics at high accuracy are required when an external parameter such as light intensity is continuously changed during the measurement of I–V characteristics.

For such purposes the channel resistance of a field-effect transistor can be varied periodically with a frequency generator at a moderate frequency (Figure 1.15(e)). The I–V characteristic is visualized with an oscilloscope. The voltage drop across the periodically variable load is applied to the x-channel of the oscilloscope and the current transformed to a voltage signal is applied to the y-channel of the oscilloscope. Periodically variable loads can be realized at relatively small size and low cost, and allow highly accurate and fast measurements and have an excellent performance, which makes them especially advantageous for teaching at any level.

1.5.4 *Measurement of the energy conversion efficiency with a pyranometer*

For the measurement of the energy conversion efficiency of a solar cell, the I–V characteristic has to be measured along with the P_{sun} at the exact moment at which the I–V characteristic is obtained. The P_{sun} is measured with a pyranometer in the unit of W/m^2 (Figure 1.17). The angles of incidence must be identical for the solar cell and for the pyranometer.

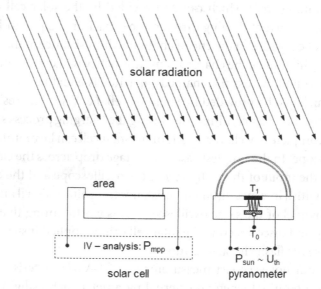

Figure 1.17. Procedure for an outdoor measurement of the solar energy conversion efficiency of a solar cell with a pyranometer.

The area of the solar cell has to be measured with a ruler for normalizing the power in the mpp of the solar cell to its area.

In the pyranometer, the blackbody absorbs the light of the complete sun spectrum. The decisive advantage of a blackbody is that the sunlight is absorbed practically over the whole relevant spectral range at identical sensitivity.

In the blackbody, the radiation energy of the sunlight is converted into heat leading to an increase of the temperature (T_1). The temperature T_1 is measured with a thermopile consisting of tens of thermocouples in series. A thermocouple is an intimate contact between two different metals. The contact potential between two different metals depends on the temperature. Therefore a thermocouple converts the temperature T_1 into a voltage signal. The signal of the thermopile is measured as the difference of the series of contact potentials between the heated blackbody and a body at the ambient or reference temperature T_0. The voltage signal of the thermopile at the output of the pyranometer is calibrated with respect to P_{sun}. It must be noted that the calibration of a pyranometer demands calibration standards and is a rather critical requirement. Pyranometers of high performance have an absolute accuracy of 1% at AM1.5, i.e. $10 \, W/m^2$.

The blackbody in the pyranometer has to be protected from variable heat transfer, i.e. from convection of surrounding air and from changes in the heat conductivity of surrounding materials. For this purpose, the blackbody is surrounded by a double-walled glass dome and embedded into heat-insulating materials. In addition, the inner part of a pyranometer is kept dry, for example, with silica gel, in order to avoid variable heat exchange due to penetrating water molecules. Furthermore, the part of the pyranometer with the blackbody kept at the reference temperature is shaded to avoid changes of T_0 during the measurement.

The advantage of the measurement of the solar energy conversion efficiency with a pyranometer is that any solar cell, PV module or string of PV modules can be characterized under given conditions at a certain moment. Therefore, accurate performance monitoring of PV power plants is coupled with real-time monitoring of P_{sun} using pyranometer(s).

There is a low-cost version of this which involves the measurement of P_{sun} with a photodiode. The point is that a certain photocurrent of a silicon photodiode can be related to a certain P_{sun}. However, in contrast to a blackbody, a silicon photodiode does not integrate over the whole relevant sun spectrum with identical sensitivity. But the sun spectrum can change depending on geography, weather conditions and time of day. Therefore, the absolute accuracy of measurements with photodiodes is lower than for measurements with a pyranometer.

1.5.5 *Measurement of the solar energy conversion efficiency with a sun simulator*

The measurement of the solar energy conversion efficiency with a pyranometer has a great disadvantage which is related to the sun. The intensity and the spectrum of sunlight change dramatically over the day and depend on various conditions. This makes the direct comparison of the solar energy conversion efficiency of solar cells measured at different places and at different times problematic. A sun simulator overcomes this serious problem of comparability of measurements. The standard test conditions (STC; AM1.5 with $1000 \, W/m^2$ and temperature of the solar cell 25°C) are standardized for the measurement of the solar energy conversion efficiency (IEC, 2008).

A sun simulator is an artificial light source with an intensity spectrum very close to that of the sun at AM1.5. The artificial reproduction of the sun spectrum at AM1.5 demands appropriate light sources, combinations of light sources and filtering of light. For example, the sun spectrum at AM0 can be approximated with a blackbody with a temperature of 5800 K. Related blackbodies are not available on earth.

A halogen lamp can be described as a blackbody with a temperature of about 3000 K (Figure 1.18). The maximum intensity of a halogen lamp is at a wavelength of about 1000 nm. The simulation of the sun spectrum with a halogen lamp is impossible due to the very low intensity in the spectral range between blue and ultraviolet light.

A xenon arc lamp is a light source with an arc temperature of about 10000 K, which is much higher than the temperature at the surface of the sun. The maximum intensity of a xenon arc lamp is at a wavelength of about 300 nm. In addition, a xenon arc lamp contains numerous spectral lines of high intensity, especially in the near-infrared range. The simulation of the sun spectrum with a xenon arc lamp is impossible due to the low intensity in the near-infrared range in comparison with ultraviolet light.

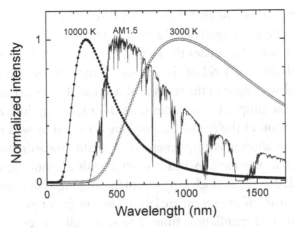

Figure 1.18. Normalized spectra of the intensity of a blackbody at 3000 K and at 10000 K in comparison with the normalized AM1.5 spectrum.

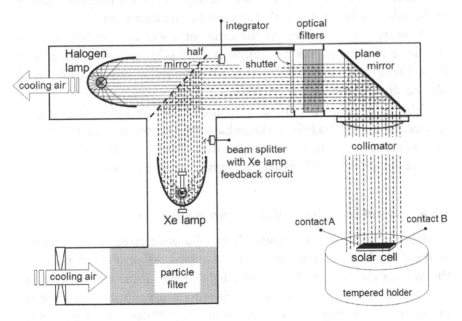

Figure 1.19. Principle set-up of a sun simulator.

The sun spectrum can be simulated with a combination of a halogen lamp providing enough intensity in the infrared and near-infrared range and a xenon arc lamp providing high intensity in the blue and violet spectral range. Figure 1.19 shows the principle set-up of a sun simulator. The light of the halogen and xenon arc lamps passes through a half mirror to a plane mirror from where the light is reflected to the collimator. The light passes the collimator and illuminates the solar cell. There are additional filters in the sun simulator to get rid of the intense infrared lines of the xenon arc lamp and to absorb light in the regions of decreased intensity in the AM1.5 spectrum. In the simplest case, water is used for filtering since a significant part of absorption in the earth's atmosphere is caused by water molecules. Furthermore, a shutter is implemented to enable measurements of I–V characteristics in the dark. A beam splitter with a xenon lamp feedback circuit is needed to adjust the lamp current for keeping a constant and highly stabilized intensity of the xenon arc lamp. The integrator gives information about accumulated irradiation time. The solar cell is kept at a tempered holder at 25°C and contacted with a source measure unit. The contact resistance is usually tested with a second contact at the anode and cathode of the solar cell before the I–V characteristics are measured.

Powerful light sources on the order of a kW are needed in a sun simulator for homogeneous illumination of solar cells with an area of $10 \times 10 \, cm^2$ and larger. The light sources should be actively cooled with filtered air.

The life of lamps is limited. Therefore, a sun simulator needs regular service. For example, a halogen lamp has to be replaced after every 50 hours of operation. In addition, the sun simulator has to be calibrated regularly with calibrated solar cells that are sensitive in different spectral ranges.

1.5.6 *Spectral dependence of the quantum efficiency*

Sunlight is white, i.e. it contains light of a wide range of wavelengths as mentioned above. The sensitivity of a solar cell strongly depends on the wavelength (λ) of the exciting light. This means that light at an equal intensity but at different wavelengths can generate very different photocurrents. The so-called spectral response (SR) describes this property. The SR is defined as the ratio between the photocurrent measured at a given wavelength ($I_{ph}(\lambda)$) and the corresponding light intensity ($P_{light}(\lambda)$).

$$SR(\lambda) = \frac{I_{ph}(\lambda)}{P_{light}(\lambda)} \tag{1.38}$$

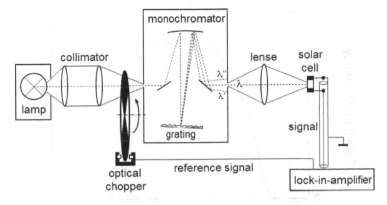

Figure 1.20. Schematic of the simplest set-up for spectral response measurements.

A simplified set-up for the measurement of the SR is shown in Figure 1.20. The heart of this set-up is a monochromator which filters required wavelengths from the spectrum of a lamp with a dispersive element such as a grating. The incoming white light is modulated with an optical chopper and the photocurrent is measured with a lock-in amplifier. The lock-in amplifier registers only that part of the photocurrent which is caused by the modulated incoming light, i.e. noise signals and the sensitivity to bias light are strongly suppressed. The signal at the lock-in amplifier is measured as a function of wavelength.

The accurate absolute calibration of a monochromatic light source demands tremendous efforts. Therefore, it is much easier to compare a measured photocurrent spectrum of an unknown solar cell ($I_{ph,1}$) with that ($I_{ph,2}$) of a solar cell where the spectral response spectrum is known (SR_2). The unknown spectral response (SR_1) is found from the following procedure.

$$SR_1(\lambda) = \frac{I_{ph,1}(\lambda)}{I_{ph,2}(\lambda)} \cdot SR_2(\lambda) \qquad (1.39)$$

A photocurrent is the photo-generated charge per time unit and the power is the energy per time unit. The photo-generated charge is the product of the number of collected photo-generated electrons ($N_e(\lambda)$) and the elementary charge (q) and the energy of light is the product of the number of photons ($N_{ph}(\lambda)$) and the photon energy h·c/λ, where h is the Planck constant and c is the velocity of light. Therefore, Equation (1.38)

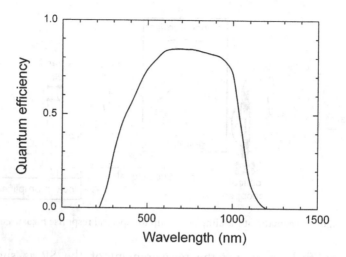

Figure 1.21. Example of a quantum efficiency spectrum of a solar cell.

can be transformed to

$$SR(\lambda) = \frac{q \cdot \lambda}{h \cdot c} \cdot \frac{N_e(\lambda)}{N_{ph}(\lambda)} = SR_{max}(\lambda) \cdot \frac{N_e(\lambda)}{N_{ph}(\lambda)} \qquad (1.40)$$

The maximum spectral response is proportional to the wavelength and amounts, for example, to 0.724 A/W at a wavelength of 900 nm.

The ratio between the number of collected photo-generated electrons and of incident photons is called quantum efficiency of a solar cell.

$$QE(\lambda) \equiv \frac{N_e(\lambda)}{N_{ph}(\lambda)} \qquad (1.41)$$

Figure 1.21 shows an example of a quantum efficiency spectrum. For the given example, the quantum efficiency is less than one over the whole spectral range and even zero at wavelengths below 220 nm and above 1200 nm. The quantum efficiency allows for obtaining information about optical and collection losses in solar cells, which will be a topic of consideration in Chapter 2. Spectral response measurements are important in the areas of research and development of solar cells.

The I_{SC} density of a solar cell can be calculated if the spectral response spectrum or the quantum efficiency spectrum of the solar cell is known. For this purpose, the product of the spectral response spectrum and the intensity spectrum of the sun or the product of the quantum efficiency spectrum and the photon flux spectrum of the sun ($\Phi_{sun}(\lambda)$) are integrated over the wavelength.

$$I_{SC} = q \cdot \int_0^\infty SR(\lambda) \cdot P_{sun}(\lambda) \cdot d\lambda \qquad (1.42')$$

$$I_{SC} = q \cdot \int_0^\infty QE(\lambda) \cdot \Phi_{sun}(\lambda) \cdot d\lambda \qquad (1.42'')$$

The values of the I_{SC} density obtained from measurements with a sun simulator and obtained from the quantum efficiency spectrum should be equal if the sun simulator and the reference solar cell for the analysis of the spectral response are well calibrated. Therefore, spectral response measurements allow for an independent correct determination of I_{SC}.

1.6 Summary

The sun emits a huge amount of power or energy flux to earth characterized by the solar constant ($1356\,W/m^2$) and by its spectral distribution with a maximum at photon energy around $1.4\,eV$. Total P_{sun} reaching the earth is about $1.3 \cdot 10^8\,GW$. For purposes of comparison, a nuclear power plant has a power of about 1 GW and a person needs on average about 0.1 kW to sustain his or her biological life. The total power received on earth can be compared with the population of mankind (more than $7 \cdot 10^9$) and the energy demand per capita for comfortable life (2 kW per capita seems sufficient). The P_{sun} reaching the earth exceeds the energy demand of mankind by several thousand times. Photovoltaic solar energy conversion is aimed at using a part of this energy for human needs in the form of electricity. The solar energy conversion efficiency is the most important parameter of solar cells and PV power plants. A high solar energy conversion efficiency combined with a low degradation rate of conversion efficiency and a low energy payback time is decisive for sustainable solar energy conversion.

A solar cell has two basic functions: the generation of a photocurrent and the generation of a photovoltage for the production of electric power. The maximum electric power of an illuminated solar cell is the product of the I_{SC}, the V_{OC} and the FF. Loads are used to extract the power from solar cells. From an illuminated solar cell, the maximum power is extracted with a load, the resistance of which has to be equal to the quotient of the potential and of the current in the mpp. The mpp of a solar cell changes with changing illumination. This has practical consequences such as mpp tracking. For worldwide comparison, solar cells and PV modules have to be characterized at STC (power at AM1.5 with $1000\,W/m^2$ and temperature of the solar cell at $25°C$).

The behavior of solar cells can be analyzed with equivalent circuits. An ideal solar cell contains only a photocurrent generator and a diode. The diode is necessary for internal charge separation, i.e. for the generation of a photovoltage. Ideal solar cells can be completely described by two fundamental parameters, the I_{SC} and the I_0, i.e. all properties of materials and combinations of materials used in a solar cell are confined in I_{SC} and I_0 (Figure 1.22).

A major issue for materials concepts in photovoltaics is to determine dependencies of I_{SC} and I_0 on limiting parameters of materials and materials combinations. The V_{OC} is a derived parameter and increases with increasing I_{SC} and/or decreasing I_0. The FF has to be obtained from analysis of the I–V and power–voltage characteristics. The solar energy conversion efficiency of ideal solar cells increases with increasing I_{SC}, i.e. under concentrated sunlight (see also Figure 1.23).

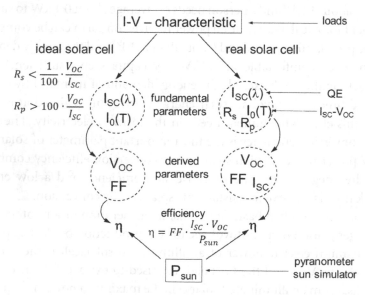

Figure 1.22. Summary of the basic characteristics of ideal and real solar cells.

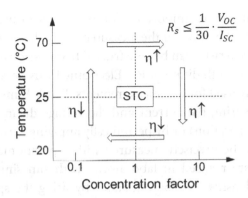

Figure 1.23. General dependence of the solar energy conversion efficiency at increasing and decreasing temperature and concentration factor of illumination.

Losses caused by resistive heating at R_s and R_p have to be considered for real solar cells under operation. An $R_{s,tol}$ of solar cells with high energy conversion efficiency should be equal or less than the quotient of V_{OC} and I_{SC} of the respective ideal solar cell divided by 100. An $R_{p,tol}$ of solar cells with high solar energy conversion efficiency should be equal or larger than the quotient of V_{OC} and I_{SC} of the respective ideal solar cell multiplied by 100. These $R_{s,tol}$ and $R_{p,tol}$ help to evaluate the potential of given materials and combinations of materials for reaching very high solar energy conversion efficiencies at certain operation conditions of solar cells and PV modules. For example, $R_{s,tol}$ are important economic factors for optimization of the amount of metals such as silver in contact grids on c-Si solar cells (see Chapter 7) and connecting cables in PV power plants.

At concentrated sunlight, the R_s limits FF and I_{SC}, whereas the R_p limits FF and V_{OC} at low light intensities. The solar energy conversion efficiency starts to reduce with increasing I_{SC} when R_s becomes larger than about a third of $R_{s,tol}$ (see also Figure 1.23). The minimum FF is 0.25. As examples, very low R_s can be realized in GaAs solar cells (see Chapter 8) and very high R_p are achieved in dye-sensitized solar cells (see Chapter 10).

The I_0 and the R_p of a solar cell can be easily obtained from I_{SC}–V_{OC} measurements. The activation energy of I_0 can be derived from the dependence of the logarithm of I_0 on the reciprocal temperature. The thermal activation of the I_0 has the consequence that V_{OC} decreases with increasing temperature, i.e. the solar energy conversion efficiency of solar cells decreases with increasing temperature (see also Figure 1.23).

Load resistances are important for reliable measurements of I–V characteristics and therefore for the measurement of the FF and R_s of a solar cell. An I–V characteristic can be constructed from various working points measured precisely with different R_L. Electronic loads or constant current sources are widely used for I–V measurements. Fast I–V measurements are possible by measuring the current and the voltage during charging of a capacitor (flash operation) or by periodically applying variable R_L.

The P_{sun} can be precisely measured with a pyranomenter. Standard test conditions are realized in laboratories with sun simulators. Short-circuit current densities can be calculated by using the spectrum of the photon flux from the sun and the spectrum of the quantum efficiency of a solar cell. For the same solar cell, the I_{SC} density measured with a sun simulator should be equal to the I_{SC} density calculated from the quantum efficiency.

1.7 Tasks

T1.1: Power installed and energy produced

Ascertain the maximum energy produced by a PV power plant of 1 MW_p installed in the south of Spain compared with a PV power plant installed in Germany during the first year of operation.

T1.2: Degradation of PV modules and sustainability

Average degradation rates of PV modules (k) of about 0.7%/year (c-Si) and 1.3%/year (thin-film) (Jordan and Kurtz, 2013) have been obtained for a large number of PV power plants.

(a) Plot the dependence of the relative decrease of the solar energy conversion efficiency and of the accumulated energy produced over a time range of up to 250 years for k = 0.5, 1.0, 2.0 and 5.0%/year.

(b) Calculate the remaining power of a 1 kW_p PV system and the maximum total energy produced with a PV power plant with 1 MW_p installed after 100 years of operation in Germany for a degradation rate of 0.5 and of 1.5%. Discuss the role of the energy payback factor (assume an energy payback time of one year).

T1.3: Photon energy in the maximum of blackbody radiation at T_S

Calculate the photon energy at which the blackbody radiation at 5800 K has a maximum.

T1.4: Wavelength of light in the maximum of blackbody radiation at T_S

Calculate the wavelength of light at which the blackbody radiation at 5800 K has a maximum.

T1.5: Maximum temperature of a solar cell under operation

(a) Calculate the maximum temperature of solar cells with solar energy conversion efficiencies of 10, 20 and 40% under illumination at $1\,kW/m^2$ on earth when convection cooling is absent compared to solar cells on a satellite.

(b) Plot the temperature of a solar cell under illumination at $1\,kW/m^2$ as a function of the solar energy conversion efficiency for a complete rooftop integration, for a free-standing PV module without finned backside and for free-standing PV modules with finned backsides increasing the surface area by four or eight times in comparison to the illuminated-by-sunlight surface area. Discuss the importance of radiative and convection cooling. An ambient temperature of 27°C is assumed.

T1.6: Air mass

Ascertain the dependence of the air mass on the angle of incidence of sunlight and discuss the validity of the expression for large angles.

T1.7: Tolerable series and parallel resistances

Estimate the $R_{s,tol}$ and $R_{p,tol}$ for a solar cell with an I_0 of $10^{-19}\,A/cm^2$ operated at a I_{SC} density of $25\,mA/cm^2$ (AM1.5) and operated at concentrated sunlight with a concentration factor of 400.

T1.8: Current–voltage characteristics

Ascertain the V_{OC} and the FFs for solar cells with an I_0 of $10^{-18}\,A/cm^2$, I_{SC} densities of 10 or $100\,mA/cm^2$ and R_p of 1 or $100\,k\Omega cm^2$.

2

Photocurrent Generation
and the Origin of Photovoltage

The most important fundamental property of a photovoltaic absorber is its forbidden band gap. The band gap limits the maximum photocurrent and it is the prerequisite for realizing a photovoltage, i.e. a difference between potential energies at which electrons may be extracted from a solar cell and at which electrons may be transferred back after passing an external load. The ultimate efficiency is analyzed as a function of the band gap. Photocurrents are reduced by reflection losses and transmission losses, which are determined by the reflectivity and by the absorption coefficient of a photovoltaic absorber. The minimization of optical losses by antireflection coatings and light trapping is discussed. The Boltzmann equations are obtained for calculating the Fermi-energies of free electrons and holes from their densities and from the effective densities of states at the valence and conduction band edges. Doping of semiconductors is explained. The condition for thermal equilibrium is described and the density of intrinsic charge carriers is introduced. The change of the Fermi-energy of minority charge carriers under illumination is a fundamental condition for a photovoltage.

2.1 Energy Gap of Photovoltaic Absorbers

2.1.1 *Light absorption and band gap of photovoltaic absorbers*

A photon disappears during an absorption event. The photon energy is transferred into the excitation energy of an electron, i.e. a photon excites an electron from the ground state (not excited) into a higher energetic state (excited).

Electrons in the excited state should be mobile for providing electric current in photovoltaic solar energy conversion. The process of photo-excitation of electrons from the ground state into the excited state (where electrons are mobile) is called the photo-generation of free electrons.

During photovoltaic solar energy conversion, photo-generated electrons are extracted from the first contact of a solar cell, pass an external load and are transferred back into the solar cell at the second contact. For the generation of electric power it must be possible to extract photo-generated electrons at a potential energy which is higher than the potential energy at which the electrons are transferred back after passing the external load. Therefore, a photovoltaic absorber suitable for solar energy conversion should be able to increase the potential energy of photo-generated electrons.

The potential energy of photo-generated electrons can only be increased if the energy of excited electrons is separated from the energy of electrons in the ground state by an energy gap. For purposes of comparison, the potential energy of water is increased by evaporation and formation of clouds. Due to raining in the mountains, water can be collected at increased potential energy. The potential energy of falling water can be converted into electricity with a turbine. The energy gap of the falling water driving the turbine is proportional to the height from which the water falls. The analogy between the energy gaps of hydro-mechanical and photovoltaic solar energy conversion is depicted in Figure 2.1.

The energy gap between electrons in ground states and free electrons in the excited states gives the upper limit for the potential between the external leads of a solar cell, i.e. the upper limit of the open-circuit voltage (V_{OC}).

Figure 2.1. Sketch displaying the analogy between hydro-mechanical and photovoltaic solar energy conversion.

The energy gap between electrons in the ground state and photo-generated electrons in the excited state is called the band gap (E_g).

A photovoltaic absorber is a material in which the energy of absorbed photons is transferred into energy of photo-generated free electrons with increased potential energy. The most important fundamental property of any photovoltaic absorber is its E_g separating energetically photo-generated electrons from electrons in the ground state.

The energetic band of all electrons in the ground state is called the valence band since it is formed by valence electrons participating in chemical bonds. The band in which excited mobile electrons can exist is called the conduction band. Electrons in the conduction band are free, i.e. they do not participate in chemical bonds but contribute to the electric conductivity.

In the dark most of the energetic states in the valence band are occupied and most of the energetic states in the conduction band are unoccupied. Under illumination, electrons (e^-) are excited from the valence band into the conduction band leaving a positive charge in the valence band. The missing electron in the valence band is called the hole (h^+). Holes are treated like mobile positive charge carriers since missing valence electrons are not fixed in chemical bonds.

Energetic bands are depicted as a function of a coordinate in so-called band diagrams. Figure 2.2 shows the band diagram of a photovoltaic absorber at photo-excitation.

A material with a conduction band separated from a valence band by a forbidden E_g is also known as a semiconductor. The E_g of a semiconductor is defined as the difference between the energies at the conduction band

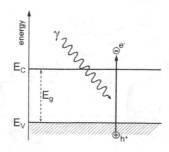

Figure 2.2.   Photo-generation of an electron-hole pair in a semiconductor. E_V, E_c and E_g denote the balance band edge, the conduction band edge and the band gap, respectively.

edge (E_C) and the valence band edge (E_V).

$$E_g = E_C - E_V \qquad (2.1)$$

The potential energy depends on the sign of electric charge. Mobile electrons have the lowest energy at the E_C. Mobile holes carry a positive charge and therefore have the lowest energy at the highest energy in the valence band, i.e. E_V.

During an absorption event, a photon with an energy equal or above E_g creates an electron-hole pair by kicking out an electron from a chemical bond state in the valence band into the conduction band. Photons with energy below E_g are not absorbed by the semiconductor.

Semiconductors can be realized with materials based on inorganic or organic matter. The most prominent inorganic semiconductors are crystalline silicon (c-Si) and gallium arsenide (GaAs) with E_g at room temperature of 1.1 and 1.42 eV, respectively. The conduction and valence bands of inorganic semiconductors are semi-infinite in relation to photon energies in the sun spectrum. The value of E_g of inorganic semiconductors decreases with increasing temperature. The temperature coefficient of E_g is of the order of half a meV per Kelvin for most crystalline semiconductors (see, for example, Chapter 8 for III–V semiconductors). The decrease of E_g with increasing temperature influences the performance of solar cells.

In organic semiconductors, the valence and conduction bands are formed by the highest occupied molecular orbital (HOMO) and lowest unoccupied molecular orbital (LUMO) bands, respectively (see Chapter 10 for details).

2.1.2 *Photon flux and maximum photocurrent of solar cells*

The part of the photon flux from the sun (Φ_{sun}) which is absorbed in the photovoltaic absorber is converted into the photocurrent of the solar cell. The Φ_{sun} provides a flow of photons with different energies ($h\upsilon$). In order to ascertain the photon flux at a certain photon energy, the power of the sun at this photon energy has to be divided by the value of the photon energy. The photon flux is given in the unit of $1/(cm^2 \cdot eV \cdot s)$.

Figure 2.3 shows the photon flux spectrum of the sun at air mass (AM) 1.5d. The maximum photon flux of $4 \cdot 10^{17}$ $1/(cm^2 \cdot eV \cdot s)$ is reached at a photon energy of about 0.9 eV.

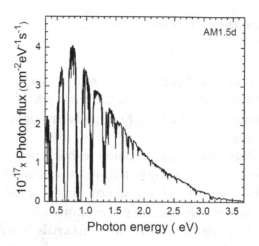

Figure 2.3.   Photon flux spectrum of the sun at AM1.5d.

Photo-generated electrons carry the photocurrent where one electron transports one elementary charge. In order to ascertain the photocurrent, the photon flux has to be integrated over all photons participating in absorption events and multiplied by the elementary charge. Photons with energy below the E_g of the photovoltaic absorber do not participate in absorption events. Therefore, the fundamental limitation of the photocurrent is given by the E_g of the photovoltaic absorber defining the lower integration boundary of the photon flux. The maximum photocurrent (I_{SC}^{max}) can be obtained if all photons with energy equal or above E_g are absorbed and if each absorbed photon generates one electron-hole pair, i.e. if the quantum efficiency is equal to 1 for photons with energy equal or larger than E_g.

$$I_{SC}^{max} = q \cdot \int_{E_g}^{\infty} \Phi_{sun}(h\nu) \cdot d(h\nu) \qquad (2.2)$$

The order of magnitude of I_{SC}^{max} can be estimated if multiplying an average photon flux of $1 \cdot 10^{17}$ 1/(cm^2·eV·s) with an integration range of 2 eV and with $q = 1.6 \cdot 10^{-19}$ As. Therefore I_{SC}^{max} is of the order of tens of mA/cm^2.

The I_{SC}^{max} at AM1.5d is plotted as a function of the E_g of the photovoltaic absorber in Figure 2.4. These values of I_{SC}^{max} have to be multiplied by a factor

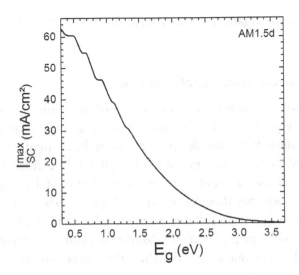

Figure 2.4. Dependence of the I_{SC}^{max} at AM1.5d on the E_g of the photovoltaic absorber.

of 1.1 to determine I_{SC}^{max} at AM1.5G. The I_{SC}^{max} decreases with increasing E_g from about 60 mA/cm^2 at $E_g = 0.5$ eV to about 1 mA/cm^2 at $E_g = 3.1$ eV. Therefore it is not useful to apply semiconductors with E_g larger than 2.8–3.0 eV for solar cells.

It is interesting to compare the short-circuit current (I_{SC}) densities of record solar cells with I_{SC}^{max}. The I_{SC} densities are 42.7, 29.47 and 35.4 mA/cm^2 at AM1.5G for record c-Si, GaAs and Cu(In, Ga)Se$_2$ solar cells (Green *et al.*, 2013). The values of I_{SC}^{max} related to the corresponding E_g of 1.1, 1.42 and 1.2 eV amount to about 44, 31 and 39 mA/cm^2, respectively. Therefore, I_{SC} of c-Si and GaAs world record solar cells is almost very close to I_{SC}^{max}, which is an important fact for the theoretical analysis of limitations in different types of solar cells.

The I_{SC} of a given solar cell increases with increasing temperature due to the decrease of the E_g of a photovoltaic absorber. The increase of the I_{SC} density can be between 1 and 2 mA/cm^2 for a strong increase of the temperature of the order of 50–70 K, depending on the absorber material (see task T2.2 at the end of this chapter).

An upper integration boundary ($E_g + \Delta E_g$) has to be considered for the calculation of I_{SC}^{max} for organic semiconductors due to the limited widths of the HOMO and LUMO bands (see Chapter 10). Therefore the values of

I_{SC}^{max} are lower for organic than for inorganic photovoltaic absorbers at the same E_g.

2.1.3 *The ultimate efficiency of solar cells*

The maximum I_{SC} density can be calculated as a function of E_g with reference to Equation (2.2). The maximum V_{OC} of a solar cell cannot exceed the value of the E_g divided by the value of the elementary charge. The upper limit of the fill factor (FF) of a solar cell is 1. The ultimate efficiency ($\eta_{ultimate}$) takes into account only the upper limits of I_{SC}, V_{OC} and FF, irrespective of whether these values are realistic for existing materials and combinations of materials suitable for photovoltaic absorbers. The $\eta_{ultimate}$ of a solar cell with one photovoltaic absorber depends only on the E_g of the absorber and can be calculated with the following equation:

$$\eta_{ultimate} = \frac{E_g \cdot \int_{E_g}^{\infty} \Phi_{sun}(h\nu) \cdot d(h\nu)}{P_{sun}} \qquad (2.3)$$

The dependence of $\eta_{ultimate}$ on E_g is presented in Figure 2.5. The $\eta_{ultimate}$ is about 48% at E_g of 0.9 and 1.1 eV. The values of $\eta_{ultimate}$ are

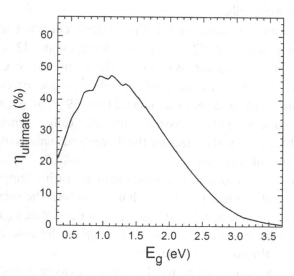

Figure 2.5.   Dependence of the ultimate efficiency of a solar cell at AM1.5d on the E_g of the photovoltaic absorber.

larger than 40% for E_g between 0.7 and about 1.6 eV. The value of $\eta_{ultimate}$ is lower than 10% for E_g larger than about 2.6 eV.

2.2 Photocurrent Limitation by Reflectivity and Transmission of Light

A part of the photon flux from the sun is lost in solar cells due to reflection and transmission of light (optical losses). The reflected and transmitted parts of the photon flux are denoted by Φ_{sun}^{refl} and Φ_{sun}^{trans}, respectively.

$$\Phi_{sun}^{abs}(h\nu) = \Phi_{sun}(h\nu) - \Phi_{sun}^{refl}(h\nu) - \Phi_{sun}^{trans}(h\nu) \tag{2.4}$$

Only the part of the photon flux which is absorbed in the photovoltaic absorber (Φ_{sun}^{abs}) can be used for photovoltaic solar energy conversion.

2.2.1 *Reflectivity and antireflection coatings*

The part of light being reflected is described by the reflection coefficient (R).

$$\Phi_{sun}^{refl}(h\nu) = R \cdot \Phi_{sun}(h\nu) \tag{2.5}$$

The reflection losses are considered for the calculation of the I_{SC} density in the following equation:

$$I_{SC} = q \cdot \int_{E_g}^{\infty} \Phi_{sun}(h\nu) \cdot [1 - R(h\nu)] \cdot d(h\nu) \tag{2.6}$$

The reflection coefficient can be calculated if the refractive indices of the absorber (n_s) and of the surrounding ambience ($n_{air} \approx 1$) are known. For normal incidence of incoming light, the reflection coefficient is (see also textbooks in optics, for example (Grimsehl, 1988):

$$R(\lambda) = \left(\frac{n_{air} - n_s(\lambda)}{n_{air} + n_s(\lambda)} \right)^2 \tag{2.7}$$

The reflection coefficient is a function of the wavelength of incoming light. For silicon and most of the crystalline semiconductors, n_s ranges between 3.5 (longer wavelengths) and 5 (shorter wavelengths). As a consequence, the reflection coefficient of crystalline semiconductors is between 0.3 and 0.4.

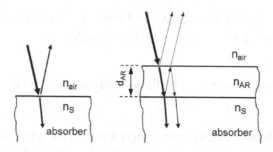

Figure 2.6. Principle of reflection and antireflection coating.

The high reflection coefficient can be reduced by using transparent antireflection coatings (Figure 2.6). An antireflection coating is a surface layer with a thickness d_{AR} and a refractive index n_{AR} the value of which is between n_s and n_{air} (see, for example, Fa. Carl Zeiss (1939)). Reflection can be suppressed by destructive interference in the antireflection coating layer. The condition of destructive interference is:

$$\frac{\lambda}{4} = n_{AR} \cdot d_{AR} \tag{2.8}$$

A minimum reflection coefficient can be obtained under condition (2.8):

$$R_{min} = \left(\frac{n_{AR}^2 - n_{air} \cdot n_s}{n_{AR}^2 + n_{air} \cdot n_s} \right)^2 \tag{2.9}$$

The reflection coefficient can be reduced to 0 for the following condition:

$$R_{min} = 0 : \quad n_{AR} = \sqrt{n_{air} \cdot n_s} \tag{2.10}$$

The refractive index of air is practically equal to 1. Many photovoltaic absorbers have a refractive index of the order of 3–5. Therefore the refractive index is about 1.7–2.2 for antireflection coatings.

The reflection coefficient can be minimized to a value close to 0 with regard to Equations (2.8)–(2.10) only for a certain wavelength. This wavelength is chosen in a range where the photocurrent generation has a maximum. The thickness of antireflection coatings is of the order of 100 nm. The reflection coefficient can be minimized at two different wavelengths by using an antireflection double layer consisting of two transparent materials with different refractive indices while the top layer has the lower refractive index.

The precise calculation of the reflection coefficient demands simulations for multi-layer systems. The effective reflection of a solar cell can be reduced to 7 ... 10 % with a single antireflection layer and even to 3 ... 4 % with a double antireflection layer. Materials for antireflection coatings are titania (TiO_2), silicon nitride (Si_3N_4), magnesium fluoride (MgF_2), silicon dioxide (SiO_2) and others.

2.2.2 *Absorption coefficient and transmission losses*

The part of the photon flux that has not been reflected decreases depending on the length of the optical path of light, which has been passed through the photovoltaic absorber. For normal incidence and planar geometry of the absorber, the length of the optical path is equal to the thickness (x) passed by the light, which will be considered in the following.

A photon flux (Φ_{ph}) reduces after passing a certain distance (x) through a material due to absorption (Figure 2.7). The change of the photon flux ($d\Phi_{ph}(x)$) absorbed in a very thin layer with a thickness dx at a depth x is proportional to the photon flux at the depth x ($\Phi_{ph}(x)$). The proportionality factor between the photon flux absorbed in the layer with thickness dx and the photon flux existing at the distance x is called the absorption coefficient (α).

The change of the photon flux at x can be expressed by the following simple differential equation:

$$\frac{d\Phi_{ph}(x)}{dx} = -\alpha \cdot \Phi_{ph}(x) \qquad (2.11)$$

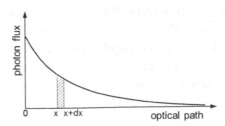

Figure 2.7. Reduction of the photon flux with increasing optical path.

The solution of Equation (2.11) is the absorption or Lambert–Beer law.

$$\Phi_{ph}(x) = \Phi_{ph}(0) \cdot exp[-\alpha(\lambda) \cdot x] \quad (2.12)$$

The photon flux at the surface ($\Phi_{ph}(0)$) is equal to the part of the photon flux that has not been reflected.

$$\Phi_{ph}(0) = \Phi_{sun} - \Phi_{sun}^{refl} \quad (2.13)$$

The product of the absorption coefficient and of the finite thickness of an absorber layer (d_{abs}) determines the part of the photon flux passing through the absorber and therefore the transmission losses. The part of the photon flux absorbed in an absorber with a thickness d_{abs} is given by the following equation if it is assumed that the optical path is equal to d_{abs}:

$$\Phi_{sun}^{abs} = \Phi_{ph}(0) \cdot [1 - e^{-\alpha(\lambda) \cdot d_{abs}}] \quad (2.14)$$

The absorption coefficient has the unit of cm^{-1}. The inverse absorption coefficient is called the absorption length (α^{-1}).

The absorption coefficient is a function of λ of the incoming light and can vary over many orders of magnitude depending on the absorber material. The absorption coefficient is equal to 0 for transparent materials and can be as large as 10^6 cm^{-1} for strongly absorbing materials.

Absorption spectra are shown as a function of photon energy for some conventional semiconductors in Figure 2.8. Absorption occurs at photon energies above E_g. For c-Si the absorption length decreases from about $100\,\mu m$ at 1.2 eV to about $10\,\mu m$ at 1.5 eV and $1\,\mu m$ at 2.2 eV. In contrast, the absorption length of CuInSe$_2$ is less than $1\,\mu m$ over nearly the complete spectral range above the E_g of CuInSe$_2$.

The behavior of Equation (2.14) is expressed in Figure 2.9 as a function of $\alpha \cdot d_{abs}$. It can be seen that more than 95% of the incoming light is absorbed if the product of the absorption coefficient and of the thickness of the absorber layer is larger than 3. This is an important design rule for solar cells. The thickness of an absorber layer should be at least three times larger than the absorption length of the absorber material.

$$d_{abs} \geq 3 \cdot \alpha^{-1} \quad (2.15)$$

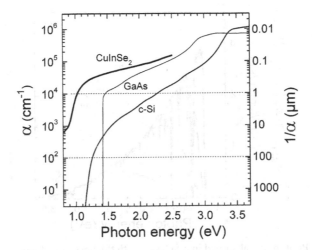

Figure 2.8. Absorption spectra of c-Si, GaAs and CuInSe$_2$.

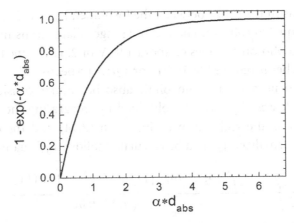

Figure 2.9. Amount of absorbed photons as a function of the product of the absorption coefficient and the absorber thickness.

At a fixed thickness of a photovoltaic absorber, the transmission losses depend on the photon energy with respect to the absorption spectrum of a given absorber (Figure 2.9).

The dependence of the portion of the photon flux absorbed in a c-Si absorber with different d_{abs} is shown in Figure 2.10. Only a very small fraction of photons with energy very close to the forbidden E_g of c-Si is lost for a thickness of the c-Si absorber of 1 cm. Already a rather significant part

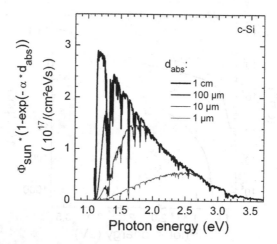

Figure 2.10. Photon flux absorbed in c-Si layers with thicknesses of 1 cm, 100 μm, 10 μm and 1 μm (increasing thickness of solid lines). The reflectivity is set to 0.

of the photons in the range between the forbidden E_g of c-Si and 1.3 eV is lost for d_{abs} equal to 100 μm. The spectral range of large transmission losses is extended to photon energies of about 1.9 eV or 2.6 eV if the thickness of the c-Si absorber is reduced to 10 μm or 1 μm, respectively.

The I_{SC} density of a photovoltaic absorber under consideration of transmission losses (I_{SC}^{abs}) can be obtained by integrating the product of the photon flux, the right term of Equation (2.14) and the elementary charge. The normalized I_{SC} can be calculated following Equation (2.16).

$$\frac{I_{SC}^{abs}}{I_{SC}^{max}} = \frac{\int_{E_g}^{\infty} \Phi_{sun}(h\nu) \cdot [1 - e^{-\alpha(h\nu) \cdot d_{abs}}] \cdot d(h\nu)}{\int_{E_g}^{\infty} \Phi_{sun}(h\nu) \cdot d(h\nu)} \qquad (2.16)$$

The dependence of I_{SC}^{abs} normalized to I_{SC}^{max} on d_{abs} is plotted in Figure 2.11 for c-Si, GaAs and CuInSe$_2$ absorbers. For a c-Si absorber, a thickness of the absorber layer larger than 200 μm is required for obtaining at least 90% of I_{SC}^{max}. This is the reason why c-Si solar cells are based on wafers, i.e. on slices of c-Si crystals with a thickness of about 200 μm or larger (see Chapter 7). More than 95% of I_{SC}^{max} are reached for GaAs absorbers with a thickness of 2 μm (see Chapter 8), whereas only a 1 μm thick CuInSe$_2$ absorber is needed to obtain 95% of I_{SC}^{max}. An absorber with the thickness of the order of 1 μm cannot be mechanically self-supporting.

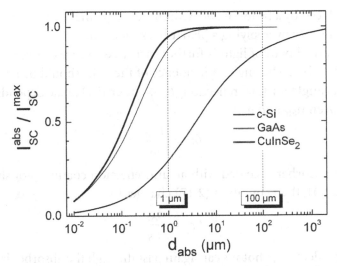

Figure 2.11. Normalized I_{SC} of c-Si, GaAs and CuInSe$_2$ absorbers as a function of the absorber thickness.

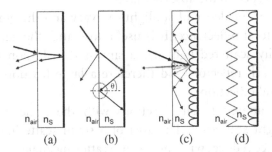

Figure 2.12. Increase of the optical path by a planar back reflector (a), total reflection (b), a scattering back reflector (c) and textured surface with scattering back reflector (d).

Photovoltaic absorbers like CuInSe$_2$ are well suited for thin-film solar cells which make use of a supporting foreign substrate (see Chapter 9).

2.2.3 *Increase of the optical path by light trapping*

Numerous measures employed in the architecture of solar cells aim to increase the optical path of light through the absorber layer. For example, a mirror at the back side of an absorber layer increases the optical path by a factor of two (Figure 2.12(a)). Therefore the thickness of an absorber layer can be reduced by a factor of two without reducing the amount of absorbed photons, merely by applying a planar back reflector.

The optical path of light is further increased in case of total reflection (Figure 2.12(b)). The angle of incidence of the light should be larger than the critical angle for total reflection (θ_C). The critical angle depends on the ratio between n_{air} and n_S.

$$\theta_C = \frac{n_{air}}{n_S} \tag{2.17}$$

If the absorber is coated with an antireflection coating (not shown in Figure (2.12)), then condition (2.17) changes to:

$$\theta_C = \frac{n_{AR}}{n_S} \tag{2.17'}$$

After reflection, photons can again pass through the absorber layer, can be reflected again at the mirror on the back side and can pass through the absorber layer a fourth time, when total reflection is achieved, and so on — photons are trapped in the absorber layer.

The angle of incidence of sunlight can vary and therefore so can the condition for total reflectance. It is useful to scatter the light at the back reflector. Ideally scattered light has a similar distribution for all angles (Lambertian back reflector) and therefore a large fraction of directions suitable for total reflection.

Light is scattered to all directions with the same probability for Lambertian light scattering. The area of a small scattering surface seen at a long distance reduces with increasing scattering angle (Figure 2.13(a)). The scattered photon flux of a small scattering surface is proportional to the area seen at a long distance from the scattering surface. For this reason the angular distribution of the flux of scattered photons is proportional to the cosine of the scattering angle (Figure 2.13(b)).

Only the light scattered into a cone with an opening angle (θ_C) can escape the absorber, i.e. this fraction of light is lost for photovoltaic solar energy conversion. For a Lambertian back reflector in a solar cell with an antireflection coating, the optical path can be increased by a factor of X_{path} (see also task T2.3 at the end of this chapter).

$$X_{path} = 2 \cdot \left(\frac{n_S}{n_{AR}} \right)^2 \tag{2.18}$$

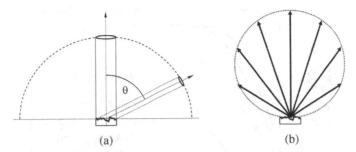

Figure 2.13. Lambertian light scattering: change of the area of a small scattering surface seen at a long distance with scattering angle θ (a) and corresponding distribution of the photon flux from a scattering surface presented as the length of arrows (b).

Equation (2.18) describes the so-called Lambertian limit of light trapping which is the limit for light trapping in ray optics. The Lambertian limit of light trapping is of the order of 8–18 for most solar cells since the ratio between the refractive indices of the absorber and of the antireflection coating is of the order of 2–3.

The normalized I_{SC}^{abs} can be calculated by adding the factor X_{path} into Equation (2.16).

$$\frac{I_{SC}^{abs}}{I_{SC}^{max}} = \frac{\int_{E_g}^{\infty} \Phi_{sun}(h\nu) \cdot [1 - e^{-\alpha(h\nu) \cdot X_{path} \cdot d_{abs}}] \cdot d(h\nu)}{\int_{E_g}^{\infty} \Phi_{sun}(h\nu) \cdot d(h\nu)} \qquad (2.19)$$

Figure 2.14 shows the dependence of the normalized I_{SC} of c-Si, GaAs and CuInSe$_2$ absorbers on d_{abs} for an optical path increased by a factor of 10. Absorber thicknesses equal to or larger than 50 μm, 200 nm and 100 nm are needed for obtaining I_{SC} densities of at least 95% of I_{SC}^{max} for c-Si, GaAs and CuInSe$_2$ absorbers, respectively. Intensive research is ongoing to develop methods to increase the optical path by values above the Lambertian limit. For example, c-Si absorbers can be reduced to 5 μm and less if an increase of the optical path by a factor of 100 can be realized for c-Si.

There are more geometries suitable for the increase of the optical path in solar cells, for example the use of textured absorber surfaces. The goal of the use of textured absorber surfaces is to increase the optical path of photons entering the absorber layer and to change the angle of incidence of incoming light in such a way that back reflected light undergoes total reflection.

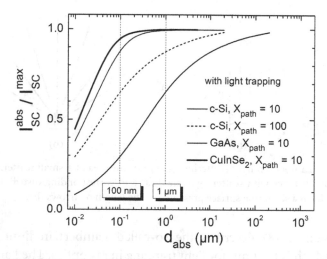

Figure 2.14. Normalized I_{SC} of c-Si, GaAs and CuInSe$_2$ absorbers as a function of the absorber thickness for an optical path increased by a factor of 10 (solid lines) and by a factor of 100 (dashed line).

Statistical ray optics (Yablonovitch, 1982) are useful for considering the surface morphology. The optical path can also be increased by implementing structures into solar cells with smaller dimensions than the wavelength of light. By related measures, the optical path can be increased above the Lambertion limit of light scattering (see Chapter 10).

The reduction of the optical path of light through absorbers is very important for several reasons. First, the amount of absorber material can be reduced, which leads to a reduction of weight and costs. Second, the increase of the optical path of light opens opportunities to increase the energy conversion efficiency of solar cells under advanced conditions, as will be shown in Chapter 4. Third, the reduction of the dimensions of photovoltaic absorbers to the range of 100 nm allows the possibility of incorporating solar cells into very thin flexible structures and membranes suitable, for example, for lightweight applications or for applications in photocatalytic water-splitting systems. Fourth, light trapping is very useful for broadening the classes of suitable photovoltaic absorbers to materials with relatively low lifetimes (for definitions, see Chapter 3). In sum, therefore, the evident benefits of light trapping provide a clear impetus for the further development of solar cells.

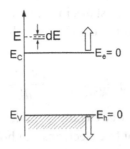

Figure 2.15. Energies of free electrons and holes in the band diagram of a semiconductor.

2.3 Free Charge Carriers in Ideal Semiconductors

2.3.1 *Densities of free charge carriers and Fermi-energy*

For the analysis of solar cells it is necessary to describe the density of free charge carriers (see, for example Seeger (2001)). In an ideal semiconductor, free electrons in the conduction band and free holes in the valence band are treated like atoms or molecules in an ideal gas while the free charge carriers are delocalized over the whole considered volume and follow the Pauli principle.

The density of free charge carriers in an interval of energy (dE, see Figure 2.15) around a considered energy value (E) is the product of dE, the density of states and the occupation probability at E. The integral of this product over energy gives the density of the free charge carriers in the respective band.

The density of free electrons and the density of free holes are denoted by n and p, respectively. The densities of states in the conduction and valence bands are denoted by $D_e(E)$ and $D_h(E)$, respectively. The occupation probability of an electron is denoted by f. The density of free electrons and the density of holes can be expressed by the following equations:

$$n = \int_{E_C}^{\infty} D_e(E) \cdot f(E) \cdot d(E) \qquad (2.20')$$

$$p = \int_{\infty}^{E_V} D_h(E) \cdot [1 - f(E)] \cdot d(E) \qquad (2.20'')$$

The density of states at a certain energy is defined as the number of states in an energy interval around this energy ($dN_{e,h}(E)$ where the indices e and h denote the number of states in the conduction and valence bands, respectively) normalized to the considered volume (V) and to the energy interval.

$$D_{e,h}(E) = \frac{1}{V} \cdot \frac{dN_{e,h}(E)}{dE} \qquad (2.21)$$

The kinetic energies of free electrons and holes are 0 at the E_C and E_V, respectively. The energies of free electrons and holes in the conduction and valence bands are considered as kinetic energies.

$$E_{kin,e} = E - E_C \qquad (2.22')$$

$$E_{kin,h} = E_V - E \qquad (2.22'')$$

The kinetic energy of a free charge carrier ($E_{kin,e,h}$) is given by its squared momentum ($k_{e,h}^2$) divided by the doubled effective mass of free electrons or holes mass ($2 \cdot m_{e,h}^*$).

$$E_{kin,e,h}(k) = \frac{k_{e,h}^2}{2 \cdot m_{e,h}^*} \qquad (2.23)$$

With regard to delocalization of free charge carriers and to the Heisenberg uncertainty principle, one state occupies the following volume in momentum space:

$$(\Delta k_{e,h})^3 = \frac{h^3}{V} \qquad (2.24)$$

The number of states at a momentum equal to or lower than $k_{e,h}$ ($N(k_{e,h})$) is the volume of a sphere with radius $k_{e,h}$ divided by $(\Delta k_{e,h})^3$ and multiplied by two due to spin degeneracy (spin up and spin down).

$$N(k_{e,h}) = \frac{4\pi}{3} \cdot |k_{e,h}|^3 \cdot \frac{2}{(\Delta k_{e,h})^3} \qquad (2.25)$$

Further, the momentum is substituted by the kinetic energy using Equation (2.23) so that the number of states becomes a function of energy and $\Delta k_{e,h}$ is substituted by transforming Equation (2.25). The kinetic energies are expressed by Equation (2.22). Following Equation (2.21), the number of states is divided by the volume and differentiated with respect to the energy. As a result, the densities of states in the conduction and valence bands are obtained:

$$D_e(E) = 4\pi \cdot \left(\frac{2 \cdot m_e^*}{h^2}\right)^{3/2} \cdot (E - E_C)^{1/2} \qquad (2.26')$$

$$D_h(E) = 4\pi \cdot \left(\frac{2 \cdot m_h^*}{h^2}\right)^{3/2} \cdot (E_V - E)^{1/2} \qquad (2.26'')$$

Equations (2.26′) and (2.26″) were obtained from first principle and contain only the effective masses of free electrons and holes as material parameters.

The effective masses of free electrons and holes do not necessarily correspond to the electron mass in free space (m_e) due to interactions of mobile charge carriers and atoms in the semiconductor. Effective masses can vary over a relatively wide range depending on the given semiconductor. For example, for c-Si m_e^* and m_h^* are $1.08 \cdot m_e$ and $0.5 \cdot m_e$, respectively (see, for example, Sze (1981)).

Equations (2.26′) and (2.26″) can be written in the following manner:

$$D_e(E) = \frac{6.7 \cdot 10^{21}}{cm^3 \cdot eV} \cdot \left(\frac{m_e^*}{m_e}\right)^{3/2} \cdot (E - E_C)^{1/2} \qquad (2.27')$$

$$D_h(E) = \frac{6.7 \cdot 10^{21}}{cm^3 \cdot eV} \cdot \left(\frac{m_h^*}{m_e}\right)^{3/2} \cdot (E_V - E)^{1/2} \qquad (2.27'')$$

Equations (2.27′) and (2.27″) describe very well the densities of free electron and hole states close to the edges of the conduction and valence bands for parabolic band for which Equation (2.22) is valid.

The occupation probabilities of free electrons and holes are given by the Fermi–Dirac statistics (Equation (2.28)), which follow from the Pauli principle. The decisive parameter in the Fermi–Dirac statistics is the Fermi-energy (E_F) at which the occupation probability is 0.5.

$$f(E) = \frac{1}{exp\left(\frac{E - E_F}{k_B \cdot T}\right) + 1} \qquad (2.28)$$

Figure 2.16 shows the density of states, the electron occupation probability and the product of both of these for free electrons and holes in energetic diagrams. The energy scale is set to 0 at E_V. The electron occupation probability is close to 1 but not equal to 1 for the valence band

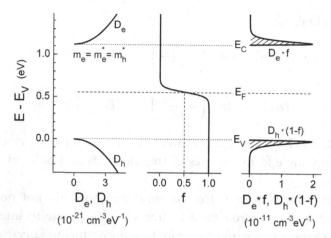

Figure 2.16. Densities of states of free electrons and holes, occupation probability of electrons and product of both for free electrons in an energy diagram for the case that the Fermi-energy is in the middle of the E_g.

and close to 0 but not equal to 0 for the conduction band. The shape of the electron density follows the exponential decrease of the occupation probability with increasing energy.

The effective densities of states at the conduction and valence band edges are obtained from the densities of states integrated from the E_C and E_V to the energies increased by $k_B \cdot T$.

$$N_{C,V} = 2 \cdot \left(2\pi \cdot k_B \cdot T \cdot \frac{m^*_{e,h}}{h^2} \right)^{3/2} \tag{2.29}$$

At room temperature, the effective densities of states at the conduction and valence band edges are about $N_C = 3 \cdot 10^{19}$ cm^{-3} and $N_V = 1 \cdot 10^{19}$ cm^{-3} for c-Si (Sze, 1981).

Taking into account the definition of N_C and N_V and the fact that the exponential in Equation (2.28) is much larger than 1, the densities of free electrons and holes are given by the so-called Boltzmann equations:

$$n = N_C \cdot exp\left(-\frac{E_C - E_{Fn}}{k_B \cdot T} \right) \tag{2.30'}$$

$$p = N_V \cdot exp\left(-\frac{E_{Fp} - E_V}{k_B \cdot T} \right) \tag{2.30''}$$

The Boltzmann statistics connect the densities of free electrons and holes with their Fermi-energies (E_{Fn} and E_{Fp}) which are equivalent to the activity or chemical potential of chemical species in chemistry. More details about free charge carriers in semiconductors can be found in textbooks about semiconductors and semiconductor devices (see, for example, Sze (1981)).

2.3.2 *Thermal equilibrium and density of intrinsic charge carriers*

Any material emits blackbody radiation depending on the temperature of the material. Blackbody radiation also contains photons with energy above E_g. Therefore free electrons and holes are generated in semiconductors by absorption of photons from blackbody radiation with photon energy above the E_g. This process is called thermal generation of free charge carriers. The thermal generation rate of free charge carriers is constant for a fixed temperature. Obviously, the densities of free electrons and holes cannot increase to infinite values. The process limiting the densities of free electrons and holes is called recombination, meaning the annihilation of free electrons and holes (see Chapter 3). Radiative recombination is defined as the annihilation of electron-hole pairs by emitting photons. The radiative recombination rate is proportional to the product of the densities of free electrons and holes since both carriers are involved in the radiative recombination process (see Chapter 3).

In thermal equilibrium, the thermal generation rate is equal to the radiative recombination rate. The densities of free electrons or holes in thermal equilibrium are denoted by n_0 or p_0, respectively. It follows from the definitions of the thermal equilibrium and of radiative recombination that the product of n_0 and p_0 is constant. The product of n_0 and p_0 can be calculated by using Equations (2.30′) and (2.30″).

The Fermi-energies of free electrons and holes can change. Therefore, the difference of the Fermi-energies of free electrons and holes must be 0 in thermal equilibrium since the product of n_0 and p_0 is constant. As a consequence, there is one common Fermi-energy of free electrons and holes in thermal equilibrium (E_{F_0}).

$$E_{F_0} = E_{Fn} = E_{Fp}: \quad n_0 \cdot p_0 = N_C \cdot N_V \cdot \exp\left(-\frac{E_g}{k_B \cdot T}\right) \qquad (2.31)$$

The densities of free electrons and holes are equal in a pure or so-called intrinsic semiconductor in thermal equilibrium. The density of intrinsic charge carriers (n_i) is defined as

$$n_i \equiv \sqrt{N_C \cdot N_V} \cdot \exp\left(-\frac{E_g}{2 \cdot k_B \cdot T}\right) \qquad (2.32)$$

The n_i is the minimum density of free carriers that can be reached in a semiconductor. The n_i increases exponentially with decreasing E_g of the semiconductor and with increasing temperature. Figure 2.17 shows the temperature dependence of n_i for GaAs, c-Si and Ge (germanium) under consideration of the temperature dependence of the band gaps.

At room temperature, the n_i of pure c-Si, GaAs and Ge are about 10^{10}, 10^6 and 10^{13} cm^{-3}, respectively. For comparison, the density of electrons in metals is of the order of 10^{22} cm^{-3}. Therefore, the conductivity of pure c-Si, GaAs and Ge is much lower than the conductivity of metals but much higher than the conductivity of insulators, which is the reason why pure c-Si, GaAs and Ge belong to the class of semiconductors.

Semiconductors with very large E_g such as diamond (5 eV) or quartz (9 eV) are insulators at room temperature due to the very low n_i. Insulators

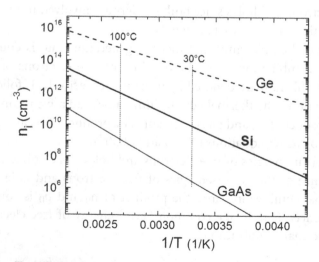

Figure 2.17. Temperature dependence of the n_i for c-Si, GaAs and Ge. The temperature dependence of E_g is considered.

become semiconducting at high temperatures due to the exponential increase of n_i with increasing temperature. *Vice versa*, all semiconductors are insulators at low temperatures.

2.3.3 *Doping, majority and minority equilibrium charge carriers*

Doping of materials is the purposeful replacement of host atoms by impurity atoms. The goal of doping semiconductors is a controlled change in conductivity. The density of free charge carriers in thermal equilibrium can be changed in semiconductors over many orders of magnitude by doping, which makes doping of semiconductors very important for electronic and photovoltaic applications.

For the doping of semiconductor crystals, host atoms are replaced by impurity atoms with a higher or with a lower valence than the valence of the host atoms (see also the example in Figure 2.18). Doping of semiconductors leads to extra electrons, which cannot be incorporated into chemical bonds if the valence of the dopant is higher than the valence of the host atom, for example, phosphorus in c-Si. The un-bonded extra electrons can be free, i.e. they can occupy states in the conduction band. On the other hand, if the valence of the dopant is lower than the valence of the host atom — for example, boron in c-Si — doping leads to electrons which are missing for the formation of chemical bonds with shared electron pairs. The missing

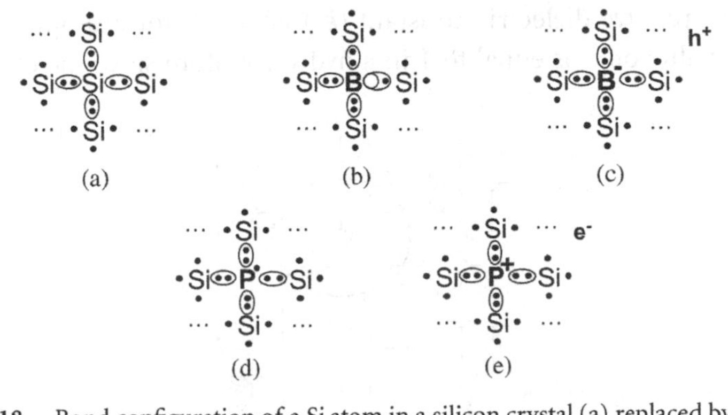

Figure 2.18. Bond configuration of a Si atom in a silicon crystal (a) replaced by boron (b, c) or phosphorus (d, e) atoms before (b, d) and after ionization (c, e).

electrons can become free holes occupying states in the valence band of the semiconductor.

Dopants causing an increase of the densities of free electrons or free holes are known as donors or acceptors, respectively. For example, Si and Ge belong to the fourth group of the periodic table of elements. Atoms of elements of the third or fifth group of the periodic table form acceptors (B,, Al, Ga, In) or donors (N, P, As, Sb) in Si or Ge crystals, respectively. Semiconductors doped with donors or acceptors are called n-type or p-type semiconductors, respectively.

Donor (D) or acceptor (A) atoms can be electrically neutral or ionized depending on whether extra or missing electrons fill states in the conduction or valence bands or not, respectively.

$$D \leftrightarrow D^+ + e^- \qquad\qquad (2.33')$$

$$A \leftrightarrow A^- + h^+ \qquad\qquad (2.33'')$$

Electrons or holes experience Coulomb attractive force with ionized donors or acceptors, similar to an electron experiencing the Coulomb attractive force with the proton in a hydrogen atom (see also Figure 2.19).

The electron in a hydrogen atom can become free if it becomes excited to an energy equal to or above the ionization energy of the hydrogen atom ($E_{ion}^H = 13.56\,eV$). With regard to the Bohr model of atoms (see, for example, Vogel (1995)), the ionization energy is proportional to the quotient of the electron mass and the squared dielectric constant. Coulomb potentials are strongly screened in inorganic semiconductors due to the quite large relative dielectric constant (ε_r) which is, for example, 11.8 for c-Si. The value of ε_r is equal to 1 in a hydrogen atom since the electron is

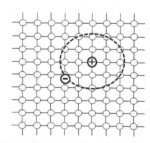

Figure 2.19. Hydrogen model of an ionized donor in a crystal.

surrounded only by vacuum. Therefore, the ionization energy of electrons or holes in doped inorganic semiconductors are more than 100 times less than the ionization energy of the hydrogen atom. The ionization energies of donors and acceptors ($E_{ion}^{D,A}$) can be estimated by taking into account $m_{e,h}^*$ (not necessarily identical with m_e), ε_r and E_{ion}^{H} (hydrogen model of donors and acceptors).

$$E_{ion}^{D,A} \approx E_{ion}^{H} \cdot \frac{m_{e,h}^*}{m_e} \cdot \frac{1}{\varepsilon_r^2} \qquad (2.34)$$

The ionization energies obtained from (2.34) are of the same order as the thermal energy at room temperature. Therefore, the thermal energy is sufficient to ionize practically all donors or acceptors in doped semiconductors at room temperature. As a consequence, the density of free electrons or holes is practically equal to the density of donor (N_D) or acceptor (N_A) atoms at room temperature. Related donors or acceptors are also called shallow donors or shallow acceptors. The so-called doping regime is relevant for solar cells. Charge neutrality gives:

$$N_D = n_0 = N_D^+ \qquad (2.35')$$

$$N_A = p_0 = N_A^- \qquad (2.35'')$$

At very low temperature the thermal energy is much lower than the ionization energy of the donors and acceptors. In this case the electrons and holes cannot escape from the region of the Coulomb potential of the ionized donors and acceptors, i.e. the electrons and holes are no longer free. This so-called freeze-out regime is not relevant for photovoltaic absorbers operated on earth or on satellites.

The density of intrinsic charge carriers increases exponentially with increasing temperature as mentioned before. At high temperatures the density of intrinsic charge carriers exceeds the density of donors or acceptors, i.e. all semiconductors become intrinsic at high temperature. This so-called intrinsic regime can become relevant for photovoltaic absorbers with low E_g at relatively high temperature.

The Fermi-energy shifts towards the E_C for n-type semiconductors and towards the E_V for p-type semiconductors (Figure 2.20).

A semiconductor is compensated if it is doped with donors and acceptors at the same time. The Fermi-energy cannot shift widely over

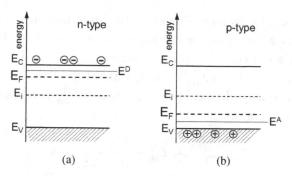

Figure 2.20. Energy diagrams for n-type (a) and p-type (b) doped semiconductors. E_i denotes the so-called intrinsic Fermi-energy of the un-doped semiconductor. E^D and E^A are the energies of the unoccupied donor and occupied acceptor states. The energy differences $E_C–E^D$ and $E^A–E_V$ correspond to the ionization energies of the donor and acceptor, respectively.

the E_g for compensated semiconductors despite possible high densities of free electrons and holes. Compensation can become important for semiconductors with amphoteric dopants such as Si which can replace Ga or As atoms in GaAs or for semiconductors with strong self-compensation such as CuInSe$_2$.

Condition (2.31) is also fulfilled for doped semiconductors in thermal equilibrium. This means that the density of the free carriers not belonging to the dopants can be calculated with respect to Equation (2.31). The density of these charge carriers is very low, and consequently they are called minority charge carriers. The free charge carriers originating from the ionized dopants are called majority charge carriers.

$$n_i^2 = n_0 \cdot p_0 \tag{2.36}$$

Free electrons are the majority charge carriers in n-type doped semiconductors, and free holes are the majority charge carriers in p-type doped semiconductors. Conversely, free holes are minority charge carriers in n-type semiconductors, and free electrons are minority charge carriers in p-type semiconductors. For example, the absorber of c-Si solar cells is usually p-type doped with a density of majority charge carriers of $p_0 = 10^{16}$ cm^{-3}. The density of minority charge carriers is $n_0 \approx 10^4$ cm^{-3} since $n_i \approx 10^{10}$ cm^{-3} for c-Si at room temperature. The density of minority charge carriers depends strongly on temperature due to the temperature dependence of n_i.

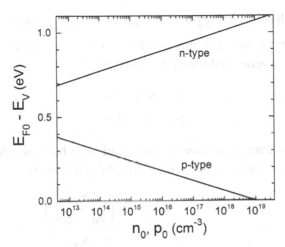

Figure 2.21. Dependence of the Fermi-energy on the density of free electrons in thermal equilibrium and density of free holes in thermal equilibrium in c-Si at 300 K.

Figure 2.21 shows an example for the dependence of Fermi-energy on n_0 and p_0 at 300 K. In the doping range the Fermi-energy can be calculated following:

$$E_{F_0} = E_C - k_B \cdot T \cdot ln\left(\frac{N_C}{n_0}\right) = E_V + k_B \cdot T \cdot ln\left(\frac{N_V}{p_0}\right) \qquad (2.37)$$

The Fermi-energy shifts towards the middle of the E_g with increasing temperature for a given density of dopants. Any doped semiconductor becomes intrinsic at high temperatures depending on E_g and doping. This limits the operation range of semiconductor devices.

The Fermi-energy shifts into the valence or conduction bands when the density of dopants becomes larger than N_V or N_C, respectively. Such highly doped semiconductors are said to be degenerated. Very high densities of donors or acceptors are sometimes denoted by $n^{+(++)}$ or $p^{+(++)}$.

The density of free charge carriers in semiconductors can be varied by doping over a range from n_i to the solubility limit of a given dopant in a given semiconductor. The solubility limit is usually of the order of 10^{20} cm^{-3}. If taking into account minority and majority charge carriers, the densities of free electrons or holes can be varied in c-Si, for example, over a range of about 20 (!) orders of magnitude.

2.3.4 *Fermi-level splitting for photo-generated charge carriers*

Non-equilibrium free electrons and holes with the densities Δn and Δp ($\Delta n = \Delta p$) are generated under illumination. Their total densities are:

$$n = n_0 + \Delta n \tag{2.38'}$$

$$p = p_0 + \Delta p \tag{2.38''}$$

The Fermi-energies of free electrons and holes become different under illumination and can be calculated if Δn or Δp is known:

$$E_{Fn} = E_C - k_B \cdot T \cdot ln\left(\frac{N_C}{n_0 + \Delta n}\right) \tag{2.39'}$$

$$E_{Fp} = E_V + k_B \cdot T \cdot ln\left(\frac{N_V}{p_0 + \Delta p}\right) \tag{2.39''}$$

The potential energy at which charge carriers can be extracted from a semiconductor is related to their Fermi-energy. As already mentioned, the Fermi energy of free charge carriers is equivalent to their chemical potential. The difference of chemical potentials between two different materials is used in batteries, for example. In photovoltaic absorbers, two different chemical potentials arise in the same material under illumination.

In solar cells, E_{Fn} and E_{Fp} are separately contacted for the same photovoltaic absorber (Figure 2.22) with, for example, a pn-homo-junction (see Chapter 4). Extracted free electrons flow from the photovoltaic absorber contacted at E_{Fn} via an external load back to the photovoltaic

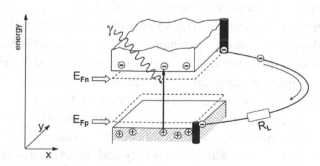

Figure 2.22. Energetic scheme of a photovoltaic absorber under illumination with separate contacts at E_{Fn} and E_{Fp} and with an external load.

absorber contacted at E_{Fp}, where the extracted electrons recombine with the free holes.

The assumption of separate contacts for E_{Fn} and E_{Fp} implies two fundamentally different functionalities: (i) separation of free charge carriers in space, for example, with a pn-junction (see Chapter 4) and (ii) transfer of free charge carriers to external leads via ohmic contacts (see Chapter 5).

The change of the Fermi-energy in equilibrium and under illumination is very different for majority and minority charge carriers. The following equations can be obtained by taking into account Equations (2.37) and (2.39)

$$E_{Fn} - E_{F0} = k_B \cdot T \cdot ln\left(\frac{n_0 + \Delta n}{n_0}\right) \qquad (2.40')$$

$$E_{F0} - E_{Fp} = k_B \cdot T \cdot ln\left(\frac{p_0 + \Delta p}{p_0}\right) \qquad (2.40'')$$

For p-type (n-type) semiconductors, p_0 (n_0) is usually much larger than Δp (Δn) so that the change of E_{Fp} (E_{Fn}) in comparison to Fermi-energy in thermal equilibrium (E_{F0}) can be neglected (see also Figure 2.23). Therefore, the illumination-induced change of the Fermi-energy of the majority charge carriers does not significantly contribute to the photovoltage of a solar cell. Conversely, the density of minority charge carriers changes over many orders of magnitude under illumination, and so the Fermi-energy of minority charge carriers increases strongly. For example, Δn is of the order of 10^{15} cm^{-3} for c-Si solar cells illuminated at

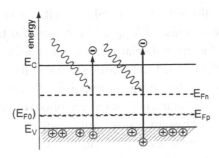

Figure 2.23. Energy diagram for p-type semiconductors under illumination.

AM1.5 which is 11 orders of magnitude larger than n_0 ($p_0 = 10^{16}\,\mathrm{cm}^{-3}$). A difference of 11 orders of magnitude in the density of free electrons corresponds to a change of E_{Fn} by 0.66 eV which is almost close to $q \cdot V_{OC}$ of the world record c-Si solar cell. Therefore, the illumination-induced change of the Fermi-energy of minority charge carriers is a fundamental reason for the photovoltage of a solar cell under operation.

The difference between the Fermi-energies of free electrons and holes (also known as Fermi-level splitting) in a photovoltaic absorber is of pivotal importance for photovoltaic energy conversion since it defines the maximum potential energy difference at which a solar cell can be operated, i.e. V_{OC} multiplied by the elementary charge.

$$E_{Fn} - E_{Fp} = k_B \cdot T \cdot ln\left(\frac{(n_0 + \Delta n) \cdot (p_0 + \Delta p)}{n_i^2}\right) \qquad (2.41)$$

The Fermi-level splitting increases with increasing Δn for a given semiconductor at fixed n_i, i.e. at fixed temperature. This is the physical reason for the increase of V_{OC} with increasing light intensity (mentioned in Chapter 1).

Equation (2.41) implements a temperature dependence of the Fermi-level splitting. Figure 2.24 shows the Fermi-level splitting under identical p_0 and Δn in Ge, Si and GaAs. The largest Fermi-level splitting is obtained for the largest value of E_g, i.e. the lowest n_i. The Fermi-level splitting decreases with increasing temperature and disappears for Ge at temperatures above 150°C.

The decrease of the Fermi-level splitting with increasing temperature is the reason for the decrease of V_{OC} and therefore for the decrease of the energy conversion efficiency of solar cells with increasing temperature. The relative decrease of the Fermi-level splitting is higher for lower E_g. Therefore the relative decrease of the energy conversion efficiency with increasing temperature is less for solar cells based on photovoltaic absorbers with higher E_g. Figure 2.25 gives an impression about the relative temperature dependence of the Fermi-level splitting for GaAs, c-Si and Ge in a temperature range relevant for most terrestrial applications of solar cells.

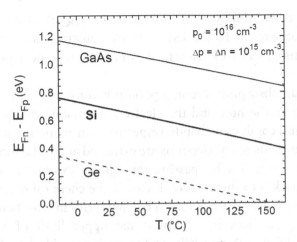

Figure 2.24. Fermi-level splitting for $p_0 = 10^{16}\,\text{cm}^{-3}$ and $\Delta n = 10^{15}\,\text{cm}^{-3}$ in photovoltaic absorbers with $E_g = 0.78$, 1.1 and 1.42 eV corresponding to the E_g of Ge, c-Si and GaAs, respectively. The temperature dependence of E_g is taken into account.

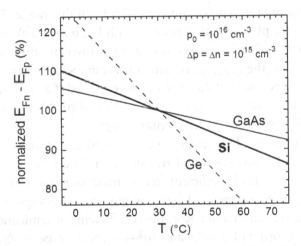

Figure 2.25. Temperature dependence of the normalized Fermi-level splitting at $30\,°\text{C}$ for $p_0 = 10^{16}\,\text{cm}^{-3}$ and $\Delta n = 10^{15}\,\text{cm}^{-3}$ in photovoltaic absorbers with $E_g = 0.78$, 1.1 and 1.42 eV for Ge, c-Si and GaAs, respectively.

2.4 Summary

Solar cells have to be made from photovoltaic absorbers or semiconductors with a forbidden E_g. The E_g of a semiconductor is decisive for photovoltaic solar energy conversion. On the one hand, the V_{OC} of a solar cell cannot exceed the value of the E_g divided by the elementary charge. On the other hand, photons with energy below the E_g cannot contribute to photocurrent generation.

During an absorption event, a photon is converted into an electron-hole pair where the hole and the electron are mobile charge carriers in the valence and conduction bands, respectively. In an illuminated solar cell with maximum efficiency, electrons are extracted at the chemical potential of free electrons and, after passing through an external load, are then transferred back into the semiconductor at the chemical potential of the free holes where the electrons then recombine with the free holes.

The $\eta_{ultimate}$ takes into account the upper limit of V_{OC} and a hypothetical FF of 1.0, which is, in theory, impossible (see Chapter 1). However, the analysis of the $\eta_{ultimate}$ gives the range of E_g suitable for very high solar energy conversion efficiency. The $\eta_{ultimate}$ is above 40% for semiconductors with E_g between 0.7 and 1.6 eV. There are numerous inorganic and organic semiconductors with E_g in this range. The $\eta_{ultimate}$ is below 10% for photovoltaic absorbers with E_g above 2.5 eV, i.e. so-called wide-gap semiconductors are not useful as photovoltaic absorbers.

According to the $\eta_{ultimate}$, realistic maximum solar energy conversion efficiencies of the order of 25–30% can be expected if assuming a realistic FF of the order of 85% (see Chapter 1) and a realistic V_{OC} of the order of 70–80% of E_g divided by the elementary charge.

The number of photons that are absorbed in the semiconductor is limited by the E_g, reflection and transmission. The reflection coefficient (R) and the absorption coefficient (α) are material parameters depending on the wavelength of incident light or the photon energy of incoming photons. The reflection and absorption coefficients of semiconductors are typically of the order of 0.3–0.4 and 10^2–10^5 cm^{-1}, respectively.

Losses in photo-generation caused by E_g cannot be minimized. Optical losses caused by reflection and transmission can be minimized by implementing antireflection coatings and by increasing the optical path of light in the photovoltaic absorber, respectively. The optical path of light can be increased by increasing the thickness of the photovoltaic absorber and/or by implementing measures for light trapping such as Lambertian light scattering combined with the total internal reflection. The optical path can be increased by about 8–18 times for this so-called Lambertian limit of light trapping, depending on the ratio of the refractive index of the photovoltaic absorber and of the antireflection coating.

The densities of states in the conduction and valence bands have been derived by taking into account only delocalization of free charge carriers and the Heisenberg uncertainty principle. The density of states increases by the square root of the difference between the energy of free charge carriers and the corresponding band edge. The density of states and the effective density of states (N_C, N_V) can be calculated if the E_g and the effective masses of free electrons and holes of the semiconductor are known. The effective mass of the electron, for example, is about $0.067 \cdot m_e$ for GaAs and $1.08 \cdot m_e$ for c-Si. The value of N_C, for example, is about $5 \cdot 10^{17}$ cm^{-3} for GaAs and $3 \cdot 10^{19}$ cm^{-3} for c-Si at room temperature.

In order to calculate the densities of free electrons and holes, the densities of states have to be multiplied by the corresponding occupation probability which is given for Fermions (following the Pauli principle meaning that two particles cannot have the same quantum numbers) by the Fermi distribution function. As a result, the Boltzmann equations are obtained for calculating the densities of free electrons and holes from their chemical potentials (which are called Fermi-energies) and from the effective densities of states at the E_C and E_V.

The n_i (intrinsic density of free charge carriers) of an ideal semiconductor follows from the absorption of blackbody radiation and radiative recombination by the semiconductor. The value of n_i^2 is proportional to the product of N_C and N_V and of the exponential of the $-E_g$ divided by the product of the Boltzmann constant and the ambient temperature. Typical values of n_i^2 are about 10^{20} cm^{-3} for c-Si and 10^{12} cm^{-3} for GaAs at room temperature. The exponential increase of n_i^2 with increasing temperature is a major limiting factor of the solar energy conversion efficiency at increased temperature.

Semiconductors can be p-type doped and n-type doped by replacing atoms in the lattice with atoms of a lower or higher valance, respectively. There is one common Fermi-energy for a semiconductor in thermal equilibrium, and the product of free electrons and free holes is equal to n_i^2. Under illumination, the Fermi-energies split into two different values (E_{Fn} and E_{Fp}) where the change of the Fermi-energy of majority charge carriers is very low in comparison to the huge change of the Fermi-energy of minority charge carriers. As a consequence, the change of the Fermi-energy of the minority charge carriers under illumination is a fundamental condition for a photovoltage.

The Fermi-level splitting increases with increasing light intensity, which gives the opportunity to increase the solar energy conversion efficiency of solar cells under concentrated sunlight.

The Fermi-level splitting decreases with increasing temperature where the relative change is larger for semiconductors with lower E_g. Therefore, V_{OC} and the solar energy conversion efficiency decrease at a greater pace with increasing temperature for solar cells made from semiconductors with lower E_g than for solar cells made from semiconductors with higher E_g.

2.5 Tasks

T2.1: I_{SC} density

Ascertain the I_{SC} density for a photovoltaic absorber with a E_g of 1.6 eV and a width of the absorption band of 0.3 eV.

T2.2: Temperature-dependent I_{SC} density

Estimate the increase of the I_{SC} density of a c-Si solar cell, the temperature of which is increased from 25°C ($E_g = 1.12$ eV) to 65°C ($E_g = 1.115$ eV).

T2.3: Lambertian limit of light trapping

Get the expression for the Lambertian limit of light trapping by taking into account a Lambertian back reflector and total internal reflection at the interface between the absorber and the antireflection coating.

T2.4: Temperature-dependent effective density of states

Solar cells are usually operated in a temperature range between 0 and 70°C. Calculate the effective density of states at the E_C of c-Si at both temperatures. The effective electron mass is $1.08 \cdot m_e$.

T2.5: Temperature-dependent density of minority charge carriers

Solar cells are usually operated in a temperature range between 0 and 70°C. Calculate the density of minority charge carriers in p-type doped c-Si at both temperatures ($p_0 = 3 \cdot 10^{16}$ cm^{-3}). The effective hole mass is $0.55 \cdot m_e$. The E_g of c-Si is 1.131 eV at 0°C and 1.113 eV at 70°C.

T2.6: Temperature-dependent Fermi-energy

Solar cells are usually operated in a temperature range between 0 and 70°C. Calculate the Fermi-energy in a p-type doped c-Si absorber ($p_0 = 3 \cdot 10^{16}$ cm^{-3}) in thermal equilibrium at both these temperatures.

T2.7: Fermi-level splitting

Calculate the Fermi-level splitting at 25°C in p-type doped c-Si with $p_0 = 3 \cdot 10^{16}$ cm^{-3} under illumination with $\Delta n = 10^{14}$, 10^{15} and 10^{16} cm^{-3}.

T2.8: Temperature-dependent Fermi-level splitting

Compare the temperature-dependent decrease of the Fermi-level splitting in a c-Si absorber ($p_0 = 3 \cdot 10^{16}$ cm^{-3}, $\Delta n = 10^{15}$ cm^{-3}) with the temperature-dependent increase of I_{SC}^{max} if the temperature is increased from 25°C to 65°C.

3

Influence of Recombination
on the Minimum Lifetime

Recombination is the annihilation of free electrons and holes. Recombination limits the density of photo-generated charge carriers and therefore the values of the photocurrent and of the photovoltage of a solar cell. The recombination rate is described by the reciprocal lifetime of free charge carriers. The minimum lifetime condition follows from the requirement of light absorption in relation to charge transport. The bulk and the interfaces of photovoltaic absorbers have to be conditioned in such a way that the lifetime of photo-generated charge carriers is equal to or longer than the minimum lifetime. The upper limits of doping of photovoltaic absorbers follow from the radiative and/or Auger recombination lifetimes. Shockley–Read–Hall and surface recombination are caused by defect states in the band gap in the bulk or at the surface of a photovoltaic absorber, respectively. The acceptable maximum densities of bulk and surface defects follow from the minimum lifetime condition. Strategies of the purposeful elimination of defect states in the forbidden band gap are illuminated for passivation of defects in the bulk and at the surface of a photovoltaic absorber.

3.1 Minimum Lifetime

3.1.1 Decay of photocurrent, recombination rate and lifetime

A photocurrent does not disappear immediately after switching off illumination or, in other words, a photocurrent still lasts for a certain time after switching off the light (see, for example, Ryvkin (1964)). To get an impression of this time one has to look at the time-dependent density of photo-generated charge carriers.

The time-dependent change of the densities of free electrons and holes is expressed by the continuity equations. In the simplest case, the densities of free electrons (n) and holes (p) depend only on generation and recombination (generation and recombination rates: G and R, respectively). This case is related to spatially homogeneously distributed densities of free electrons and holes when gradients of densities and electric fields are absent, i.e. diffusion and drift of free charge carriers can be neglected.

$$\frac{\partial n}{\partial t} = G - R \tag{3.1'}$$

$$\frac{\partial p}{\partial t} = G - R \tag{3.1''}$$

The densities of free electrons and holes are the sums of the densities of free electrons or holes in thermal equilibrium (n_0 and p_0) and of the density of photo-generated electrons (Δn) or density of photo-generated holes (Δp). Similarly, the generation rate is the sum of the thermal generation rate (G_0) and the photo-generation rate (ΔG).

A photocurrent is proportional to the density of photo-generated charge carriers. Under stationary illumination, i.e. at constant ΔG, a stationary photocurrent is reached, i.e. Δn is constant (Figure 3.1).

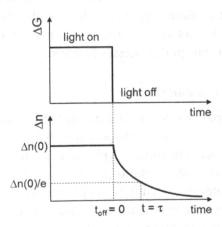

Figure 3.1. Photo-generation rate (top) and density of photo-generated free electrons (bottom) before and after switching off illumination at $t_{off} = 0$.

The decay of photocurrent transients depends on the recombination rate of photo-generated free charge carriers (ΔR), which is in turn proportional to Δn. The proportionality factor between ΔR and Δn has the unit of inverse time. The reciprocal proportionality factor between ΔR and Δn is called the lifetime of photo-generated charge carriers (τ). Therefore:

$$\Delta R \equiv \frac{\Delta n}{\tau} \qquad (3.2)$$

The continuity Equation (3.1') of photo-generated electrons can be written after switching off illumination ($\Delta G = 0$) and taking into account definition (3.2) as:

$$\frac{\partial(\Delta n)}{\partial t} = -\frac{\Delta n}{\tau} \qquad (3.3)$$

The differential Equation (3.3) has the following solution:

$$\Delta n(t) = \Delta n(0) \cdot exp\left(-\frac{t}{\tau}\right) \qquad (3.4)$$

Equation (3.4) describes an exponential decay with the exponential decay time constant τ, i.e. the photocurrent or the density of photo-generated free charge carriers decreases for a constant τ to the value 2.718 times less than the stationary value ($\Delta n(0)$) after passing a time interval of decay equal to τ. It is interesting to note that the lifetime is not necessarily constant but can depend on Δn and even on time, depending on the recombination and transport processes involved.

3.1.2 *Diffusion length and lifetime*

Photo-generated free charge carriers are generated in the photovoltaic absorber and collected at the corresponding charge-selective electron or hole contacts. In order to be collected, photo-generated free charge carriers have to pass from the place of photo-generation to the place of collection. The passage of photo-generated free charge carriers to the corresponding charge-selective electron or hole contacts takes some time, during which the photo-generated free charge carriers should not recombine. Only those photo-generated free charge carriers can be collected that can pass to the corresponding charge-selective electron or hole contacts within a time shorter than their lifetime.

The passage of photo-generated free charge carriers to the corresponding charge-selective electron or hole contacts can be driven either by diffusion or drift. Concentration gradients and electric fields are the origin for diffusion and drift of photo-generated free charge carriers, respectively. Diffusion drives the passage of photo-generated free charge carriers to the corresponding charge-selective electron or hole contacts (for example, for crystalline silicon (c-Si) see Chapter 7). Diffusion will be considered in the following. In a diffusion process the squared average displacement of a diffusing species is proportional to the considered time interval.

Photo-generated free charge carriers can move until they recombine, i.e. they can diffuse only within their lifetime. The average distance over which free charge carriers can diffuse within their lifetime is known as the diffusion length (L). The squared diffusion length is proportional to their lifetime. The proportionality factor between the squared diffusion length and the lifetime is called the diffusion coefficient (D). The diffusion length can be calculated if the diffusion coefficient and the lifetime are known, and *vice versa* the lifetime can be calculated for a given diffusion length.

$$L = \sqrt{D \cdot \tau} \qquad (3.5)$$

Incidentally, the diffusion coefficient of crystalline semiconductors is practically constant for densities of free charge carriers below about 10^{17} cm^{-3} and decreases for larger densities of free charge carriers.

3.1.3 *Minimum lifetime condition for photovoltaic absorbers*

Photo-generated free charge carriers can be collected at their corresponding charge-selective contact if they are photo-generated closer to the contact than L. It was shown in Chapter 2 (see Equation (2.15)) that more than 95% of incoming light is absorbed if the thickness of the photovoltaic absorber (d_{abs}) is equal to or larger than three times the absorption length (α^{-1}). However, absorption is only useful if the photo-generated charge carriers can be collected at their corresponding charge-selective contacts, i.e. if not only d_{abs} but also L is equal to or larger than $3 \cdot \alpha^{-1}$. Therefore the following condition for the minimum lifetime of a photovoltaic absorber (τ_{min}) can be written:

$$\tau_{min} = \frac{9 \cdot \alpha^{-2}}{D} \tag{3.6}$$

Equation (3.6) is the minimum lifetime condition and connects the optical properties with the electronic properties of a photovoltaic absorber. It is, after Equation (2.15), the second important demand for photovoltaic absorbers. The minimum lifetime condition has to be fulfilled for reaching reliable solar energy conversion efficiencies, and therefore defines the demands on those electronic properties, which limit the lifetime of photo-generated free charge carriers for a photovoltaic absorber with a given absorption length and diffusion coefficient.

With regard to Figure 2.8, the absorption lengths of c-Si and the absorption length of GaAs (gallium arsenide) are of the order of 100 μm and 1 μm, respectively. The diffusion coefficients are about 15 and 60 cm^2/s for c-Si and GaAs, respectively. Therefore, the following minimum lifetimes are obtained for photovoltaic absorbers based on c-Si or GaAs:

$$\tau_{min}(c - Si) \approx 100 \, \mu s \tag{3.7'}$$

$$\tau_{min}(GaAs) \approx 10 \, ns \tag{3.7''}$$

These conditions are rather harsh with respect to doping and the purity of the materials, as will be shown in the following. It is worth noting that, since the electronic properties of photovoltaic absorbers sensitively depend on their production technology, the minimum lifetime condition also defines the demands for suitable technologies and therefore the limits for basic cost factors of photovoltaic absorbers.

The minimum lifetime condition can be used for the estimation of the density of photo-generated charge carriers from the maximum short-circuit current (I_{SC}) density given by Equation (2.2). A current is defined as the charge (ΔQ) passing through a contact within a certain time interval (Δt). If the time interval is limited by the minimum lifetime, the charge that can pass through the contact is roughly the areal density of charge carriers photo-generated within a layer of thickness of the diffusion length, i.e. Δn

multiplied by the diffusion length.

$$I = \frac{\Delta Q}{\Delta t} \approx q \cdot \frac{\Delta n \cdot L}{\tau_{min}} \tag{3.8}$$

With respect to the minimum lifetime condition, the following equation can be applied to estimate Δn:

$$\Delta n \approx \frac{\tau_{min}}{q \cdot 3 \cdot \alpha^{-1}} \cdot I_{SC}^{max} \tag{3.9}$$

For a photovoltaic absorber based on c-Si, I_{SC}^{max} (maximum photocurrent) is about $44\,\text{mA/cm}^2$ (air mass (AM)1.5). Therefore, Δn is about $1 \cdot 10^{15}\,\text{cm}^{-3}$ for solar cells based on c-Si and operated under illumination at AM1.5.

The diffusion length and τ_{min} can be reduced if increasing the optical path of light in the photovoltaic absorber by light trapping (see, for example, Equation (2.18) in Chapter 2 of this book). For example, the increase of the optical path of light by a factor of ten would allow a reduction of the minimum lifetime of a photovoltaic absorber, referring to Equation (3.6), by a factor of 100 (!). The reduction of the minimum lifetime due to application of light trapping opens opportunities for technologies with softer requirements, for example, for maximum defect densities.

3.2 Radiative Recombination

Radiative recombination (see, for example, van Roesbroeck and Shockley (1954)) is the annihilation of an electron-hole pair under emission of a photon (Figure 3.2), i.e. a free electron (e^-) falls from the conduction band back into the valence band by filling the place of a missing electron (a hole, h^+) and by giving the excess energy to a photon (γ', Equation (3.10)). Radiative recombination cannot be avoided in photovoltaic absorbers.

$$e^- + h^+ \rightarrow \gamma' \tag{3.10}$$

The radiative recombination rate (R_{rad}) is proportional to the densities of free electrons and holes since both of them are involved in the radiative recombination process:

$$R_{rad} = B \cdot n \cdot p \tag{3.11}$$

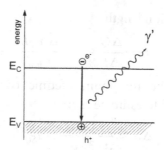

Figure 3.2.   Energy scheme for radiative recombination.

The proportionality factor (B) is the radiative recombination rate constant and amounts to about $3 \cdot 10^{-15}$ cm^3/s for c-Si (Varshni, 1967a) and about $2 \cdot 10^{-10}$ cm^3/s for GaAs (Strauss *et al.*, 1993). The large difference between the radiative recombination rate constants of c-Si and GaAs is caused by the large difference in the absorption coefficients of both semiconductors.

Since the densities of photo-generated electrons and holes are equal, Equation (3.11) can be written in the following form by using Equation (2.38):

$$R_{rad} = B \cdot n_0 \cdot p_0 + B \cdot \Delta n \cdot (n_0 + p_0 + \Delta n) \qquad (3.11')$$

With reference to definition (3.2) and Equation (3.11'), the radiative recombination lifetime of photo-generated free charge carriers (τ_{rad}) can be written as:

$$\tau_{rad} = \frac{1}{B \cdot (n_0 + p_0 + \Delta n)} \qquad (3.12)$$

The density of majority charge carriers is much higher than the density of minority charge carriers. At AM1.5 the density of photo-generated electrons and holes is usually much higher than the densities of minority or intrinsic free charge carriers but much smaller than the density of majority charge carriers. High injection is reached when the density of photo-generated charge carriers becomes higher than the density of majority charge carriers. Equation (3.12) can be simplified for n-type and p-type

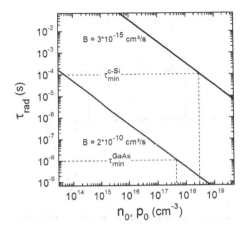

Figure 3.3. Dependence of the radiative recombination lifetime on the density of majority charge carriers in c-Si and GaAs.

semiconductors and for high injection as:

$$\tau_{rad} = \begin{cases} (B \cdot p_0)^{-1} & (p - type) \\ (B \cdot n_0)^{-1} & (n - type) \\ (B \cdot \Delta n)^{-1} & (\Delta n \gg n_0, p_0) \end{cases} \qquad (3.12')$$

Equation (3.12') shows that τ_{rad} depends on the density of majority charge carriers in photovoltaic absorbers. Figure 3.3 shows the dependence of τ_{rad} on the densities of the majority charge carriers for c-Si and GaAs absorbers.

The minimum lifetimes are reached for c-Si and GaAs at densities of majority charge carriers of about $3 \cdot 10^{18}$ cm^{-3} and $5 \cdot 10^{17}$ cm^{-3}, respectively. This means that, under the assumption that radiative recombination is the only limiting recombination process, photovoltaic absorbers based on c-Si or GaAs can be doped with majority carriers up to these densities without losses in the solar energy conversion efficiency. Doping at higher densities of majority charge carriers is not useful as a result of collection losses.

3.3 Auger Recombination

Auger recombination (Auger, 1923) is the annihilation of an electron-hole pair under excitation of a third free charge carrier up to a higher

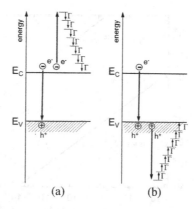

Figure 3.4. Energy schemes for Auger recombination with n ≫ p (a) and n ≪ p (b).

kinetic energy. In a photovoltaic absorber, the kinetic energy is dissipated by a cascade of phonons (denoted by Γ) emission, i.e. heating of the photovoltaic absorber (Figure 3.4). The energy of the recombining electron can be transferred to another free electron or to a second free hole (Equation (3.13)). Auger recombination cannot be avoided in photovoltaic absorbers.

$$2 \cdot e^- + h^+ \to e^- + i \cdot \Gamma \qquad (3.13')$$

$$e^- + 2 \cdot h^+ \to h^+ + j \cdot \Gamma \qquad (3.13'')$$

The Auger recombination rate (R_{Auger}) is proportional to the product of the squared density of free electrons and the density of free holes or to the product of the density of free electrons and the squared density of free holes since a third free charge carrier is involved in the Auger recombination process.

$$R_{Auger} = C_{A,e} \cdot n^2 \cdot p \qquad (3.14')$$

$$R_{Auger} = C_{A,h} \cdot n \cdot p^2 \qquad (3.14'')$$

The proportionality factors ($C_{A,e}$ and $C_{A,h}$) are the Auger recombination rate constants. The Auger recombination rate constant does not strongly depend on the band structure of a given photovoltaic absorber since the third free charge carrier is excited deep into the conduction or valence bands where the density of states is very large. Therefore, there is practically no need to distinguish between C_A for the process when the energy is transferred to a second free electron or to a second free hole ($C_A = C_{A,e} \approx C_{A,h}$). C_A is similar for most semiconductors and amounts to about $2 \cdot 10^{-30}$ cm^6/s for c-Si (Yablonovitch and Gmitter, 1986a); Wang and Macdonald, 2012), and $7 \cdot 10^{-30}$ cm^6/s (Strauss *et al.*, 1993) or $1.3 \cdot 10^{-30}$ cm^6/s (Picozzi *et al.*, 2002) for GaAs.

As already mentioned, the densities of photo-generated electrons and holes are equal and are much higher than the densities of minority free charge carriers. Therefore, the following equations can be written for the Auger recombination lifetimes (τ_{Auger}) with respect to Equations (2.38), (3.2) and (3.14) in the cases of p-type and n-type semiconductors under illumination at AM1.5 and at high injection.

$$\tau_{Auger} = \begin{cases} (C_A \cdot (p_0)^2)^{-1} & (p - type) \\ (C_A \cdot (n_0)^2)^{-1} & (n - type) \\ (C_A \cdot (\Delta n)^2)^{-1} & (\Delta n \gg n_0, p_0) \end{cases} \qquad (3.15)$$

Equation (3.15) shows that τ_{Auger} depends on the squared density of majority charge carriers in photovoltaic absorbers. Figure 3.5 shows the dependence of τ_{Auger} on the densities of the majority charge carriers.

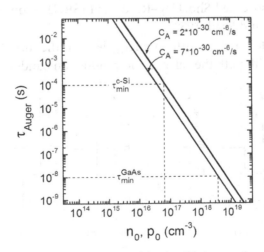

Figure 3.5. Dependence of the Auger recombination lifetime on the density of majority charge carriers.

The minimum lifetimes are reached for c-Si and GaAs at densities of majority charge carriers of about $6 \cdot 10^{16}$ cm^{-3} and $4 \cdot 10^{18}$ cm^{-3}, respectively. The density of majority charge carriers at which the minimum lifetime is reached for c-Si is less for Auger recombination than for radiative recombination. On the other side the density of majority charge carriers at which the minimum lifetime is reached for GaAs is larger for Auger recombination than for radiative recombination. Therefore, under the assumption that radiative and Auger recombination are the only limiting recombination processes, photovoltaic absorbers based on c-Si or GaAs can be doped with majority carriers up to about $5 \cdot 10^{16}$ cm^{-3} for c-Si (limited by Auger recombination) and $5 \cdot 10^{17}$ cm^{-3} for GaAs (limited by radiative recombination).

3.4 Shockley–Read–Hall Recombination

3.4.1 *Elementary processes at a trap state*

An ideal photovoltaic absorber does not have electronic states in the forbidden band gap. However, real photovoltaic absorbers have localized defect states in the forbidden band gap due to imperfections, arising, for example, in a crystalline lattice during the growth process. Defect states in the forbidden band gap of a semiconductor are also known as trap states and cause the so-called Shockley–Read–Hall (SRH) recombination (Hall, 1952; Shockley and Read, 1952).

A trap or defect state with an energy level in the forbidden band gap (E_t) interacts with both the valence and conduction bands (see Figure 3.6).

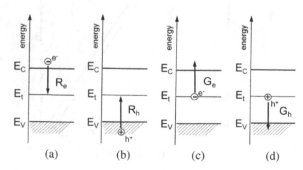

Figure 3.6. Elementary processes at a trap state: electron trapping (a), hole trapping (b), electron emission (c) and hole emission (d).

These interactions lead to four elementary processes at a trap state. A free electron can be captured from the conduction band at the unoccupied defect state. This process is called electron trapping. The trapped electron is no longer mobile, but localized at the trap state.

An electron can be emitted from an occupied trap state into the conduction band, a process which is known as electron de-trapping or electron emission. Similarly, a free hole can be trapped at an occupied trap state (hole trapping). The trapped hole is also localized. A trapped hole can be emitted from an unoccupied trap state into the valence band (de-trapping or emission of a hole). A defect state is characterized by the rates of all four processes: capture rate of an electron (R_e), capture rate of a hole (R_h), emission rate of an electron (G_e) and emission rate of a hole (G_h). An electron-hole pair recombines at a trap state if both the free electron and the free hole are trapped at the trap state at the same time.

The density of occupied traps (n_t) can be calculated by using the Boltzmann statistics. Traps are defects localized in space. The density of defects in the bulk is denoted by N_t:

$$n_t = N_t \cdot exp\left(-\frac{E_t - E_F}{k_B \cdot T}\right) \tag{3.16}$$

For SRH recombination, the rate Equations (3.1) have to be extended by the rate equation of the density of occupied trap states and G, G_e, G_h, R_e and R_h have to be taken into account.

$$\frac{\partial n}{\partial t} = G - R_e + G_e \tag{3.17$'$}$$

$$\frac{\partial p}{\partial t} = G - R_h + G_h \tag{3.17$''$}$$

$$\frac{\partial n_t}{\partial t} = R_e - R_h - G_e + G_h \tag{3.17$'''$}$$

The equations under (3.17) are not independent of each other, as they are coupled by the five rates. In a solar cell, trapped electrons and trapped holes can be emitted by excitation with photons originating from thermal emission or from sunlight, similar to the generation of free electrons and holes due to excitation of electrons from the valence into the conduction band.

A trap state can be charged differently depending on the nature of the trap and on occupation. An acceptor-like trap is negatively charged when it is occupied and neutral when it is unoccupied. Conversely, a donor-like trap is neutral when it is occupied and positively charged when it is unoccupied. This has to be considered for charge neutrality of a photovoltaic absorber.

3.4.2 *Emission and capture rates of electrons and holes*

The electron emission rate is proportional to the density of occupied traps whereas the hole emission rate is proportional to the density of unoccupied traps, i.e. the difference between the density of traps and the density of occupied traps. The proportionality factors are the electron and hole emission rate constants, β_e and β_h, respectively.

$$G_e = \beta_e \cdot n_t \tag{3.18'}$$

$$G_h = \beta_h \cdot (N_t - n_t) \tag{3.18''}$$

The capture rate of a free electron is proportional to the density of free electrons, to the density of unoccupied traps and to the thermal velocity (v_{th}). The capture rate of a free hole is proportional to the density of free holes (p), to the density of occupied traps (n_t) and to the thermal velocity. The thermal velocity (about 10^7 cm/s at room temperature) has influence on the probability of a charge carrier to meet a localized trap. The proportionality factors have the unit of the area and are the electron and hole capture cross sections, σ_e and σ_h, respectively. The capture cross sections of traps can be understood as the extension of a trap perpendicular to the direction of motion of a free charge carrier. Capture cross sections are of the order of 10^{-15} cm^2 for trap states close to the middle of the forbidden band gap.

$$R_e = \sigma_e \cdot v_{th} \cdot (N_t - n_t) \cdot n \tag{3.19'}$$

$$R_h = \sigma_h \cdot v_{th} \cdot n_t \cdot p \tag{3.19''}$$

The emission rates are not independent parameters. They can be obtained if assuming steady state, i.e. when the densities of free holes, free electrons and occupied traps are constant, and in the absence of photo-generation, i.e. Δn is 0. In this case the electron or hole emission rate is equal to the electron or hole capture rate, respectively. The following equations are obtained for β_e and β_h by considering Equations (3.18) and (3.19), and by taking into account the Boltzmann statistics for free electrons and holes and occupied traps:

$$\beta_e = \sigma_e \cdot v_{th} \cdot N_C \cdot exp\left(-\frac{E_C - E_t}{k_B \cdot T}\right) \tag{3.20$'$}$$

$$\beta_h = \sigma_h \cdot v_{th} \cdot N_V \cdot exp\left(-\frac{E_t - E_V}{k_B \cdot T}\right) \tag{3.20$''$}$$

Equations (3.20$'$) and (3.20$''$) show that the emission rates depend exponentially on the energy position of the trap state and on temperature. The electron or hole emission rates are large for traps close to the conduction or valence band edges, respectively.

3.4.3 *Minimum Shockley–Read–Hall recombination lifetimes*

The maximum capture rates are reached for electrons or holes when all trap states are unoccupied or occupied, respectively (Sze, 1981). The minimum SRH lifetimes are obtained for the conditions of the maximum capture rates with respect to Equations (3.2) and (3.19):

$$\tau_{e,SRH}^{min} = \frac{1}{\sigma_e \cdot v_{th} \cdot N_t} \tag{3.21$'$}$$

$$\tau_{h,SRH}^{min} = \frac{1}{\sigma_h \cdot v_{th} \cdot N_t} \tag{3.21$''$}$$

The recombination lifetime of free electrons and holes is the sum of the SRH electron and hole lifetimes since both an electron and a hole have to be captured during the recombination of an electron-hole pair.

Figure 3.7 shows the dependence of the minimum SRH lifetimes as a function of the density of traps. The minimum lifetimes are reached for c-Si and GaAs at densities of traps in the forbidden gap of about 10^{12} cm^{-3} and 10^{16} cm^{-3}.

The condition for a minimum density of traps in the band gap of a photovoltaic absorber, which is established in Figure 3.7, is very harsh for c-Si. It means that only one defect per 10^{11} host Si atoms is acceptable for high-efficiency c-Si solar cells. Extremely pure silicon crystals can be produced by using distillation of precursor molecules and float zone crystal

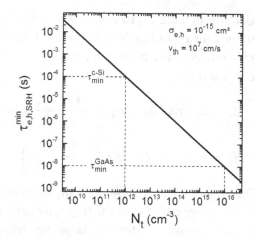

Figure 3.7. Dependence of the minimum SRH lifetime on the density of defects in the bulk under the assumption of equal capture cross sections for electrons and holes with $\sigma_e = \sigma_h = 10^{-15}$ cm^2.

growth (see Chapter 7). For this reason, for example, a highly enriched ^{28}Si crystal has been proposed as a new standard to determine the Avogadro constant and the kilogram definition (Becker *et al.*, 2007).

3.4.4 *Shockley–Read–Hall recombination rate*

The SRH recombination rate can be calculated for a single state in accordance to its energetic position in the forbidden band gap in the case of steady state. In steady state two equations with two unknown variables, n_t and G, are obtained from Equations (3.17):

$$G = R_e - G_e = R_h - G_h \tag{3.22}$$

The density of occupied traps is obtained if the equation containing the capture and emission rates of electrons and holes is solved in accordance with Equations (3.22), (3.18), (3.19) and (3.20):

$$n_t = \frac{\sigma_e \cdot N_t \cdot n + \sigma_h \cdot N_t \cdot N_V \cdot exp\left(-\frac{E_t - E_V}{k_B \cdot T}\right)}{\sigma_e \cdot \left[n + N_C \cdot exp\left(-\frac{E_C - E_t}{k_B \cdot T}\right)\right] + \sigma_h \cdot \left[p + N_V \cdot exp\left(-\frac{E_t - E_V}{k_B \cdot T}\right)\right]} \tag{3.23}$$

The SRH recombination rate (R_{SRH}) is equal to the generation rate in steady state so that R_{SRH} can be calculated considering Equations (3.22), (3.21) and the Boltzmann statistics for intrinsic free charge carriers (2.28):

$$R_{SRH} = \frac{n \cdot p - n_i^2}{\tau_{h,SRH}^{min} \cdot \left[n + N_C \cdot exp\left(-\frac{E_C - E_t}{k_B \cdot T}\right)\right] + \tau_{e,SRH}^{min} \cdot \left[p + N_V \cdot exp\left(-\frac{E_t - E_V}{k_B \cdot T}\right)\right]} \quad (3.24)$$

Equation (3.24) (Sze, 1981) is rather complex. The energy of the trap state and the densities of majority and photo-generated free charge carriers are the most important parameters in Equation (3.24).

In the following, identical capture cross sections ($\sigma_e = \sigma_h = 10^{-15}$ cm^2) are assumed in order to facilitate a simplified analysis. As an example, a p-type c-Si absorber is chosen and the ratio between the density of photo-generated charge carriers and the density of majority charge carriers is varied. The SRH recombination lifetime is calculated with reference to Equation (3.24) and the definition of the recombination rate. The dependence of the SRH lifetime is plotted in Figure 3.8(a) as a function

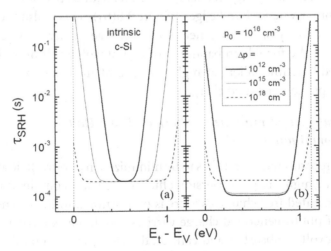

Figure 3.8. Shockley–Read–Hall recombination lifetime as a function of the energy of the trap state in the forbidden band gap for intrinsic (a) and for p-type c-Si with $p_0 = 10^{16}$ cm^{-3} (b) at densities of photo-generated charge carriers of 10^{12}, 10^{15} and 10^{18} cm^{-3} (thick solid line, thin solid line and dashed line, respectively).

of E_t for intrinsic c-Si at three different densities of photo-generated free charge carriers.

For intrinsic c-Si, the minimum SRH lifetime is the doubled value of $\tau_{e,h,SRH}^{min}$. The lowest SRH recombination lifetime is reached for trap states with energies in the middle of the forbidden band gap independent of the density of majority charge carriers and/or of the density of photo-generated charge carriers. Trap states around the middle of the band gap are formed in c-Si, for example, by copper impurities ($E_t - E_V = 0.52\,eV$), gold impurities ($E_t - E_V = 0.54\,eV$) or iron impurities ($E_t - E_V = 0.55\,eV$) (Sze and Irvine, 1968). Incidentally, copper, gold and iron atoms also form other states in addition to the states around the middle of the band gap. Impurities forming trap states around the middle of the band gap of a photovoltaic absorber are also called "lifetime killers" since they limit — and do so very efficiently — the lifetime of photo-generated free charge carriers.

The SRH recombination lifetime increases when E_t shifts towards the valence or conduction band edges, i.e. decreasing $E_t - E_V$ or with decreasing $E_C - E_t$, respectively. This means that defects with trap states closer to the conduction or valence band edges are less recombination active. However, the SRH recombination lifetime strongly decreases with the increasing density of majority charge carriers for defects with trap states close to the conduction or valence band edges (Figure 3.8(b)). This is also the case for increasing density of photo-generated free charge carriers. Therefore, any trap state in the forbidden band gap of a photovoltaic absorber becomes a so-called "lifetime killer" at high densities of majority charge carriers and/or of photo-generated charge carriers.

3.4.5 *Strategies for minimizing Shockley–Read–Hall recombination*

The SRH recombination rate has to be minimized in photovoltaic absorbers for reaching high energy conversion efficiencies. The recombination rate of photo-generated free charge carriers is proportional to their density. The density of photo-generated charge carriers should be as large as possible in a photovoltaic absorber and cannot, therefore, serve to minimize SRH recombination.

The capture cross section of a trap state is given by the nature of a given defect and cannot be changed. Therefore, the most efficient way to reduce the SRH recombination rate is the reduction of the density of intrinsic defects and of defects introduced by unwanted impurities. Intrinsic defects can be, for example, host atoms in a crystal, which are displaced from a lattice position to an interstitial position, or atoms which are part of extended structural defects such as dislocations or grain boundaries in crystalline or polycrystalline photovoltaic absorbers, respectively.

The generation of intrinsic defects is thermally activated and requires high activation energies of the order of the band gap in semiconductor crystals. For example, massive displacement of silicon atoms in c-Si occurs only at temperatures very close to the melting point of c-Si. The density of intrinsic defects in a c-Si lattice at room temperature can be estimated if it is assumed that there is an activation energy of about 1.1 eV and a density of host atoms of about $5 \cdot 10^{22}$ cm^{-3} (with respect to the density of c-Si equal to 2.33 g/cm^3 and a molar mass of Si atoms of 28 g/mol). The expected density of intrinsic point defects in a c-Si lattice is about 19 orders of magnitude less than the density of host atoms. This is extremely small in comparison to the density of free charge carriers.

So-called defect passivation technologies have been developed for numerous cases when the incorporation of unwanted impurities (important, for example, in c-Si) or the generation of intrinsic defects (important, for example, in amorphous silicon) cannot be suppressed sufficiently during the production of the photovoltaic absorber. Defect passivation means the transformation of a recombination active defect into an electronically passive defect, i.e. a defect that does not have electronic states in the forbidden band gap. Passivation strategies are explained in Figure 3.9 for bulk defects in photovoltaic absorbers.

The first strategy of defect passivation aims to modify the chemical nature of bulk defects. A chemical reaction between a defect and a passivating atom leads to the removal of trap states from the band gap. Simultaneously, electronic bonding and anti-bonding states are formed, the energy of which is inside the valence and conduction bands or very close to the valence and conduction band edges of the photovoltaic absorber, respectively (Figure 3.9(a)). Defects can be passivated, for example, with hydrogen (hydrogen passivation), which is decisive for passivation of amorphous silicon absorbers (see Chapter 9). Second, the local density of defects can be reduced by so-called gettering (Figure 3.9(b)). A getterer is additionally introduced into a photovoltaic absorber. Getterers serve as

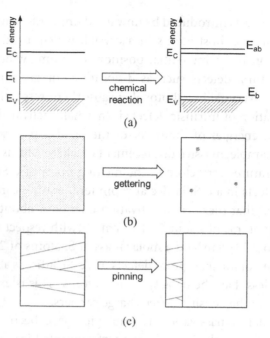

Figure 3.9. Passivation of defects in photovoltaic absorbers: (a) removal of trap states from the band gap by chemical reaction, E_b and E_{ab} denote the energies of the bonding and anti-bonding states, respectively, (b) decrease of the local density of defects by gettering and (c) pinning of dislocations in a layer outside the photovoltaic absorber.

sinks for diffusing impurity atoms which form defects with electronic states in the band gap. Related sinks can be formed by atoms, complexes of atoms, precipitates or structural defects. The application of getterers allows the lifetime and therefore the diffusion length of photo-generated free charge carriers to increase in a certain region of a photovoltaic absorber. Gettering based on SiO_2 precipitates is applied, for example, in c-Si produced by Chrochralski growth (see Chapter 7).

Recombination active defect states are usually formed along extended structural defects such as dislocations. Third, extended structural defects such as dislocations can be stopped by pinning them, for example, in defect-rich layers outside a photovoltaic absorber layer (Figure 3.9(c)). The pinning of dislocations is a very efficient method to reduce defects in photovoltaic absorbers based especially on III–V semiconductors (see Chapter 8).

3.5 Surface Recombination

3.5.1 *Surface recombination velocity*

Surfaces and interfaces are part of solar cells. Well-defined interfaces are required, for example, for charge separation or extraction of charge carriers to external leads. The bond configuration changes abruptly at the surface of a photovoltaic absorber. The abrupt change of the bond configuration at surfaces and interfaces is the reason for tremendous differences between electronic states in the bulk of a photovoltaic absorber and at its surface. The density of surface states can be very large since the density of surface bonds with changed configuration is of the order of 10^{14}–10^{15} cm^{-2} depending on the surface structure. Therefore surfaces and interfaces can be regions with very high densities of trap states in the forbidden gap of the photovoltaic absorber. Free electrons and holes can be captured at surface states in the forbidden gap similarly to the SRH recombination in the bulk of a photovoltaic absorber. The annihilation of electron-hole pairs via surface states in the forbidden band gap is called surface recombination.

Surface states usually have a broad distribution due to disorder. Disorder is caused by variations in bond angles and bond lengths. The broad distribution of surface states is different to sharply distributed trap states in crystalline semiconductors where trap states belong to well-defined bond configurations with fixed structure.

Disorder at interfaces and abrupt changes of bond configurations between semiconductors can be minimized or even avoided. Defects at interfaces are avoided, for example, at pn-homo-junctions which are formed between an n-type and a p-type doped region of the same semiconductor crystal (see Chapter 4). The pn-homo-junction is the most perfect of interfaces in solar cells. The density of interface states is negligible at hetero-junctions between two different semiconductors with the same lattice constants grown by epitaxy on each other (decisive for solar cells with III–V semiconductors, see Chapter 8).

In analogy to SRH, surface states are characterized by the energy and capture and emission rates of electrons and holes (Figure 3.10). Surface states with energies around the middle of the band gap are highly recombination active since the capture rates for both free electrons and

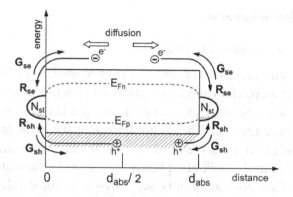

Figure 3.10. Schematic of surface recombination for a symmetric photovoltaic absorber.

free holes are large. Surface states can be donor- or acceptor-like depending on whether the occupied state is neutral and the unoccupied state is positively charged (donor-like) or whether the unoccupied state is neutral and the occupied state is negatively charged (acceptor-like).

Photo-generated free electrons and holes have to reach the surface where they can recombine. This means that the surface recombination rate decreases with increasing distance between the surface and the place where the photo-generated free charge carriers are photo-generated. Therefore the capture rates of free electrons (R_{se}) and holes (R_{sh}) at surface states are proportional to the half of the reciprocal thickness of a homogeneous photovoltaic absorber layer with two equivalent surfaces.

The capture rates of free electrons or holes are proportional to the densities of free electrons (n_s) or holes (p_s) at the surface, respectively, and to the product of the density of surface defects (N_{st}) and thermal velocity (v_{th}). The proportionality factors are, in analogy to SRH recombination, the surface capture cross section for electrons (σ_{se}) and holes (σ_{sh}):

$$R_{se} = \sigma_{se} \cdot v_{th} \cdot N_{st} \cdot \frac{n_s}{d/2} \qquad (3.25')$$

$$R_{sh} = \sigma_{sh} \cdot v_{th} \cdot N_{st} \cdot \frac{p_s}{d/2} \qquad (3.25'')$$

The product of the surface capture cross section, v_{th} and N_{st} has the unit of velocity. The electron and hole surface recombination velocities (s_n and

s_p, respectively), which are actually capture velocities, are defined as:

$$s_n = \sigma_{se} \cdot v_{th} \cdot N_{st} \tag{3.26'}$$

$$s_p = \sigma_{sh} \cdot v_{th} \cdot N_{st} \tag{3.26''}$$

The capture rates of free electrons and holes at surface states are traditionally written with the surface recombination velocities as the proportionality factors (see, for details, Eades and Swanson (1985)).

The density of photo-generated charge carriers is lower near the surface than in the bulk of a photovoltaic absorber due to surface recombination. The surface serves as a sink for photo-generated free electrons and holes, i.e. the Fermi-level splitting is reduced near the surface.

Surface recombination has to be considered in the boundary conditions for suitable models describing the distribution of photo-generated free charge carriers in space. If the access of photo-generated free charge carriers is limited by diffusion, then the following boundary conditions can be written:

$$-D_n \cdot \left.\frac{\partial n}{\partial x}\right|_{x=0,d_{abs}} = s_n \cdot n|_{x=0,d_{abs}} \tag{3.27'}$$

$$-D_p \cdot \left.\frac{\partial p}{\partial x}\right|_{x=0,d_{abs}} = s_p \cdot p|_{x=0,d_{abs}} \tag{3.27''}$$

3.5.2 *Surface recombination lifetime*

The surface recombination lifetimes of photo-generated free electrons and holes are obtained by using the definition of the lifetime of photo-generated free charge carriers and Equations (3.26):

$$\tau_{se} = \frac{d_{abs}}{2 \cdot s_n} \tag{3.28'}$$

$$\tau_{sh} = \frac{d_{abs}}{2 \cdot s_p} \tag{3.28''}$$

The surface recombination lifetimes described by Equations (3.28') and (3.28'') have the sense of the minimum surface recombination lifetime in analogy to the minimum SRH recombination lifetime. The surface recombination lifetimes are plotted as a function of N_{st} in Figure 3.11

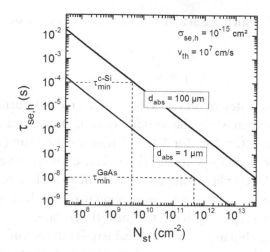

Figure 3.11. Surface recombination lifetimes as a function of the density of surface defects for $\sigma_{s,e} = \sigma_{s,h} = 10^{-15}$ cm^2 and for two thicknesses (1 and 100 μm) of the photovoltaic absorber.

for identical surface capture cross section for free electrons and holes and for two thicknesses of the absorber.

The minimum lifetime condition is reached for c-Si absorbers with a thickness of 100 μm for densities of surface states between 10^9 and 10^{10} cm^{-2}, which corresponds to surface recombination velocities between 10 and 100 cm/s. A density of surface states below 10^{10} cm^{-2} is very low and demands sophisticated technologies for surface and interface treatment. For a GaAs absorber with thickness of 1 μm, the minimum lifetime condition is reached for N_{st} below 10^{12} cm^{-2}.

Light trapping can be applied to reduce the thickness of photovoltaic absorbers as mentioned earlier. However, surface recombination becomes more crucial with a reduction of d_{abs} and the density of surface states must be reduced in accordance with the reduction of d_{abs} (see Chapter 4).

3.5.3 *Strategies for minimizing surface recombination*

The purposeful reduction of the surface recombination rate is called surface passivation. Strategies of surface passivation are directed to the reduction of the density of free charge carriers at the surface as well as of the density of surface states. Surface passivation plays a very important role for solar cells with high solar energy conversion efficiency.

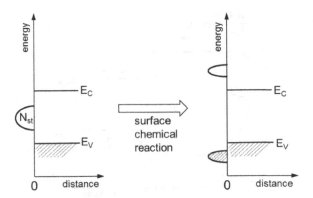

Figure 3.12. Surface passivation by surface chemical reactions removing surface states from the region around the middle of the band gap of the photovoltaic absorber.

A very effective method for surface passivation is the reduction of the density of surface defects with electronic states in the forbidden band gap of the photovoltaic absorber (see Figure 3.12).

The density of surface defects can be strongly reduced if surface atoms participate in chemical bonds with electronic states in the valence or conduction band or at least very close to the valence and conduction band edges of the photovoltaic absorber. For example, surface atoms can be partially oxidized (see, for example, the Si/SiO_2 interface in Chapter 7) or unsaturated bonds at interfaces can be hydrogen passivated.

The total surface recombination rate is proportional to the product of the densities of free electrons and holes at the surface. Therefore the surface recombination rate can be strongly reduced if one of the densities of free electrons or holes is very low at the surface. The reduction of n_S and/or p_S is possible by incorporating barriers repelling free charge carriers from the surface. Barriers can be realized, for example, with doping profiles at the side of the photovoltaic absorber (back surface field, see Chapter 7) or with fixed electrostatic charge in an additional layer close to the absorber surface (see, for example, Chapter 7).

Concepts of surface passivation for solar cells are based on the incorporation of so-called passivation layers (Figure 3.13). The aims of a passivation layer are (i) the reduction of the density of surface states at the interface between the passivation layer and the photovoltaic absorber and (ii) the formation of barriers for photo-generated charge carriers between

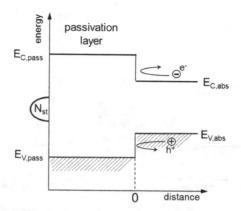

Figure 3.13. Surface passivation with a surface passivation layer avoiding defect states at the interface between the passivation layer and the absorber, and avoiding the penetration of free electrons and holes from the photovoltaic absorber to the external surface of the passivation layer.

the photovoltaic absorber and the external surface of the passivation layer. The first aim is achieved by choosing the right chemical bond configuration at the interface between the passivation layer and the photovoltaic absorber. The second aim is achieved by realizing so-called type I hetero-junctions (see Chapter 8 for details). The passivation layer should be much thicker than the tunneling length (for details on tunneling see Chapter 5).

The barriers for free electrons and holes at interfaces between a passivation layer and photovoltaic absorber are given by the differences between the valence and conduction band edges of the passivation layer ($E_{V,pass}$ and $E_{C,pass}$, respectively) and of the photovoltaic absorber ($E_{V,abs}$ and $E_{C,abs}$, respectively). The following condition should be fulfilled for effective surface passivation:

$$E_{C,pass} - E_{C,abs} \gg k_B \cdot T \qquad (3.29')$$

$$E_{V,abs} - E_{V,pass} \gg k_B \cdot T \qquad (3.29'')$$

Historically, the most prominent passivation layer is SiO_2 grown on c-Si (see Chapter 7). Practically perfect surface passivation can be reached by epitaxial growth of a passivation layer having the same lattice constant as the crystal of the photovoltaic absorber (for example, passivation of a GaAs surface with an $Al_xGa_{1-x}As$ layer, see Chapter 8).

3.6 Summary

The density of charge carriers that are photo-generated in a photovoltaic absorber is limited by recombination processes. The recombination rate increases with decreasing lifetime of minority charge carriers in a photo-voltaic absorber. The minimum lifetime condition has been introduced as a criterion with which to analyze materials requirements for photovoltaic absorbers.

The minimum lifetime follows from the demands of strong absorption and of sufficient carrier collection. Strong absorption means that d_{abs} should be equal to or larger than $3 \cdot \alpha^{-1}$. The thickness of the photovoltaic absorber can be reduced by a factor X_{path} giving the ratio by which the optical path is increased under light trapping (see Chapter 2). Sufficient carrier collection means that photo-generation takes place at a distance equal to or shorter than the diffusion length (L) of minority charge carriers. Under the conditions of strong absorption and sufficient carrier collection the diffusion length is given by the diffusion coefficient (D) and the minimum lifetime (τ_{min}):

$$\frac{3 \cdot \alpha^{-1}}{X_{path}} = \sqrt{D \cdot \tau_{min}} \qquad (3.30)$$

The density of majority charge carriers should be as high as possible since the density of minority charge carriers in thermal equilibrium should be as low as possible for reaching a high Fermi-level splitting in an illuminated photovoltaic absorber (see Chapter 2). However, the maximum useful density of equilibrium majority charge carriers is limited by radiative and/or Auger recombination, which cannot be avoided in photovoltaic absorbers.

Shockley–Read–Hall and surface recombination are proportional to the densities of bulk and surface defects which have to be minimized by technological measures. Passivation strategies are directed to the reduction of SRH and surface recombination rates by the elimination of defects. For this purpose, bond configurations with defect states in the forbidden band gap can be transformed by chemical reactions into bond configurations without states in the band gap. Further, regions with very low densities of defects can be obtained in photovoltaic absorbers by gettering

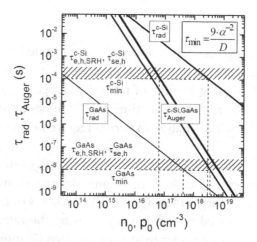

Figure 3.14. Summarized lifetimes for the various recombination mechanisms as a function of the density of dopants.

atoms forming defect states or by pinning extended defects outside the photo-active part of the absorber.

Figure 3.14 summarizes the lifetimes for the various recombination mechanisms as a function of the density of dopants for c-Si and GaAs absorbers.

The minimum lifetimes are about 10 ns for GaAs and 100 μs for c-Si absorbers. For a GaAs absorber the density of majority charge carriers is limited by radiative recombination to about $5 \cdot 10^{17}$ cm^{-3}. For a c-Si absorber the density of majority charge carriers is limited by Auger recombination to about $2 - 5 \cdot 10^{16}$ cm^{-3}. For c-Si absorbers the density of defects in the bulk should be equal to or less than 10^{11} cm^{-3} and the density of surface defects should be less than 10^{10} cm^{-2}.

The total lifetime follows from the lifetime of all recombination processes in a photovoltaic absorber. The reciprocal total lifetime (τ_{total}) is the sum of the reciprocal lifetimes of all recombination processes.

$$\frac{1}{\tau_{total}} = \frac{1}{\tau_{rad}} + \frac{1}{\tau_{Auger}} + \frac{1}{\tau_{SRH}} + \frac{2 \cdot s_{n,p}}{d} \tag{3.31}$$

The total lifetime of photo-generated free charge carriers is limited by the recombination process with the shortest lifetime. The total lifetime of a given photovoltaic absorber has to be longer than the minimum lifetime obtained from the minimum lifetime condition.

Strategies and technological measures to minimize SRH and surface recombination are under permanent development in order to reach new record energy conversion efficiencies and in order to further reduce the energy required for production of solar cells without losses in the energy conversion efficiency.

3.7 Tasks

T3.1: Minimum lifetime and light trapping

Ascertain the minimum lifetimes for c-Si absorbers with a single pass of light, a planar back reflector and a Lambertian back reflector for light trapping.

T3.2: Density of photo-generated charge carriers in a GaAs absorber

Estimate the density of photo-generated charge carriers in a GaAs absorber with and without a back reflector under illumination at AM1.5G with regard to the minimum lifetime condition. Compare the result with the density of photo-generated charge carriers in a c-Si absorber without a back reflector.

T3.3: Density of majority charge carriers in a c-Si absorber

Ascertain the maximum densities of majority charge carriers and the diffusion lengths of minority charge carriers for c-Si absorbers with lifetimes of 7 ms and 7 μs.

T3.4: Maximum limit of lifetime

Is it realistic to produce a silicon crystal with a lifetime of minority charge carriers of 157 s? If so, why?

T3.5: Intrinsic carrier lifetime

The intrinsic carrier lifetime is related to the hypothetical lifetime, which one would obtain for an absolutely defect-free un-doped semiconductor. Calculate the intrinsic carrier lifetimes for c-Si and GaAs.

T3.6: Lifetime at the solubility limit of doping

The solubility limit is of the order of 10^{20} cm^{-3} for many dopants. Calculate the lifetime of c-Si and GaAs absorbers at the solubility limit. Get the density of defects in the bulk which may be accepted at the solubility limit with respect to the lifetime, and estimate the related diffusion length.

T3.7: Total lifetime

Calculate the total lifetime of photo-generated electrons in a p-type doped c-Si absorber ($p_0 = 3 \cdot 10^{16}$ cm^{-3}, thickness $100\,\mu$m) with a density of defects in the bulk of 10^{12} cm^{-3} ($\sigma_e = 10^{-15}$ cm^2) and a surface recombination velocity of 10 cm/s at both sides of the absorber.

4

Charge Separation Across pn-junctions

The charge-selective contact is the heart of each solar cell since charge separation at the charge-selective contact provides the driving force for directed motion of photo-generated charge carriers. At a charge-selective contact, photo-generated charge carriers are separated towards the side of the contact with lower potential energy. Junctions between n- and p-type doped semiconductors (pn-junctions) are the most prominent charge-selective contacts since they can be realized as nearly ideal charge-selective contacts. Electrostatic properties such as the width of the space charge regions and the diffusion potential of a pn-junction are introduced. Charge-selective contacts are characterized by the diode saturation current density (I_0), which gives the fundamental limitation of the open-circuit voltage of an illuminated solar cell. The dependence of I_0 on the density of intrinsic charge carriers, on the density of donors at the n-type doped and on the density of acceptors at the p-type doped sides of the pn-junction as well as the dependence of I_0 on the diffusion coefficients and diffusion lengths of minority charge carriers is obtained. The role of Shockley–Read–Hall recombination at a pn-junction is illuminated and considered in the two-diode model with recombination in the neutral regions of the absorber and with recombination in the region of the charge-selective contact. The influence on I_0 of surface recombination in addition to the thickness of the n-type or p-type doped semiconductor layers is analyzed. Principles to achieve very high solar energy conversion efficiencies of solar cells are elucidated.

4.1 Concept of the Ideal Charge-Selective Contact

The splitting of the Fermi-energies of electrons and holes in an illuminated photovoltaic absorber is the first prerequisite for obtaining a photovoltage in a solar cell (see Chapter 2). The second prerequisite is the ability to extract photo-generated electrons from an illuminated photovoltaic absorber at a high chemical potential (Fermi-energy of electrons, E_{Fn}) and to transfer the electrons back to the photovoltaic absorber at a reduced chemical potential (Fermi-energy of holes, E_{Fp}) after passing through an external load. The electrons transferred back to the photovoltaic absorber recombine with holes.

The extraction of electrons at E_{Fn} is realized by electron separation with charge-selective electron contacts, where holes are extracted by separation with charge-selective hole contacts at E_{Fp}.

Charge separation in a solar cell is a directed transfer of photo-generated charge carriers from a photovoltaic absorber across an interface into electronic states of a contacting material (see Figure 4.1). The driving force for a directed charge transfer is given by the decrease of the potential energy of transferred mobile charge carriers in their final state, i.e. transferred photo-generated charge carriers have a lower potential energy after the charge transfer than before.

Transferred photo-generated electrons or holes should remain mobile in the contacting material for getting them out to external leads. Mobile

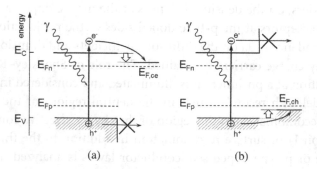

Figure 4.1. Schematic band diagrams of ideal charge-selective electron (a) and hole (b) contacts acting as sinks for photo-generated electrons and holes, respectively. The Fermi-energies E_{Fn} and $E_{F,ce}$ (a) and E_{Fp} and $E_{F,ch}$ (b) are aligned at charge-selective electron and hole contacts. The transfer of photo-generated holes or electrons is blocked from entering the contacting material of charge-selective electron or hole contacts, respectively.

electrons and holes in the contacting material are characterized by their Fermi-energies ($E_{F,ce}$ and $E_{F,ch}$). The nature of electronic states in the contacting material is not important. Contacting materials can be inorganic or organic semiconductors, metals and electrolytes.

Charge-selective electron or hole contacts act as sinks for photo-generated electrons or holes, respectively. In contrast, the transfer of photo-generated holes or electrons should be blocked from the photovoltaic absorber into the contacting materials of charge-selective electron or hole contacts, respectively. The transfer of photo-generated holes or electrons is blocked by barriers.

In ideal charge-selective contacts the (i) potential energy of transferred photo-generated electrons (holes) is reduced, (ii) transfer of photo-generated holes (electrons) is blocked and (iii) $E_{F,ce}$ ($E_{F,ch}$) is aligned with E_{Fn} (E_{Fp}).

Electronic defect states in the forbidden band gap must be avoided at the contact for realizing charge-selective contacts close to ideal. In practicality, this is only possible with pn-junctions made in one semiconductor crystal with an extremely low density of defects such a crystalline silicon (c-Si) or gallium arsenide (GaAs) (pn-homo-junction or pn-junction). Charge-selective contacts can also be realized with hetero-junctions between two different semiconductors (hetero-junction) or between semiconductors and metal or electrolyte contacts (Schottky contacts). The most important charge-selective contact is the pn-junction.

4.2 The Ideal Semi-Infinite pn-junction in Thermal Equilibrium

4.2.1 *Formation of a pn-junction*

A pn-junction is formed at the contact between n-type and p-type doped semiconductors. If the n-type and p-type doped semiconductors have the same forbidden band gap, the pn-junction is defined as a homo-junction. A homo-junction is considered as metallurgically sharp if inter-diffusion of dopants across the junction can be neglected. In the following, an ideally flat and metallurgically sharp pn-junction will be analyzed following Sze (1981).

The density of free electrons is much higher at the n-type doped side than at the p-type doped side, and similarly the density of free holes is

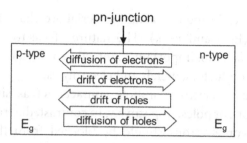

Figure 4.2. Diffusion and drift of electrons and holes across a pn-junction in thermal equilibrium.

much higher at the p-type doped side than at the n-type doped side of the pn-junction. Therefore gradients of the densities of free electrons and holes exist at a pn-junction.

The gradients of the densities of free electrons and holes cause diffusion of free electrons from the n-type to the p-type doped and of free holes from the p-type to the n-type doped sides of the pn-junction. There are permanent diffusion currents of free electrons and holes across the pn-junction (see Figure 4.2). Free electrons and holes diffusing to the p-type and n-type doped sides of the pn-junction, respectively, recombine with the majority charge carriers.

The disappearance of majority charge carriers from the pn-junction is called depletion. The ionized donors and acceptors at the n-type and p-type doped sides of the pn-junction are left in the so-called depletion regions. The charge of ionized donors or acceptors is not compensated by free electrons or holes in the depletion regions at the n-type and p-type doped sides of the pn-junctions, respectively. Therefore the diffusion of free electrons and holes to opposite sides of the pn-junction leads to electrostatic charging of the pn-junction. The depletion regions at the n-type and p-type sides of the pn-junction are also known as space charge regions for this reason.

As a result of electrostatic charging, electric fields are formed at a pn-junction. Free electrons or holes thermally generated in the space charge regions are accelerated in the electric field towards the n-type and p-type doped sides of the pn-junction, respectively. These drift currents have the opposite sign of the diffusion currents. The drift currents compensate the diffusion currents so that the net current is 0 in thermal equilibrium.

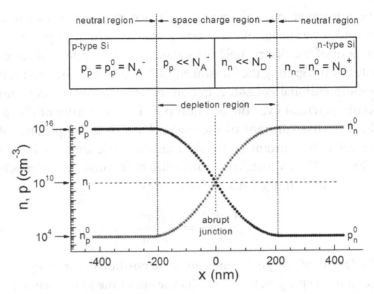

Figure 4.3. Densities of free electrons and holes across a pn-junction of c-Si at room temperature for densities of majority carriers of 10^{16} cm^{-3} and extensions of the respective neutral, depletion and space charge regions. The symbols p_p^0, n_p^0, n_n^0 and p_n^0 denote the densities of free holes and electrons at the p-type doped side and of the free electrons and holes at the n-type doped sides, respectively, in thermal equilibrium.

Figure 4.3 illustrates the densities of free electrons and holes across a pn-junction in the case of c-Si for equal densities of donors and acceptors at the n-type and p-type sides, respectively. The densities of mobile charge carriers and ionized dopants are equal in the neutral regions. Majority charge carriers become minority charge carriers at the pn-junction. The densities of free electrons and holes are equal to the intrinsic density of free charge carriers at the pn-junction.

4.2.2 *Space charge regions of a pn-junction*

The densities of free charge carriers decrease strongly between the neutral and space charge regions within a very short distance, and so the densities of free charge carriers are much lower than the densities of ionized dopants practically over the entirety of the space charge regions. Therefore the space charge regions can be approximated by layers with certain thicknesses and with constant densities of charge equal to the densities of donors and acceptors at the n-type and p-type sides of the pn-junction, respectively.

The uncompensated charge in the space charge regions causes an electric field and a change of the electric potential across a pn-junction. The distributions of the electric field (ε) and of the electric potential (φ) can be calculated by integrating the Poisson equation once and twice, respectively. The one-dimensional Poisson equation connects the second derivative of the electric potential over the position (x) with the density of charge (ρ) and the dielectric constant of the material which is the product of the relative dielectric constant (ε_r) and of the dielectric constant in vacuum ($\varepsilon_0 = 8.8 \cdot 10^{-14}$ As/Vcm). The relative dielectric constant ranges between 10 and 12 for most inorganic semiconductors.

$$\frac{\partial^2 \varphi(x)}{\partial x^2} = -\frac{\rho(x)}{\varepsilon_r \cdot \varepsilon_0} \tag{4.1}$$

The density of charge depends upon the position of the charge carriers in relation to the pn-junction. The thicknesses of the space charge regions at the p-type and n-type doped sides of the pn-junction are denoted by x_p and x_n, respectively. The density of charge is equal to 0 in the neutral regions, i.e. where x is less than $-x_p$ or larger than x_n. The density of charge is equal to the negative product of the elementary charge and the density of acceptors in the space charge region at the p-type doped side of the pn-junction, i.e. for x ranging between $-x_p$ and 0. In the space charge region at the n-type doped side of the pn-junction, i.e. for x ranging between 0 and x_n, the density of charge is equal to the product of the elementary charge and the density of donors (see also Figure 4.4(a)).

$$x < -x_p: \quad \rho = 0 \tag{4.2}$$

$$x \in (-x_p, 0): \quad \rho = -q \cdot N_A \tag{4.2'}$$

$$x \in (0, x_n): \quad \rho = q \cdot N_D \tag{4.2''}$$

$$x > x_n: \quad \rho = 0 \tag{4.2'''}$$

The electric field is equal to 0 in the neutral regions, and the density of charge is constant in the space charge regions. Therefore, integration of the Poisson equation leads to a linear dependence of the electric field on the position in the space charge regions.

$$x < -x_p: \quad \varepsilon = 0 \tag{4.3}$$

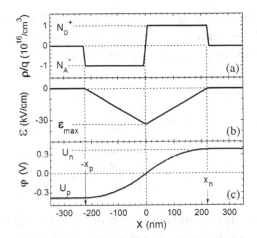

Figure 4.4. Distributions of the density of charge (a), electric field (b) and electric potential (c) across a pn-junction in the case of c-Si with $N_D = N_A = 10^{16}$ cm^{-3}.

$$x \in (-x_p, 0): \quad \varepsilon = -\frac{q \cdot N_A}{\varepsilon_r \cdot \varepsilon_0} \cdot (x + x_p) \qquad (4.3')$$

$$x \in (0, x_n): \quad \varepsilon = \frac{q \cdot N_D}{\varepsilon_r \cdot \varepsilon_0} \cdot (x - x_n) \qquad (4.3'')$$

$$x > x_n: \quad \varepsilon = 0 \qquad (4.3''')$$

The electric potential is a smooth function. It is reasonable to set the electric potential to 0 at the pn-junction. The integration constants are set to $-U_p$ and U_n in the neutral regions at the p-type and n-type doped sides of the pn-junction, respectively. The values of U_p and U_n correspond to the potential drops across the space charge regions at the p-type and n-type sides of the pn-junction, respectively. The second integration of the Poisson equation leads to the distribution of the electric potential.

$$x < -x_p: \quad \varphi = -U_p \qquad (4.4)$$

$$x \in (-x_p, 0): \quad \varphi = -\frac{q \cdot N_A}{\varepsilon_r \cdot \varepsilon_0} \cdot \left(\frac{x^2}{2} + x_p \cdot x\right) \qquad (4.4')$$

$$x \in (0, x_n): \quad \varphi = \frac{q \cdot N_D}{\varepsilon_r \cdot \varepsilon_0} \cdot \left(\frac{x^2}{2} - x_n \cdot x\right) \qquad (4.4'')$$

$$x > -x_n: \quad \varphi = U_n \qquad (4.4''')$$

The thicknesses of the space charge regions at the p-type and n-type doped sides of the pn-junction can be calculated from Equations (4.4′) and (4.4″) if the voltage drops across the corresponding space charge regions are known.

$$x_p = \sqrt{\frac{2 \cdot \varepsilon_r \cdot \varepsilon_0}{q \cdot N_A}} \cdot U_p \tag{4.5′}$$

$$x_n = \sqrt{\frac{2 \cdot \varepsilon_r \cdot \varepsilon_0}{q \cdot N_D}} \cdot U_n \tag{4.5″}$$

Charge neutrality demands that the product of the density of acceptors and the thickness of the space charge region at the p-type side is equal to the product of the density of donors and the thickness of the space charge region at the n-type side of the pn-junction.

$$N_A \cdot x_p = N_D \cdot x_n \tag{4.6}$$

The thicknesses of the space charge regions and the potential drops across the space charge regions are not independent from each other (see task T4.1 at the end of this chapter).

Figures 4.4(b) and 4.4(c) give an example of the distributions of the electric field and of the electric potential across a pn-junction in the case of c-Si for equal densities of donors and acceptors (10^{16} cm^{-3}). The space charge regions are extended over about 200 nm for the given example. The electric field is negative since it is directed from the negative to the positive charge. The maximum electric field is reached directly at the pn-junction. The maximum electric field is identical at the p-type and n-type doped sides of the pn-junction (see task T2). For the given example, the maximum electric field is about -34 kV/cm. This is a typical value for electric fields across charge-selective contacts in solar cells. Across the space charge regions the electric potential increases continuously from $-U_p$ in the neutral region of the p-type doped side to U_n in the neutrality region at the n-type doped side of the pn-junction.

The thickness of the space charge region decreases with increasing density of donors or acceptors with regard to Equations (4.5′) and (4.5″).

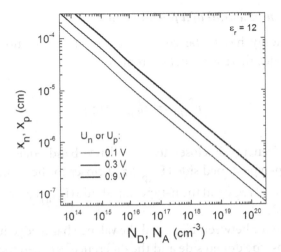

Figure 4.5. Dependence of the thickness of the space charge region on the density of donors and density of acceptors for a semiconductor with a relative dielectric constant equal to 12 and for different values of the potentials drops across the space charge region.

Depending on the densities of dopants at both sides of the pn-junction, the potential drop across a space charge region can vary significantly.

Figure 4.5 shows the dependence of the thickness of the space charge region as a function of the density of donors or acceptors for a semiconductor with a relative dielectric constant of 12 and for different values of U_p or U_n.

The densities of donors or acceptors amount to the order of $5 \cdot 10^{16}$ cm^{-3} or $5 \cdot 10^{17}$ cm^{-3} in solar cells based on c-Si or GaAs absorbers, respectively (see Chapter 3). Therefore the thickness of the space charge region is of the order of 100–200 nm or 30–50 nm for pn-junctions in solar cells based on c-Si or GaAs absorbers, respectively. These thicknesses are much less than the corresponding absorption length of c-Si and the absorption length of GaAs (see Chapter 2). It can be concluded that more than 90% or 80% of the incoming photons are not absorbed within the space charge regions in solar cells based on c-Si or GaAs absorbers, respectively.

The thickness of the space charge region decreases to values of the order of only 2–4 nm for highly doped semiconductors. Space charge regions with thicknesses this small are not suitable charge-selective contacts, but they are very important for the formation of ohmic contacts (see Chapter 5).

4.2.3 Diffusion potential of a pn-junction

The Fermi-energy has to be constant across a pn-junction in thermal equilibrium since there is no net current flow:

$$E^0_{F(p)} = E^0_{F(n)} = E_{F0} \tag{4.7}$$

The Fermi-energy is closer to the valence band edge in the neutral region at the p-type doped side ($E^0_{F(p)}$) and closer to the conduction band edge in the neutral region at the n-type doped side ($E^0_{F(n)}$) of the pn-junction (see Figure 4.6).

The difference between $E^0_{F(p)}$ and the valence band edge in the neutral region at the p-type doped side and the difference between the conduction band edge and $E^0_{F(n)}$ at the n-type doped side of the pn-junction are calculated by the Boltzmann equations (2.30′) and (2.30″). The Boltzmann equations for the densities of majority charge carriers (p^0_p and n^0_n) in the neutral regions are:

$$p^0_p = N_V \cdot \exp\left(-\frac{E_{F0} - E_{V(p)}}{k_B \cdot T}\right) \tag{4.8′}$$

$$n^0_n = N_C \cdot \exp\left(-\frac{E_{C(n)} - E_{F0}}{k_B \cdot T}\right) \tag{4.8″}$$

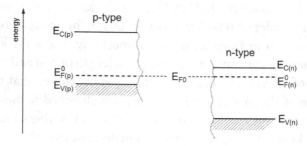

Figure 4.6. Alignment of the Fermi-energy of contacted p-type ($E^0_{F(p)}$) doped and n-type ($E^0_{F(n)}$) doped semiconductors in thermal equilibrium. The valence and conduction band edges at the p-type and n-type doped sides are denoted by $E_{V(p)}$, $E_{C(p)}$, $E_{V(n)}$ and $E_{C(n)}$, respectively.

The respective densities of the minority charge carriers in the neutral regions (n_p^0 and p_n^0) can be written as:

$$p_n^0 = N_V \cdot \exp\left(-\frac{E_{F0} - E_{V(n)}}{k_B \cdot T}\right) \qquad (4.9')$$

$$n_p^0 = N_C \cdot \exp\left(-\frac{E_{C(p)} - E_{F0}}{k_B \cdot T}\right) \qquad (4.9'')$$

The energy of an electron in the conduction band in the neutral region at the p-type doped side is higher than the energy of an electron in the neutral region at the n-type doped side of the pn-junction due to the fact that the Fermi-energy is constant in thermal equilibrium. The kinetic energy of free electrons is much smaller than the forbidden band gap of the semiconductor. The kinetic energy of free electrons is equal in the neutrality regions at the p-type and n-type doped sides of the pn-junction. Therefore the difference between the energies of the conduction band edge in the neutrality regions at the p-type and n-type doped sides of the pn-junction corresponds to the difference of the potential energy of free electrons at both sides.

The potential energy of an electron decreases with increasing electrostatic potential due to the definition of the direction of the electric field. The potential energy of a free electron in the conduction band therefore follows exactly the negative electrostatic potential across a pn-junction (see Figure 4.7).

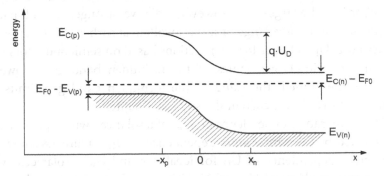

Figure 4.7. Schematic band diagram of a pn-junction in the thermal equilibrium. U_D denotes the diffusion potential of a pn-junction.

The difference between the conduction band edges in the neutrality regions at the p-type ($E_{C(p)}$) and n-type ($E_{C(n)}$) doped sides of the pn-junction is caused by the diffusion of majority charge carriers during the formation of the pn-junction. The diffusion potential (U_D) of a pn-junction is defined as the difference of $E_{C(n)}$ and $E_{C(p)}$ divided by the elementary charge. The value of U_D is equal to the negative sum of the potentials drops across the space charge regions of a pn-junction.

$$U_D \equiv \frac{E_{C(p)} - E_{C(n)}}{q} = -(U_p + U_n) \tag{4.10}$$

The diffusion potential of a pn-junction can be obtained if the forbidden band gap of the semiconductor and the densities of the acceptors and donors at the p-type and n-type sides of the pn-junction are known. According to Figure 4.7:

$$U_D = \frac{1}{q} \cdot [E_g - (E_{F0} - E_{V(p)}) - (E_{C(n)} - E_{F0})] \tag{4.11}$$

Equation (4.11) can be transformed by using the Boltzmann equations (2.30′) and (2.30″) and the definition of the density of intrinsic charge carriers (Equation (2.32)) and by considering complete ionization of donors and acceptors, i.e. p_p^0 and n_n^0 are equal to N_A and N_D, respectively.

$$U_D = \frac{k_B \cdot T}{q} \cdot \ln\frac{N_A \cdot N_D}{n_i^2} \tag{4.12}$$

The open-circuit voltage (V_{OC}) of a solar cell cannot exceed the U_D, i.e. U_D is the upper limit for V_{OC}. Equation (4.12) shows that the higher N_A and/or N_D, the larger U_D. However, radiative or Auger recombination limit the useful doping range of a photovoltaic absorber (see Chapter 3). Furthermore, U_D is larger for pn-junctions based on semiconductors with lower n_i, i.e. with larger values of the forbidden band gap. However, as already mentioned, the value of the forbidden band gap limits the maximum short-circuit current density (see Chapter 2).

The temperature dependence of V_{OC} of a solar cell with a pn-junction is dominated by the temperature dependence of U_D. Figure 4.8 shows the temperature dependence of U_D for ideal c-Si and GaAs solar cells with two different products of N_A and N_D. The U_D of a pn-junction decreases with increasing temperature. For a given semiconductor, the value of U_D

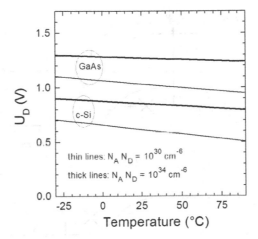

Figure 4.8. Temperature dependence of the U_D for ideal c-Si and GaAs solar cells with $N_A \cdot N_D$ equal to 10^{30} and 10^{34} cm^{-3}.

is less reduced with increasing temperature if the product of N_A and N_D is increased.

4.3 Collection of Photo-Generated Charge Carriers at a pn-junction

4.3.1 *pn-junction as an ideal charge-selective electron and hole contact*

Photo-generation takes place at the n-type doped and p-type doped sides of a solar cell with a pn-junction as the charge-selective contact. Electrons are minority charge carriers at the p-type doped side. The potential energy of an electron photo-generated at the p-type doped side (Figure 4.9(a)) is higher than the potential energy of electrons at the n-type doped side where the electrons are majority charge carriers. As a consequence, an electron photo-generated at the p-type doped side is transferred to the n-type doped side where the Fermi-energy of the charge-selective electron contact corresponds to the Fermi-energy of electrons at the n-type doped side of the pn-junction. The potential energy of a hole photo-generated at the p-type doped side is lower than the potential energy of holes at the n-type doped side. This potential barrier rejects the transfer of holes photo-generated at the p-type doped side to the n-type doped side. Therefore the

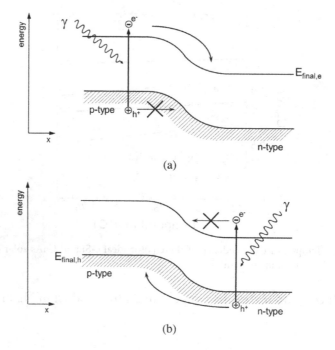

Figure 4.9. Separation of photo-generated electrons and holes at a pn-junction.

n-type doped side of a pn-junction is a charge-selective contact for electrons photo-generated at the p-type doped side.

The potential energy of a hole photo-generated at the n-type doped side of the pn-junction is higher than the potential energy of holes at the p-type doped side. Therefore a hole photo-generated at the n-type doped side can be easily transferred to the p-type doped side of the pn-junction (Figure 4.9(b)). In contrast, the potential energy of an electron photo-generated at the n-type doped side of the pn-junction is lower than the potential energy of electrons at the p-type doped side, i.e. the transfer of an electron photo-generated at the n-type doped side to the p-type doped side is blocked. Consequently, the p-type doped side is a charge-selective contact for holes photo-generated at the n-type doped side of the pn-junction.

The majority charge carriers electrically contact the photo-generated minority charge carriers at the opposite side of the pn-junction. Therefore the majority charge carriers probe the densities of the minority charge carriers at the opposite side of the pn-junction, which is equivalent to the contact of the chemical potential of the minority charge carriers, i.e. their Fermi-levels. Ideal pn-junctions are ideal charge-selective contacts since the potential energy of the blocking barrier is much larger than the thermal energy of photo-generated charge carriers.

4.3.2 Charge carriers photo-generated in a space charge region

A proportion of photons from sunlight are absorbed in space charge regions of charge-selective contacts. Mobile charge carriers photo-generated in a space charge region are driven by the electric field. The maximum transit time of a mobile charge carrier photo-generated in the electric field of a pn-junction can be estimated by taking into account the potential drop across a space charge region, the extension of a space charge region and the drift mobility of photo-generated charge carriers. For example, the electron drift mobility (μ_e) is of the order of 1000 cm^2/Vs for c-Si (Jacoboni *et al.*, 1977).

The electron drift mobility is defined as the proportionality factor between the electron drift velocity (v_e) and the electric field.

$$v_e \equiv \mu_e \cdot \varepsilon \tag{4.13}$$

In the simplest case, the drift velocity can be estimated as the thickness of the space charge region divided by the transit time (t_{tr}). The value of $t_{tr,e}$ can be estimated for electrons photo-generated in the space charge region at the p-type doped side of the pn-junction by using the following equation:

$$t_{tr,e} \approx \frac{x_p^2}{\mu_e \cdot U_p} \tag{4.14}$$

The resulting value of $t_{tr,e}$ as well as of $t_{tr,h}$ is about 1 ps for c-Si solar cells and even less for GaAs solar cells. The lifetimes of photo-generated free charge carriers are much longer than 1 ps in c-Si or GaAs. Therefore all charge carriers photo-generated in the space charge region of a pn-junction are collected.

4.3.3 Charge carriers photo-generated in a neutral region

The space charge region of a pn-junction is usually much thinner than the absorption length for most photons in the spectrum of sunlight. Therefore the vast majority of excess charge carriers is photo-generated in the neutral regions of a solar cell. For example, less than 10% or 20% of the maximum photocurrent can be photo-generated in a c-Si or GaAs solar cell within a space charge region with a thickness of 100 or 30 nm, respectively. Therefore, the collection of photo-generated charge carriers is mainly limited by diffusion of minority charge carriers.

Minority charge carriers photo-generated in the neutral regions diffuse towards the pn-junction and disappear when reaching the space charge region. The relation between the absorption length (α^{-1}) and the diffusion lengths of minority charge carriers ($L_{n,p}$) is crucial for the collection of photo-generated charge carriers and has been considered for the definition of the minimum lifetime condition (see Chapter 3). However, the minimum lifetime condition does not give information about losses of carrier collection.

The dependence of the photocurrent on the ratio between α^{-1} and $L_{n,p}$ will be derived for dominant light absorption in the neutral region at the p-type doped side of a pn-junction (see also Gärtner (1959)). The thickness of a photovoltaic absorber is given by the thicknesses of the p-type (H_p) and n-type (H_n) doped regions. It is assumed in the following that H_n as well as x_n (or transparent n-type doped side) and x_p are much thinner than α^{-1} and that H_p is much thicker than α^{-1} (see Figure 4.10).

The density of photo-generated electrons is denoted by Δn. The stationary diffusion equation of electrons photo-generated in the neutrality region of the p-type doped side of the pn-junction and the boundary

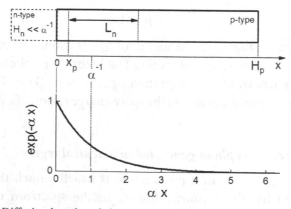

Figure 4.10. Diffusion length and absorption length at a pn-junction with a transparent n-type doped side.

conditions can be written in the following way:

$$0 = D_n \cdot \frac{\partial^2(\Delta n)}{\partial x^2} + G(x) - R_n \qquad (4.15)$$

$$\Delta n|_{x=x_p} = 0 \qquad (4.16')$$

$$\Delta n|_{x \to \infty} = 0 \qquad (4.16'')$$

The photo-generation rate G(x) at the depth x is proportional to the photon flux from the sun (Φ_{sun}), to the part of the photons that are not reflected, and to the amount of transmitted photons:

$$G(x) = \Phi_{sun} \cdot (1 - R) \cdot \alpha \cdot \exp(-\alpha \cdot x) \qquad (4.17)$$

The recombination rate (R, please do not confuse this with the reflectivity which is denoted by R in Chapter 3) is given by definition (3.2) and the product of the electron diffusion coefficient (D_n) and the electron lifetime can be replaced by the electron diffusion length (L_n). Equation (4.15) can be rewritten as:

$$0 = \frac{\partial^2(\Delta n)}{\partial x^2} + \frac{\Phi_{sun} \cdot (1 - R) \cdot \alpha \cdot \exp(-\alpha \cdot x)}{D_n} - \frac{\Delta n}{L_n^2} \qquad (4.18)$$

The approach for solving this differential equation is:

$$\Delta n = A1 \cdot \exp(-\alpha \cdot x) + A2 \cdot \exp\left(-\frac{x}{L_n}\right) + A3 \cdot \exp\left(\frac{x}{L_n}\right) \qquad (4.19)$$

The constant A3 is equal to 0 with respect to boundary condition (4.16''). The constant A2 is obtained from the boundary condition (4.16'):

$$A2 = -A1 \cdot \exp(-\alpha \cdot x_p) \cdot \exp\left(\frac{x_p}{L_n}\right) \qquad (4.20)$$

Constant A1 has to be calculated by taking into account Equation (4.20) and by taking the second derivative of Δn (density of photo-generated electrons) from Equation (4.18):

$$A1 = \frac{\Phi_{sun} \cdot (1 - R) \cdot \alpha}{D_n} \cdot \left(\frac{1}{\alpha^2 - \frac{1}{L_n^2}}\right) \qquad (4.21)$$

The density of photo-generated electrons is:

$$\Delta n = \frac{\Phi_{sun} \cdot (1 - R) \cdot \alpha}{D_n} \cdot \left(\frac{1}{\alpha^2 - \frac{1}{L_n^2}} \right)$$

$$\cdot \left[\exp(-\alpha \cdot x) - exp\left(-\alpha \cdot x_p + \frac{x_p - x}{L_n} \right) \right] \qquad (4.22)$$

The diffusion current density (I_n) of photo-generated electrons is defined as:

$$I_n = -q \cdot D_n \cdot \frac{\partial(\Delta n)}{\partial x} \qquad (4.23)$$

The diffusion current density is calculated following Equation (4.23), by taking into account Equation (4.22) and setting x equal to x_p:

$$I_n = q \cdot \Phi_{sun} \cdot (1 - R) \cdot \exp(-\alpha \cdot x_p) \cdot \left(\frac{1}{\frac{\alpha^{-1}}{L_n} + 1} \right) \qquad (4.24)$$

Incidentally, solution (4.24) is the part of the diffusion current density related to the neutral region in depletion layer photoeffects (Gärtner, 1959). The last term in Equation (4.24) describes the dependence of the diffusion current density on the ratio between the absorption length (α^{-1}) and the diffusion length of photo-generated electrons in the neutral region the p-type doped side of the pn-junction (L_n). The last term in Equation (4.24) is very important since it reflects the collection losses of photo-generated charge carriers for any type of solar cell. It shows, for example, that only half of the possible photocurrent is reached at the minimum carrier lifetime condition.

The dependence of the last term in Equation (4.24) is plotted as a function of α^{-1}/L_n in Figure 4.11. It can be seen that the diffusion length of photo-generated charge carriers should exceed the absorption length by a factor of 10 or 100 in order to achieve photocurrents above 90% or 98% of the possible photocurrent, respectively. Therefore, c-Si wafers with a diffusion length of even more than 1 mm are needed for the realization of c-Si solar cells with very high solar energy conversion efficiency.

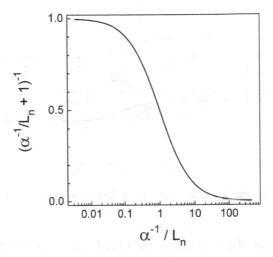

Figure 4.11. Plot of the reciprocal sum of 1 and the ratio between the absorption length (α^{-1}) and the diffusion length of photo-generated minority charge carriers (L_n) as a function of α^{-1}/L_n.

4.4 Diode Saturation Current Density of a pn-junction

4.4.1 *Fermi-level splitting and densities of minority charge carriers*

Fermi-level splitting occurs under illumination at both the n-type doped and p-type doped sides of a pn-junction. For the minority charge carriers the Fermi-energies change strongly, in contrast to the very small change of the Fermi-energies of the majority charge carriers. The Fermi-energies of the minority electrons and holes at the p-type doped and n-type doped sides of the pn-junction are denoted by $E_{Fn(p)}$ and $E_{Fp(n)}$, respectively. The Fermi-energies of the majority electrons and holes at the n-type doped and p-type doped sides of the pn-junction are denoted by $E_{Fn(n)}$ and $E_{Fp(p)}$, respectively. The Fermi-energies of the majority charge carriers contact the Fermi-energies of the minority charge carriers:

$$E_{Fn(n)} = E_{Fn(p)} \qquad (4.25')$$

$$E_{Fp(p)} = E_{Fp(n)} \qquad (4.25'')$$

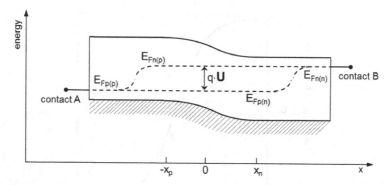

Figure 4.12. Schematic band diagram of an illuminated pn-junction with external contacts.

External leads of a solar cell contact the Fermi-energies of the majority charge carriers (Figure 4.12). For ideal ohmic contacts (see Chapter 5) the Fermi-energies in the external leads are identical to the Fermi-energies of the majority charge carriers. The difference between the Fermi-energies in the two external leads is measured as a potential energy, i.e. as the voltage drop between two wires (U) multiplied by the elementary charge. Therefore the photovoltage measured between the external leads of an illuminated solar cell with an ideal pn-junction can be written as:

$$q \cdot U \equiv E_{Fn(n)} - E_{Fp(p)} \qquad (4.26)$$

The Fermi-energies of the minority charge carriers can be expressed by the Fermi-energies of the majority charge carriers and U by using Equations (4.25) and (4.26):

$$E_{Fn(p)} = E_{Fp(p)} + q \cdot U \qquad (4.27')$$

$$E_{Fp(n)} = E_{Fn(n)} - q \cdot U \qquad (4.27'')$$

The densities of the minority charge carriers in the illuminated pn-junction are, according to Equations (4.9) and (4.27):

$$n_p = n_p^0 \cdot exp\left(\frac{q \cdot U}{k_B \cdot T}\right) \qquad (4.28')$$

$$p_n = p_n^0 \cdot exp\left(\frac{q \cdot U}{k_B \cdot T}\right) \qquad (4.28'')$$

Therefore the densities of the photo-generated minority charge carriers are:

$$\Delta n_p = n_p^0 \cdot \left[exp\left(\frac{q \cdot U}{k_B \cdot T} \right) - 1 \right] \tag{4.29'}$$

$$\Delta p_n = p_n^0 \cdot \left[exp\left(\frac{q \cdot U}{k_B \cdot T} \right) - 1 \right] \tag{4.29''}$$

4.4.2 Diode saturation current density of a semi-infinite pn-junction

An illuminated solar cell is operated under steady state condition. The diffusion coefficients and the recombination rates of the minority charge carriers are denoted by D_n, D_p, R_p and R_n, respectively. Here the steady state diffusion equations are considered for minority charge carriers injected at an external forward potential:

$$0 = D_p \cdot \frac{\partial^2 (\Delta p_n)}{\partial x^2} - R_p \tag{4.30'}$$

$$0 = D_n \cdot \frac{\partial^2 (\Delta n_p)}{\partial x^2} - R_n \tag{4.30''}$$

The diffusion Equations (4.30') and (4.30'') can be transformed by taking into account the definitions of the recombination rate (3.2) and of the diffusion length (3.5).

$$\frac{\Delta p_n}{L_p^2} = \frac{\partial^2 (\Delta p_n)}{\partial x^2} \tag{4.31'}$$

$$\frac{\Delta n_p}{L_n^2} = \frac{\partial^2 (\Delta n_p)}{\partial x^2} \tag{4.31''}$$

The density of diffusing minority charge carriers becomes 0 behind the pn-junction since the pn-junction acts as an ideal sink. The diffusion Equations (4.31') and (4.31'') can be solved with the following approach:

$$\Delta p_n = B_p \cdot exp\left(-\frac{x}{L_p} \right) \tag{4.32'}$$

$$\Delta n_p = B_n \cdot exp\left(-\frac{x}{L_n} \right) \tag{4.32''}$$

The constants B_p and B_n in Equations (4.32') and (4.32'') are obtained from the densities of minority charge carriers at the pn-junction at $x = 0$, and Δp_n and Δn_p are given by Equations (4.29') and (4.29''), respectively.

The diffusion current density of the electrons (I_n) is given by Equation (4.23) and that of the holes (I_p) is given by:

$$I_p = -q \cdot D_p \cdot \frac{\partial(\Delta p_n)}{\partial x} \tag{4.23'}$$

The diffusion current densities of the electrons (I_n) and holes (I_p) are obtained from Equations (4.23) and (4.23') by introducing the first derivative of Equations (4.32') and (4.32'') and using Equations (4.29') and (4.29''). Furthermore, it is considered that the diffusion current is independent of x and that the net current density (I) is the sum of I_p and I_n. In addition, the density of minority charge carriers in thermal equilibrium can be replaced by the density of majority charge carriers in thermal equilibrium. The densities of the majority charge carriers in equilibrium are equal to the densities of donors and acceptors if assuming complete ionization. The following final expression is obtained for the current density across the ideal pn-junction under forward bias:

$$I = I_0 \cdot \left[exp \left(\frac{q \cdot U}{k_B \cdot T} \right) - 1 \right] \tag{4.33}$$

Equation (4.33) is the diode equation of the ideal pn-junction with I_0 given by:

$$I_0 = I_{0,n} + I_{0,p} = q \cdot n_i^2 \cdot \frac{D_n}{L_n \cdot N_A} + q \cdot n_i^2 \cdot \frac{D_p}{L_p \cdot N_D} \tag{4.34}$$

The I_0 limits V_{OC} (see Chapter 1) and should therefore be as low as possible. Equation (4.34) is the simplest model for a I_0 of a solar cell and already contains eight parameters: temperature and band gap, densities of donors and acceptors at the n-type doped and p-type doped sides of the pn-junction and the diffusion coefficients and diffusion length of photo-generated minority charge carriers at the n-type doped and p-type doped sides of the pn-junction. The parameters are not independent of each other. For example, an increase of N_A or N_D leads to a decrease of L_n or L_p, respectively. The interdependence of other parameters makes the minimization of I_0 yet more complex.

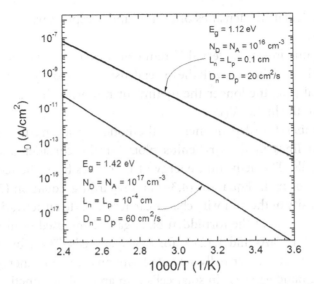

Figure 4.13. Arrhenius plots of the I_0 of two solar cells with fixed band gaps of 1.12 and 1.42 eV. Densities of donors and acceptors of 10^{17} cm^{-3} or 10^{16} cm^{-3}, diffusion lengths of photo-generated charge carriers of 1 μm or 1 mm and diffusion coefficients of 60 cm^2/s or 20 cm^2/s were assumed for the solar cell with the band gap of 1.42 eV or 1.12 eV, respectively.

Figure 4.13 gives an impression about possible orders of magnitude of diode saturation current densities for GaAs and c-Si solar cells. The diode saturation current densities of a semi-infinite pn-junction are plotted following Equation (4.34) as a function of the reciprocal temperature for two approximate sets of parameters equal for both electrons and holes.

The diffusion lengths of minority charge carriers are kept at only 1 μm for the considered GaAs solar cell. The value of I_0 at room temperature is about $3 \cdot 10^{-18}$ A/cm^2 for the considered GaAs solar cell, which is larger by more than one order of magnitude in comparison to the I_0 estimated for a record GaAs solar cell (see Chapter 1).

Relatively long diffusion lengths of minority charge carriers of 1 mm are chosen for the considered c-Si solar cell while the densities of donors and acceptors are kept at only 10^{16} cm^{-3}. The chosen parameters result in an I_0 at room temperature of about $3 \cdot 10^{-13}$ A/cm^2 for the considered c-Si solar cell, which is about three times larger than the estimated value of I_0 of the world record c-Si solar cell (see Chapter 1). The value of I_0 would be three times less if the densities of donors and acceptors were increased by three times.

In pn-junctions with very different densities of donors and acceptors, the I_0 is limited by the side with the lower density of majority charge carriers. As a general rule, the longer the lifetime of minority charge carriers, the lower I_0 and the higher V_{OC}.

The temperature dependencies of the diode saturation current densities are straight lines in the Arrhenius plots for the considered GaAs and c-Si solar cells. The activation energy of I_0 follows from the temperature dependence of n_i^2 in Equation (4.34) and regarding Equation (2.32). The squared density of the minority charge carriers leads to an activation energy equal to the value of the forbidden band gap of the used semiconductor, i.e. 1.42 eV for GaAs and 1.12 eV for c-Si solar cells. Therefore the diode saturation current is thermally activated with an activation energy $E_A^{ideal\text{-}pn}$ equal to the band gap (E_g) in solar cells with an ideal pn-junction.

$$E_A^{ideal,pn} = E_g \tag{4.35}$$

It is worth noting that the conclusion expressed in (4.35) follows from the condition that the thermal generation rate is only balanced by radiative recombination in the thermal equilibrium (see Chapter 2).

4.4.3 *Shockley–Read–Hall recombination in a pn-junction*

Photovoltaic absorbers are not perfect, they can have defect states around the middle of the forbidden band gap for which Shockley–Read–Hall (SRH) recombination rates are large (see Chapter 3). The SRH recombination rate is given by Equation (3.24). For the analysis of the influence of SRH recombination on the I_0 of a pn-junction some simplifying assumptions are introduced following Goetzberger *et al.* (1997) (see Figure 4.14).

First, it is assumed that the energy of the defect states is equal to the Fermi-energy of the un-doped intrinsic semiconductor.

$$E_t = E_i \tag{4.36}$$

Second, similar effective densities of states at the valence and conduction band edges are assumed.

$$N_C = N_V \tag{4.37}$$

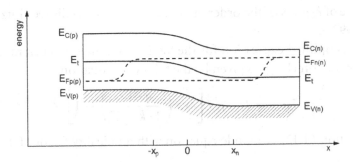

Figure 4.14. Schematic energy band diagram of a pn-junction with defect states at the middle of the forbidden band gap in non-equilibrium.

Third, it is assumed that the minimum SRH lifetimes are equal for electrons and holes.

$$\tau_{e,SRH}^{min} = \tau_{h,SRH}^{min} = \tau_0 \tag{4.38}$$

The following expression is obtained for the SRH recombination rate (R_{SRH}) if taking into account Equations (3.24), the simplifications (4.36)–(4.38), the Boltzmann equation (2.30) and the definition of the density of intrinsic charge carriers (2.32).

$$R_{SRH} = \frac{(n_0 + \Delta n) \cdot (p_0 + \Delta n) - n_i^2}{\tau_0 \cdot (n_0 + p_0 + 2 \cdot \Delta n + 2 \cdot n_i)} \tag{4.39}$$

The maximum SRH recombination rate at the pn-junction is obtained from Equation (4.39) by transforming it with the binomial equation:

$$R_{SRH,pn}^{max} = \frac{n - n_i}{2 \cdot \tau_0} \tag{4.40}$$

The density of photo-generated electrons at the pn-junction corresponds to half of the Fermi-level splitting under illumination. Therefore Equation (4.40) can be transformed to:

$$R_{SRH,pn}^{max} = \frac{n_i}{2 \cdot \tau_0} \cdot \left[exp \left(\frac{q \cdot U}{2 \cdot k_B \cdot T} \right) - 1 \right] \tag{4.41}$$

The SRH recombination rate leads to a current density which is the product of the SRH recombination rate, the elementary charge and the thickness of the region where the SRH recombination rate is large (δ_{SRH}).

The value of δ_{SRH} is of the order of 10 nm (for more details see Götzberger *et al.* (1997)).

The simplified expression for the SRH recombination current density of a pn-junction (I_{SRH}) is given by the following equation:

$$I_{SRH} = I_{0,SRH} \cdot \left[exp\left(\frac{q \cdot U}{2 \cdot k_B \cdot T} \right) - 1 \right] \qquad (4.42)$$

Equation (4.42) is a diode equation with an ideality factor of 2 and an I_0 given by:

$$I_{0,SRH} = \frac{q \cdot \delta_{SRH}}{2 \cdot \tau_0} \cdot n_i \qquad (4.43)$$

The diode saturation current density for SRH recombination is of the order of 10^{-12}–10^{-8} A/cm^2 for c-Si solar cells with respective SRH lifetimes in the c-Si absorber (τ_0 between 1 μs and 10 ms). The value of $I_{0,SRH}$ can even be much larger if the charge-selective contact is formed, for example, at a hetero-junction with a very high density of defects at the interface. Further $I_{0,SRH}$ is proportional to n_i and therefore has an activation energy of:

$$E_A^{SRH,pn} = \frac{E_g}{2} \qquad (4.44)$$

Incidentally, defect states causing SRH recombination are important for any charge-selective contact independent on whether they are formed by pn-junctions or hetero-junctions. Equations (4.42) and (4.43) describe a rather simplified case. SRH recombination can be distinguished experimentally by analyzing the ideality factor and the thermal activation energy of the I_0. In reality the ideality factors of c-Si solar cells, for example, are usually between 1.1 and 1.3.

4.4.4 *The two-diode model of an illuminated pn-junction*

A part of photo-generated charge carriers recombines in the n-type and p-type doped neutral regions while the other part of photo-generated charge carriers recombines via deep defect states at the pn-junction of a solar cell. Therefore both recombination processes are considered in the equivalent circuit of a solar cell by two diodes connected in parallel (Figure 4.15). The diode saturation current densities of the two-diode model of a solar cell are limited by so-called bulk recombination in the

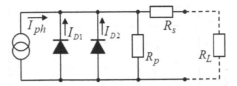

Figure 4.15. Equivalent circuit of the two-diode model of a solar cell.

neutrality regions (I_0) given by Equation (4.34) for a semi-infinite pn-junction and by SRH recombination at the charge-selective contact ($I_{0,SRH}$), given in a simplified form by Equation (4.43) for a pn-junction.

$$-I = I_0 \cdot \left[exp\left(\frac{q \cdot (U - I \cdot R_s)}{k_B \cdot T} \right) - 1 \right]$$

$$+ I_{0,SRH} \cdot \left[exp\left(\frac{q \cdot (U - I \cdot R_s)}{2 \cdot k_B \cdot T} \right) - 1 \right]$$

$$+ \frac{U - I \cdot R_s}{R_p} - I_{SC} \qquad (4.45)$$

Figure 4.16 shows examples of the influence of SRH recombination on current–voltage characteristics for a solar cell with I_0 equal to 10^{-10} A/cm^2 (a) and 10^{-12} A/cm^2(b). The current–voltage characteristics remain practically unchanged for very low values of $I_{0,SRH}$.

The fill factor (FF) and V_{OC} of a solar cell decrease with increasing $I_{0,SRH}$. The decrease of FF and V_{OC} sets in at lower values of $I_{0,SRH}$ for solar cells with lower I_0. This is shown in more detail in Figure 4.17 where the dependencies of V_{OC} and FF on $I_{0,SRH}$ are given for solar cells with I_0 equal to 10^{-9} and 10^{-13} A/cm^2. The V_{OC} of the solar cell with I_0 equal to 10^{-13} A/cm^2 starts to decrease at values of $I_{0,SRH}$ of about 10^{-8} A/cm^2 while the decrease of FF sets in already at values below 10^{-9} A/cm^2. The value of V_{OC} decreases from 0.65 V at $I_{0,SRH}$ equal to 10^{-7} A/cm^2 to 0.54 V at $I_{0,SRH}$ equal to 10^{-6} A/cm^2. At the same time the FF decreases only from about 75% to 70% in the same range of $I_{0,SRH}$.

Solar cells with large I_0 are less sensitive to recombination at the charge-selective contact. For a solar cell with I_0 equal to 10^{-9} A/cm^2 the value of V_{OC} starts to reduce remarkably for $I_{0,SRH}$ larger than 10^{-6} A/cm^2 while FF

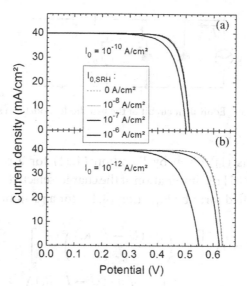

Figure 4.16. Current–voltage characteristics for solar cells with two diodes ($I_0 = 10^{-10}$ and 10^{-12} A/cm^2 (a) and (b), respectively, and $I_{SRH,0} = 0$, 10^{-8}, 10^{-7} and 10^{-6} A/cm^2, dotted, thin, middle and thick solid lines, respectively), very low R_S and very large R_p.

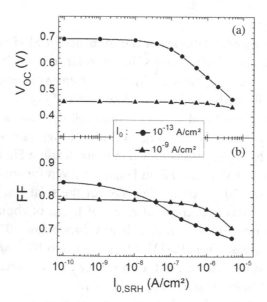

Figure 4.17. Dependence of V_{OC} (a) and FF (b) on the I_0 of SRH recombination for solar cells with I_0 equal to 10^{-9} (triangles) and 10^{-13} (circles) A/cm^2, very low R_S and very large R_p.

already reduces from 78% at $I_{0,SRH}$ equal to 10^{-7} A/cm^2 to 76% at $I_{0,SRH}$ equal to 10^{-6} A/cm^2.

4.4.5 *Consideration of surface recombination*

Solar cells have a finite thickness. Photo-generated charge carriers can reach the surfaces opposite to the pn-junction at the p-type and n-type doped sides of a solar cell. These surfaces are often known as front surface and back surface, depending on the direction of incoming sunlight or external surfaces of the absorber.

Photo-generated charge carriers can recombine at the front and back surfaces of a solar cell. The thickness of the n-type and p-type doped sides of the solar cell (H_n and H_p, respectively) and the corresponding surface recombination velocities of the minority charge carriers (s_p and s_n at the external surface of the n-type and p-type doped regions of the solar cell, respectively) have to be considered for the calculation of the I_0 of a solar cell in addition to the forbidden band gap, temperature, densities of donors and acceptors, diffusion constants and diffusion lengths of minority charge carriers (Figure 4.18).

The stationary diffusion equations (Equations (4.31)) have to be solved for minority charge carriers for the following boundary conditions:

$$\Delta n_p|_{x=x_p} = 0 \qquad (4.46')$$

$$\Delta p_n|_{x=x_n} = 0 \qquad (4.46'')$$

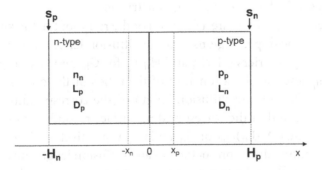

Figure 4.18. Scheme of a pn-junction with limited thickness of the p-type and n-type doped regions and given surface recombination velocities of the minority charge carriers.

and

$$\Delta n_p|_{x=H_p} \cdot s_n = -D_n \cdot \frac{\partial(\Delta n_p)}{\partial x} \tag{4.47'}$$

$$\Delta p_n|_{x=H_n} \cdot s_p = -D_p \cdot \frac{\partial(\Delta p_n)}{\partial x} \tag{4.47''}$$

The boundary conditions (4.47) take into account the limitation of surface recombination by the access of minority charge carriers to the surface (Chapter 3). The densities of minority charge carriers can be obtained by taking into account negative and positive arguments in the exponentials of the approach for solving Equations (4.47) which brings in the functions of sinh and cosh. The diffusion currents of the minority charge carriers can be calculated following Equations (4.23) and (4.23') for photo-generated electrons and holes, respectively. The solutions are given for the current densities for minority electrons and holes (Götzberger *et al.*, 1997):

$$I_n = I_{0,n} \cdot G_{fn} \cdot \left[exp\left(\frac{q \cdot U}{k_B \cdot T}\right) - 1 \right] \tag{4.48'}$$

$$I_p = I_{0,p} \cdot G_{fp} \cdot \left[exp\left(\frac{q \cdot U}{k_B \cdot T}\right) - 1 \right] \tag{4.48''}$$

The diode saturation current densities are a product of the diode saturation current densities of the semi-infinite pn-junction (see Equation (4.34)) and of the so-called geometry factors (G_{fn} and G_{fp}). Equations (4.48') and (4.48'') show that the I_0 of a solar cell can increase or decrease depending on the behavior of the geometry factor.

The geometry factors are a function of the ratios of the thickness of the p-type or n-type doped regions and the diffusion length of the respective minority charge carriers (H_p/L_n and H_n/L_p for G_{fp} and G_{fn}, respectively). Further, G_{fp} and G_{fn} are a function of the ratios of the quoteients of the diffusion coefficient and diffusion length of the corresponding minority charge carriers and of the corresponding surface recombination velocities (($D_p/L_p)/s_p$ and $(D_n/L_n)/s_n$ for G_{fp} and G_{fn}, respectively). The quotients of the corresponding diffusion coefficient and diffusion length are denoted by $s_{\infty p}$ and $s_{\infty n}$ and can be treated as recombination velocities in the p-type and n-type doped regions. The ratios $H_{p(n)}/L_{n(p)}$ and $s_{n(p)}/s_{\infty n(p)}$ have an influence on the diode saturation current densities and therefore play an important role for the value of V_{OC}.

The values of the geometry factors can be calculated by using the following equations (Götzberger *et al.*, 1997):

$$G_{fn} = \frac{\cosh\left(\frac{H_p}{L_n}\right) + \frac{s_{\infty n}}{s_n} \cdot \sinh\left(\frac{H_p}{L_n}\right)}{\frac{s_{\infty n}}{s_n} \cdot \cosh\left(\frac{H_p}{L_n}\right) + \sinh\left(\frac{H_p}{L_n}\right)} \qquad (4.49')$$

$$G_{fp} = \frac{\cosh\left(\frac{H_n}{L_p}\right) + \frac{s_{\infty p}}{s_p} \cdot \sinh\left(\frac{H_n}{L_p}\right)}{\frac{s_{\infty p}}{s_p} \cdot \cosh\left(\frac{H_n}{L_p}\right) + \sinh\left(\frac{H_n}{L_p}\right)} \qquad (4.49'')$$

The geometry factors G_{fn} and G_{fp} are plotted in Figure 4.19 as a function of the ratios H_p/L_n and H_n/L_p, respectively. The geometry factors are independent of surface recombination, i.e. equal to 1, if the n-type or p-type doped regions are thicker than the diffusion length of the corresponding minority charge carriers or if the ratios $s_{\infty p}/s_p$ or $s_{\infty n}/s_n$ are equal to 1.

There are two very different regimes if H_n or H_p are thinner than L_p or L_n, respectively. The geometry factors are larger than 1 for large

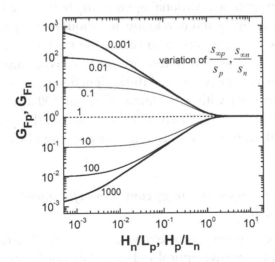

Figure 4.19. Dependence of the geometry factors on the ratio of the thickness of the doped region and the diffusion length of the corresponding minority charge carriers for different ratios of the surface recombination velocity and the bulk recombination velocity of the corresponding minority charge carriers.

surface recombination velocities, i.e. when $s_{\infty p}/s_p$ or $s_{\infty n}/s_n < 1$. This is the case, for example, for surface recombination velocities with values above 1000 cm/s or 10 cm/s for $D_{n(p)}$ of about 10 cm^2/s and $L_{n(p)}$ of about 100 μm or 1 cm, respectively. Incidentally, surface recombination velocities of even more than 10^6 cm/s are required for ohmic contacts (see Chapter 5). The geometry factors can increase by more than two orders of magnitude if, for example, H_p/L_n or H_n/L_p are about 0.01 and $s_{\infty p}/s_p$ or $s_{\infty n}/s_n$ are about 0.001. An increase of the I_0 by a factor of 100, for example, leads to a decrease of V_{OC} by 0.12 V. Therefore there is no sense, with respect to a maximum of V_{OC}, in reducing H_p/L_n or H_n/L_p to values below 1 if surface recombination cannot be controlled in a way that $s_{\infty p}/s_p$ or $s_{\infty n}/s_n$ are larger than 1, i.e. a strong increase of the diffusion length to values much larger than α^{-1} can improve carrier collection but reduce V_{OC}. In addition, advanced light trapping is only useful if it is combined with advanced surface passivation.

The ratios $s_{\infty p}/s_p$ and $s_{\infty n}/s_n$ are larger than 1 in the regime of low surface recombination. In this regime the geometry factors can be reduced by more than one order of magnitude if H_p/L_n or H_n/L_p are less than 0.1–0.01 and $s_{\infty p}/s_p$ or $s_{\infty n}/s_n$ are larger than 10–100. The reduction of the geometry factor gives an additional opportunity to increase V_{OC}. This can be achieved by applying and developing techniques for sophisticated light trapping and surface passivation to reduce the thickness of the absorber and therefore $H_{p(n)}$ and to reduce the surface recombination velocity, respectively. For example, electron diffusion lengths of 1 mm can be realized in p-type doped c-Si with a thickness of about 100 μm and a surface recombination velocity of 10 cm/s. In this case H_p/L_n and $s_n/s_{\infty n}$ are about 0.1 and 0.05, respectively, leading to a decrease of G_{fn} by about ten times.

4.4.6 *The way to very high energy conversion efficiencies of solar cells*

Very high solar energy conversion efficiencies can be reached for solar cells if all losses including resistive, optical and recombination losses are reduced to a minimum. Resistive and optical losses can already be reduced to a very few percent of the overall energy conversion efficiency. Recombination losses are important for carrier collection, i.e. for I_{SC} (short-circuit current), and for the I_0, i.e. for V_{OC}. The reduction of recombination losses to a minimum is challenging. The general way to minimize recombination losses in solar cells is (i) the increase of the diffusion lengths of minority charge carriers, (ii) the reduction of the surface recombination velocity and (iii) the decrease of the absorber thickness by light trapping.

The application of Equations (4.26), (4.37), (4.43), (4.45), (4.48) and (4.49) opens the opportunity of analyzing general trends in the dependencies of solar cell parameters on parameters of materials and combinations of materials. However, detailed simulations of the complete solar cell including the local morphology and the architecture of the solar cell are required for the optimization of a certain type of solar cell under given conditions taking into account the detailed parameters of implemented materials, combinations of materials and technologies.

Figure 4.20 gives an example for the dependence of I_{SC} and V_{OC} on the surface recombination velocity for different diffusion lengths of electrons

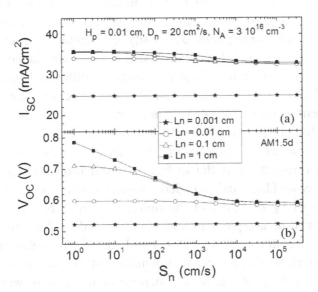

Figure 4.20. Dependence of I_{SC} and V_{OC} at air mass (AM)1.5d on the surface recombination velocity for a c-Si solar cell with a very thin n-type and a 100-μm thick p-type doped region for different diffusion lengths of electrons photo-generated in the p-type doped region. Resistive losses and losses due to reflection are not considered. An optical path equal to the absorber thickness is taken into account.

photo-generated in the p-type doped absorber at fixed H_p and very thin H_n. Light trapping is not taken into account. The values of I_{SC} and V_{OC} are 25 mA/cm^2 and 0.522 V, respectively, independent of s_n for a diffusion length of only 10 μm. A diffusion length of 100 μm is significantly larger or close to the absorption length for most photons in the sun spectrum with energy above the forbidden band gap of c-Si. An increase of L_n to 100 μm leads to an increase of I_{SC} and V_{OC} to 34 mA/cm^2 and 0.6 V for low values of s_n and of about 33 mA/cm^2 and 0.58 V for high values of s_n, respectively, i.e. the influence of surface recombination becomes remarkable but is still weak. It is worth noting that the energy conversion efficiency is about 16% for c-Si solar cells with related parameters.

An increase of L_n to 1 mm leads to saturation of I_{SC} at 36 mA/cm^2 for low values of s_n while I_{SC} remains at about 33 mA/cm^2 for high values of s_n. At the same time V_{OC} increases only slightly to about 0.59 V for high values of s_n. The V_{OC} reaches 0.62 V for s_n equal to 10^3 cm/s and L_n equal to 1 mm. This means that an increase of the L_n to values above 100 μm is not useful for c-Si solar cells if the surface recombination velocity cannot be reduced to values below 1000 cm/s. A surface recombination velocity of 10^3 cm/corresponds to a density of recombination active surface defects of the order of 10^{11} cm^{-2} if taking into account Equation (3.26') and Figure 3.11. The value of V_{OC} increases for a diffusion length of 1 mm to 0.70 and 0.71 V for s_n equal to 10 and 1 cm/s, respectively, corresponding to the present world record (Zhao *et al.*, 1998).

The value of V_{OC} should be increased to reach a new world record of the energy conversion efficiency of c-Si solar cells. This may be reached by a further reduction of the surface recombination velocity and by a further increase of L_n. This is very demanding, but seems possible. However, an increase of L_n to values of the order of 1 cm demands a reduction of the density of acceptors to values of the order of $3 \cdot 10^{15}$ cm^{-3} if taking into account Equations (3.12) and (3.15). But a decrease of N_A causes an increase of I_0 according to Equation (4.34) and therefore causes a reduction of V_{OC}. A significant increase of the present world record of the energy conversion efficiency of c-Si solar cells can therefore only be achieved by reducing H_p to values of the order of only 10 μm. This is possible, in principle, by implementing strategies of advanced light trapping into solar cells (see Figure 2.14). However, one has to keep in mind that related measures will demand a *de facto* revolution in production technologies.

4.5 Summary

A photo-generated charge carrier can pass an ideal charge-selective contact only in the direction in which the potential energy is reduced for the given charge carrier. This condition should hold for both photo-generated electrons and holes, which are separated to the opposite sides of an ideal charge-selective contact. Contacts at which the potential energy is reduced only for one kind of photo-generated charge carriers are not ideal charge-selective.

At a contact between a p-type and n-type doped semiconductor layer (pn-junction), the potential energy is reduced at the side where the charge carriers are the majority. This means that photo-generated electrons are separated from the p-type doped region into the n-type doped region, and similarly the photo-generated holes are separated from the n-type doped region into the p-type doped region. Therefore ideal charge-selective contacts can be formed by pn-junctions.

Charge-selective contacts such as pn-junctions are sinks for photo-generated minority charge carriers, which reach the sink from neutral regions of the photovoltaic absorber by diffusion. Space charge regions are formed at pn-junctions and the corresponding potential drop across the space charge region is the driving force for the diffusion of minority charge carriers towards the pn-junction. Therefore the potential drop across a pn-junction is the diffusion potential of a pn-junction. The diffusion potential is the upper limit of V_{OC}.

Photo-generated minority charge carriers are collected at a pn-junction. All charge carriers photo-generated in a space charge region of a pn-junction are collected since the drift time is much shorter than the lifetime of photo-generated charge carriers. For charge carriers photo-generated in the neutral bulk of a photovoltaic absorber, the efficiency of the collection depends on the distance between the pn-junction and the region in which the minority charge carriers were photo-generated in relation to the diffusion length of minority charge carriers. Therefore, in first approximation, the quantum efficiency of a solar cell is limited by the reciprocal sum of 1 and the quotient of the absorption length and the diffusion length. Investigation into the spectral dependence of the quantum efficiency gives the opportunity of obtaining information about the diffusion length of photo-generated minority charge carriers.

The Fermi-energy of photo-generated minority charge carriers is contacted with the Fermi-energy of the majority charge carriers at the opposite side of a pn-junction, and the Fermi-energy of majority charge carriers is contacted with external leads in solar cells. Therefore the electric potential between the two external leads of a solar cell corresponds to the difference between the Fermi-energies of the minority charge carriers in the p-type and n-type doped layers forming the pn-junction.

The most important characteristic of a charge-selective contact is the I_0 which limits V_{OC}. A diode equation related to bulk recombination follows from the dependence of the density of minority charge carriers on an external potential and diffusion of minority charge carriers towards the pn-junction. The value of I_0 depends on the squared density of intrinsic charge carriers (i.e. on temperature, band gap and effective masses of electrons and holes), on the densities of acceptors and donors at the p-type and n-type doped regions of the pn-junction and on the diffusion constants and diffusion lengths of minority charge carriers in the p-type and n-type doped regions. The ideality factor of a diode limited by bulk recombination is equal to 1.

Shockley–Read–Hall recombination via deep defect states at a pn-junction also limits the current density across a pn-junction. The dependence of the SRH recombination current on the external potential is described by a diode equation with a diode saturation current density $I_{0,SRH}$. The value of $I_{0,SRH}$ is proportional to the density of intrinsic charge carriers, the extension of the region with a high SRH recombination rate and the inverse SRH lifetime. The ideality factor of a diode limited by SRH recombination at the pn-junction is equal to 2.

The two-diode model considers recombination in general in the bulk of the photovoltaic absorber and SRH recombination at the charge-selective contact. Two diodes with diode saturation current densities I_0 and $I_{0,SRH}$ and with corresponding ideality factors of 1 and 2 are connected in parallel in the two-diode model. Both recombination processes can be distinguished by the thermal activation energy of the I_0, which is equal to the band gap of the photovoltaic absorber for bulk recombination and to half of the band gap for SRH recombination at the charge-selective contact.

Surface recombination at the external surface of the p-type and n-type doped layers of a solar cell can lead additionally to a reduction of the density of minority charge carriers if the diffusion length of minority charge carriers is equal or less than the thickness of the corresponding p-type or n-type doped layer. The influence of surface recombination is described by the geometry factor multiplied by I_0. The geometry factor depends on the ratio between the thickness of the p-type or n-type doped layer and the diffusion length of the corresponding minority charge carriers, as well as on the ratio between the surface recombination velocity and the diffusion coefficient divided by the diffusion length of the corresponding minority charge carriers.

Very high solar energy conversion efficiencies can be reached only for solar cells with low thicknesses of photovoltaic absorbers where the diffusion length of minority charge carriers is very high and the surface recombination velocities are extremely low. Related conditions can only be reached with GaAs (see Chapter 8) and c-Si (see Chapter 7) solar cells.

4.6 Tasks

T4.1: Diffusion potential

Calculate the U_D of a pn-junction with (a) $N_A = 10^{16}$ and $N_D = 10^{19}$ cm^{-3} for c-Si and (b) $N_A = 10^{18}$ and $N_D = 10^{17}$ cm^{-3} for GaAs at room temperature.

T4.2: Space charge regions of a pn-junction

Calculate the thickness of the space charge regions for a pn-junction with (a) $N_A = 10^{16}$ and $N_D = 10^{19}$ cm^{-3} for a c-Si solar cell and (b) $N_A = 10^{18}$ and $N_D = 10^{17}$ cm^{-3} for a GaAs solar cell at room temperature.

T4.3: Maximum electric field at a pn-junction

Ascertain the maximum electric field at a pn-junction with (a) $N_A = 10^{16}$ and $N_D = 10^{19}$ cm^{-3} for a c-Si solar cell and (b) $N_A = 10^{18}$ and $N_D = 10^{17}$ cm^{-3} for a GaAs solar cell at room temperature.

T4.4: Saturation current density of a semi-infinite pn-junction in the limit of Auger recombination

Ascertain the diode I_0 of a semi-infinite pn-junction of a c-Si solar cell with $N_A = 10^{16}$ and $N_D = 10^{19} \, cm^{-3}$ at room temperature and compare with a c-Si record solar cell.

T4.5: Diode saturation current density of a semi-infinite pn-junction in the limit of radiative recombination

Ascertain the I_0 of a semi-infinite pn-junction of a GaAs solar cell with $N_A = 10^{18}$ and $N_D = 10^{17} \, cm^{-3}$ at room temperature and compare with a record GaAs solar cell.

T4.6: Temperature dependence of V_{OC} for c-Si solar cells

Ascertain the temperature dependence of V_{OC} for a c-Si solar cell with the I_0 in the Auger limit.

T4.7: Diode saturation current density for SRH recombination at the pn-junction

Maximum densities of defects with electronic states in the middle of the band gap were obtained with regard to the minimum lifetime condition. Estimate the $I_{0,SRH}$ for recombination via defects at a pn-junction for a c-Si solar cell with (a) $N_t = 10^{12} \, cm^{-3}$ and (b) $N_t = 10^{16} \, cm^{-3}$ at room temperature.

T4.8: Dependence of V_{OC} on SRH recombination at the pn-junction

Ascertain the dependence of V_{OC} on the diode saturation current densities considering bulk recombination (I_0) and SRH recombination ($I_{0,SRH}$) at a pn-junction and plot V_{OC} as a function of $I_{0,SRH}$ for $I_0 = 10^{-26}$, 10^{-20}, 10^{-14} and $10^{-8} \, A/cm^2$ ($I_{SC} = 0.04 \, A/cm^2$).

T4.9: Geometry factor of solar cells with a pn-junction

Estimate the geometry factor for the electron I_0 of a c-Si solar cell with $N_A = 10^{16}$ and $N_D = 10^{19} \, cm^{-3}$. The thickness of the p-type doped region and of the surface recombination velocity are $H_p = 150 \, \mu m$ and $s_n = 10 \, cm/s$, respectively.

5

Ohmic Contacts for Solar Cells

Ohmic contacts connect a photovoltaic absorber from the two sides of the charge-selective contact of a solar cell with metal leads connecting to external load. This chapter is devoted to the concept of ideal ohmic contacts and to realization principles of ohmic contacts at metal–semiconductor and semiconductor–semiconductor interfaces. Barriers are formed at metal–semiconductor interfaces depending on the metal work function, the electron affinity, doping of the semiconductor and interfacial defects. High densities of interface defects cause Fermi-level pinning. Tunneling of free charge carriers through very thin barrier layers at highly doped semiconductors is decisive for ohmic contacts in solar cells with very high solar energy conversion efficiencies. The transmission probability is derived for a triangular barrier the width and height of which are given by the extension of the space charge region and by the barrier at the contact, respectively. The tunneling or contact resistance is obtained as a function of the barrier height and of the density of dopants in a highly doped semiconductor. Consequences for the tolerable series resistance and design of solar cells are discussed. Furthermore, tunneling contacts are considered between a highly doped n-type and a highly doped p-type semiconductor for integrated series connection in multi-junction solar cells. Ohmic contacts may be realized as well by defect-assisted tunneling through barrier layers.

5.1 Concept of Ideal Ohmic Contacts

Metal leads connect a solar cell with an external load. A metal–semiconductor contact is formed at the interface between a metal lead

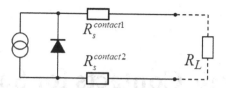

Figure 5.1. Equivalent circuit of a solar cell with contact resistances and a load.

and a photovoltaic absorber. Metal–semiconductor contacts are crucial for minimizing losses in a solar cell.

Ideal metal–semiconductor contacts are expressed by two ohmic contacts in the equivalent circuit of a solar cell (Figure 5.1). Ohmic means that the voltage drop across the contact is proportional to the current flowing through the contact independent of the direction of the current, i.e. the function of charge-selectivity of a solar cell is not influenced by ohmic metal–semiconductor contacts. As a consequence, ideal ohmic metal–semiconductor contacts do not introduce potential barriers either for mobile electrons or holes. The proportionality factor between the voltage drop across an ohmic contact and the flowing current is known as contact resistance ($R_s^{contact1}$, $R_s^{contact2}$).

For ideal ohmic contacts of a solar cell, the Fermi-energies of the two metal leads (E_{Fm1} and E_{Fm2}) are aligned with the Fermi-energies of the majority charge carriers ($E_{Fn(n)}$ and $E_{Fp(p)}$) at the two sides of the photovoltaic absorber, which are separated by the charge-selective contact (see Chapter 4). Therefore, the potential difference between the external leads of a solar cell (U) is equivalent to the difference between $E_{Fn(n)}$ and $E_{Fp(p)}$. With regard to Ohm's law and Equation (4.26), one can write the following condition for ideal ohmic contacts of a solar cell:

$$U = I \cdot R_L = \frac{E_{Fn(n)} - E_{Fp(p)}}{q} - I \cdot (R_s^{contact1} + R_s^{contact2}) \qquad (5.1)$$

Equation (5.1) becomes equal to Equation (4.26) if the values of $R_s^{contact1,2}$ are so low that the voltage drops across the ohmic contacts can be neglected. For reaching very high solar energy conversion efficiencies, the values of $R_s^{contact1}$ and $R_s^{contact2}$ should be lower than the tolerable series resistance of a solar cell (see also Chapter 1).

Ohmic contacts are not charge-selective. This means, in contrast to charge-selective contacts, that a splitting of Fermi-energies is impossible

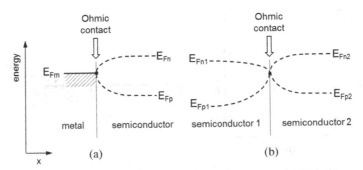

Figure 5.2. Band diagrams of ideal ohmic contacts between a metal and a semiconductor (a) and between two semiconductors (b).

at ohmic contacts. Therefore, the Fermi-energy of the metal (E_{Fm}) is equal to the Fermi-energies of free electrons and holes at ohmic metal–semiconductors contacts (Figure 5.2(a)). The definition of ideal ohmic contacts with n-type or p-type doped semiconductors is given as:

$$E_{Fm} \equiv E_{Fn(n)} \equiv E_{Fp(n)} \qquad (5.2')$$
$$E_{Fm} \equiv E_{Fn(p)} \equiv E_{Fp(p)} \qquad (5.2'')$$

As a consequence of definition (5.2), the surface recombination velocity is infinitely high at ideal ohmic contacts. In reality, the surface recombination velocity (see Equations (3.26') and (3.26'')) does not exceed the thermal velocity of free charge carriers since there is no physical meaning if the product of the capture cross section and the density of surface states becomes larger than 1. Therefore, the surface recombination velocity at ideal ohmic contacts is about 10^7 cm/s.

Ohmic contacts can also be realized between two semiconductors (Figure 5.2(b)). Ohmic contacts between two semiconductors are called recombination contacts. Recombination contacts are decisive for integrated series connections in multi-junction solar cells (see also Chapters 6 and 8).

Recombination losses at ohmic contacts have to be minimized despite the fact that the surface recombination velocity cannot be reduced. The recombination rate at ohmic contacts is proportional to the density of photo-generated charge carriers. Therefore, recombination losses at ohmic contacts can be reduced by minimizing the density of photo-generated

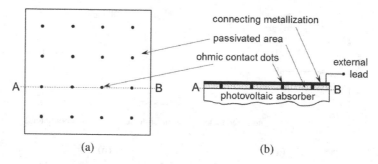

(a)   (b)

Figure 5.3. Bottom view of a solar cell with ohmic contact dots which connect the metallization and passivated areas (a) and a cross section along line A–B (b).

minority charge carriers in front of the ohmic contact. The density of photo-generated minority charge carriers at ohmic contacts is reduced by implementing potential barriers, which reject minority charge carriers in the photovoltaic absorber in front of the ohmic contact. Related barriers, for example, can be achieved with a so-called back surface field (see Chapter 7).

Another measure for reducing recombination losses at ohmic contacts in solar cells is related to the drastic reduction of the area of the ohmic contact by a factor of the order of 1000 in combination with the excellent surface passivation of the part of the semiconductor surface (see Chapter 3) that is not covered by an ohmic contact (Figure 5.3).

The ohmic contact area is minimized, for example, by using interconnected grid lines at illuminated front contacts and by applying contact dots at back contacts. The distance between ohmic contact dots has to be larger than the thickness of the photovoltaic absorber and/or the diffusion length of minority charge carriers. In this case, the probability of photogenerated minority charge carriers reaching the charge-selective contact is much higher than the probability of recombination at the ohmic contact. The ohmic contact dots are connected with a metallization layer at back contacts. The reduction of the ohmic contact area is decisive for reaching very high solar energy conversion efficiencies, especially with crystalline silicon (c-Si) solar cells (see Chapter 7).

The minimization of the ohmic contact area as well as the concentration of sunlight demand very low resistances of metal–semiconductor contacts in solar cells of the order of 10^{-4}–10^{-5} Ωcm^2 or even less.

5.2 Ohmic Metal–Semiconductor Contacts

The behavior of metal–semiconductor contacts has been described in numerous textbooks, such as Rhoderick (1978) or Sze (1981). Here only the principal aspects that are important for solar cells are considered.

5.2.1 *Work function, electron affinity and ionization energy*

Fermi-energies of metals (E_{Fm}) in relation to the energies of valence and conduction band edges of semiconductors (E_V and E_C) are important for charge transport across metal–semiconductor interfaces. All characteristic energies can be compared to one common reference energy — the vacuum level (E_{vac}). This is expressed in Figure 5.4.

The work function of a metal (Φ_m) is defined as the difference between the vacuum level and the Fermi-energy of the metal.

$$\Phi_m \equiv E_{vac} - E_{Fm} \tag{5.3}$$

The work function of a metal corresponds to the energy, which is needed to replace a free electron from the metal into the vacuum, for example, by photoemission. In an ultra-high vacuum, the work function of pure metal surfaces depends on the nature of the metal and on the facet from which electrons are emitted into the vacuum. For example, the values of Φ_m range between 5.1 and 5.47 eV for gold surfaces, between 4.52 and 4.74 eV for silver surfaces, and between 4.06 and 4.26 eV for aluminum surfaces. Calcium has a low work function of 2.87 eV (for more details see Sze (1981) and the references therein).

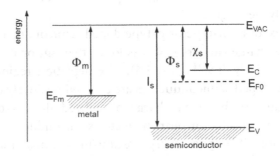

Figure 5.4. Energy diagram of the metal work function (Φ_m) and of the electron affinity (χ_s), the work function (Φ_s) and the ionization energy (I_s) of a semiconductor.

The work function of a semiconductor (Φ_s) is defined as the difference between the vacuum level and the Fermi-energy of the semiconductor in thermal equilibrium (E_{F0}). The Fermi-energy of a semiconductor in thermal equilibrium — and therefore also Φ_s — depend on the doping of the semiconductor. The ionization energy of a semiconductor (I_s) is the energy needed to transfer an electron from a bond state into vacuum, i.e. the difference between the vacuum level and the valence band edge. The electron affinity of a semiconductor (χ_s) is defined as the difference between the conduction band edge and the vacuum level. The work function of a semiconductor can be calculated if the ionization energy or the electron affinity and the doping of the semiconductor (see Equation (2.37)) are known:

$$\Phi_s = I_S - (E_{F0} - E_V) = \chi_s + (E_C - E_{F0}) \tag{5.4}$$

The electron affinity, for example, is 4.05 eV for c-Si and 4.07 eV for gallium arsenide (GaAs) (Cowley and Sze, 1965).

5.2.2 Formation of metal–semiconductor contacts

When a metal and a semiconductor are brought into intimate contact, electrons flow from the side with the higher Fermi-energy to the side with the lower Fermi-energy until equilibrium is reached, i.e. electrons are transferred from the side with the lower work function to the side with the higher work function.

Majority charge carriers are accumulated at the surface of an n-type or p-type doped semiconductor if Φ_m is larger than Φ_s or if Φ_m is lower than Φ_s, respectively (see Figure 5.5).

The surface of an n-type or p-type doped semiconductor is depleted from majority charge carriers if Φ_m is lower than Φ_s or if Φ_m is higher than Φ_s, respectively (see Figure 5.6). The depletion regions of the n-type or p-type doped semiconductors are charged positively or negatively, respectively, due to the fixed charge of ionized donors or acceptors. Charging the semiconductor near the metal–semiconductor contact leads to the formation of a space charge region in a similar manner to the formation of a space charge region at a pn-junction (see Chapter 4). The distribution of the electric field and of the electrostatic potential in the

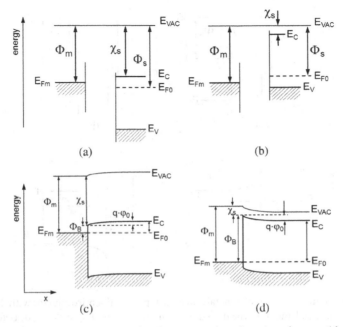

Figure 5.5. Band diagrams of a metal and an n-type doped semiconductor with $\Phi_m < \Phi_s$ (a, c) and of a metal and a p-type doped semiconductor with $\Phi_m > \Phi_s$ (b, d) before (a, b) and after (c, d) contact formation. Majority charge carriers are accumulated at the semiconductor surfaces (c, d).

space charge region can be calculated by integrating the Poisson equation (4.1) once or twice, respectively.

The transfer of electrons from the metal into the semiconductor requires energy equal to or larger than the difference between the electron affinity of the semiconductor and the metal work function. The energetic barrier for electron transfer across metal–semiconductor contacts (Φ_B) is given by Φ_m and χ_s (the Schottky limit):

$$\Phi_B = \Phi_m - \chi_s \tag{5.5}$$

The conduction or valence bands of depleted surfaces of n-type or p-type doped semiconductors bend upwards or downwards, respectively. The so-called surface band bending (φ_0) is positive or negative for depleted surfaces of n-type or p-type doped semiconductors, respectively. The surface band bending is equal to the difference between Φ_B and the difference between the conduction band edge and the Fermi-energy in

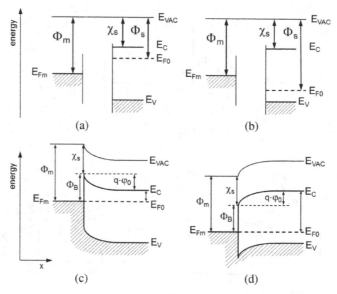

Figure 5.6. Band diagrams of a metal and an n-type doped semiconductor with $\Phi_m > \Phi_s$ (a, c) and of a metal and a p-type doped semiconductor with $\Phi_m < \Phi_s$ (b, d) before (a, b) and after (c, d) contact formation. The semiconductor surfaces are depleted from majority charge carriers (c, d).

the bulk semiconductor (E_C-E_{F0}).

$$q \cdot \varphi_0 = \Phi_B - (E_C - E_{F0}) \qquad (5.6)$$

For ohmic contacts, the potential energy of majority charge carriers should decrease towards the contact such that majority charge carriers can flow from the semiconductor into the metal. This is the case for metal–semiconductor contacts, which are in accumulation (Figure 5.5(c, d)). The surface band bending at metal–semiconductor contacts that are in depletion corresponds to the increase of the potential energy of majority charge carriers moving from the bulk of the semiconductor towards the metal–semiconductor contact (Figure 5.6(c, d)). Metal–semiconductor contacts with a barrier for majority charge carriers, i.e. $q \cdot \varphi_0$, are called Schottky contacts (Schottky, 1938).

Metals with a relatively low work function close to the electron affinity of the semiconductor are required for the formation of an ohmic contact with a given n-type doped semiconductor. In contrast, metals with a much larger metal work function close to the ionization energy of the semiconductor are needed for the formation of an ohmic contact with the same semiconductor, p-type doped.

The barrier height only depends on the nature of the contacting metal and semiconductor in the Schottky limit. However, interfacial layers at metal–semiconductor contacts can strongly influence the value of Φ_B. Interfacial layers at metal–semiconductor contacts are caused, for example, by inter-diffusion of atoms across the metal–semiconductor interface, by partial replacement of atoms at the semiconductor surface with metal atoms or by residual contamination, which was not removed before contact formation. The thickness of related interfacial layers can be of the order of several mono-layers of atoms.

Interfacial layers introduce defect states at the surface of the semiconductor very close to the metal contact (see Figure 5.7). An electric dipole is formed at the interface dipole layer disturbing the dependence of Φ_B on Φ_m:

$$\Phi_B = \Phi_m - \chi_s - \Delta \qquad (5.7)$$

Defect states are part of metal–semiconductor contacts. Occupied donor interface defect states are neutral, whereas unoccupied donor interface defect states are positively charged. Occupied or unoccupied acceptor interface defect states are negatively charged or neutral, respectively.

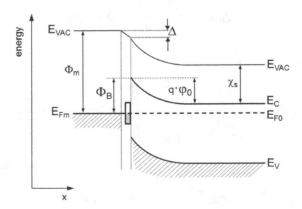

Figure 5.7. Band diagram of a metal–semiconductor contact with an interfacial layer and defects at the semiconductor surface.

5.2.3 *Fermi-level pinning at metal–semiconductor contacts*

A variation of the metal work function causes a shift of the Fermi-energy at the semiconductor surface (E_{FS}) and therefore a change of the occupation of defect states, i.e. the electrostatic charge at the semiconductor surface is changed due to a change of E_{FS}. The change of the electrostatic charge at surface states is compensated by a change of the electrostatic charge in the depletion region of the semiconductor.

Defect states with a high density can pin the Fermi-energy at a semiconductor surface. Fermi-level pinning means that E_{FS} cannot be shifted in relation to the conduction or valence band edges at the surface of the semiconductor (E_{CS} and E_{VS}, respectively). In this case, the barrier height at a metal–semiconductor contact is independent of the metal work function (the so-called Bardeen limit; Bardeen (1947)). In the following, the Fermi-level pinning will be analyzed for a free semiconductor surface with donor and acceptor surface states (Figure 5.8).

With regard to charge neutrality at a free semiconductor surface, the sum of the charge in the space charge region (Q_{sc}), of the charge in donor ($Q_{st,d}$) and of the charge in acceptor ($Q_{st,a}$) states is equal to 0.

$$0 = Q_{sc} + Q_{st,d} + Q_{st,a} \tag{5.8}$$

Following Equations (4.5) and (4.6), the fixed charge in the space charge region, for example, of an n-type doped semiconductor is:

$$Q_{sc,n} = \sqrt{\frac{2 \cdot \varepsilon_r \cdot \varepsilon_0}{q} \cdot N_D \cdot \varphi_0} \tag{5.9'}$$

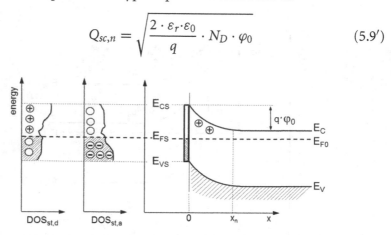

Figure 5.8.   Band diagram and schematic densities of surface states at a semiconductor surface with occupied and unoccupied donor and acceptor surface states.

The band bending at the surface and the space charge of a p-type doped semiconductor in depletion are negative:

$$Q_{sc,p} = -\sqrt{\frac{2 \cdot \varepsilon_r \cdot \varepsilon_0}{q} \cdot N_A \cdot |\varphi_0|} \tag{5.9''}$$

In thermal equilibrium, the charge in occupied acceptor and unoccupied donor states follows from integration over the densities of acceptor ($DOS_{st,a}$) and donor ($DOS_{st,d}$) surface states from the valence band edge at the surface (E_{VS}) to the Fermi-energy at the surface (E_{FS}), and from E_{FS} to the conduction band edge at the surface (E_{CS}), respectively.

$$Q_{st,a} = -\int_{E_{VS}}^{E_{FS}} DOS_{st,a}(E) \cdot dE \tag{5.10'}$$

$$Q_{st,d} = \int_{E_{FS}}^{E_{CS}} DOS_{st,d}(E) \cdot dE \tag{5.10''}$$

For qualitative analysis, it is sufficient to assume constant distributions of $DOS_{st,a}$ and $DOS_{st,d}$. The densities of acceptor- and donor-like surface defects are $N_{st,a}$ and $N_{st,d}$, respectively.

$$Q_{st,a} = -N_{st,a} \cdot \frac{E_{FS} - E_{VS}}{E_g} \tag{5.11'}$$

$$Q_{st,d} = N_{st,d} \cdot \frac{E_{CS} - E_{FS}}{E_g} \tag{5.11''}$$

The differences E_{FS}–E_{VS} and E_{CS}–E_{FS} can be expressed by the surface band bending and by the Fermi-energy of majority charge carriers in the bulk of an n-type doped semiconductor:

$$E_{FS} - E_{VS} = E_g - (E_C - E_{F0}) - q \cdot \varphi_0 \tag{5.12'}$$

$$E_{CS} - E_{FS} = (E_C - E_{F0}) + q \cdot \varphi_0 \tag{5.12''}$$

Similarly, the differences E_{FS}–E_{VS} and E_{CS}–E_{FS} can be expressed for a p-type doped semiconductor by:

$$E_{FS} - E_{VS} = -[(E_V - E_{F0}) + q \cdot \varphi_0] \tag{5.12'''}$$

$$E_{CS} - E_{FS} = E_g + (E_V - E_{F0}) + q \cdot \varphi_0 \tag{5.12''''}$$

The differences between the conduction and valence band edges and the Fermi-energy in the bulk of the semiconductor can be calculated by using Equation (2.37) and by taking into account complete ionization of donors or acceptors.

$$E_C - E_{F0} = k_B \cdot T \cdot ln\frac{N_C}{N_D} \tag{5.13'}$$

$$-(E_V - E_{F0}) = k_B \cdot T \cdot ln\frac{N_V}{N_A} \tag{5.13''}$$

Equation (5.8) can be transformed for n-type doped semiconductors.

$$\sqrt{\frac{2 \cdot \varepsilon_r \cdot \varepsilon_0 \cdot N_D}{q}} \cdot \sqrt{\varphi_0} + \frac{q \cdot \varphi_0 + (E_C - E_{F0})}{q} \cdot N_{st,d}$$

$$- \frac{E_g - q \cdot \varphi_0 - (E_C - E_{F0})}{q} \cdot N_{st,a} = 0 \tag{5.14}$$

An equivalent expression can be obtained for p-type doped semiconductors. Equation (5.14) has the following form:

$$0 = \varphi_0^2 - (2 \cdot B + A) \cdot \varphi_0 + B^2 \tag{5.15}$$

with

$$A = \frac{2 \cdot \varepsilon_r \cdot \varepsilon_0 \cdot N_D}{q} \cdot \left[\frac{E_g}{q \cdot (N_{st,a} + N_{st,d})}\right]^2 \tag{5.15'}$$

and

$$B = \frac{N_{st,a}}{N_{st,a} + N_{st,d}} \cdot \frac{E_g}{q} - \frac{E_C - E_{F0}}{q} \tag{5.15''}$$

The solution of Equation (5.15) in a physical meaning is:

$$\varphi_0 = \frac{2 \cdot B + A}{2} - \sqrt{\left(\frac{2 \cdot B + A}{2}\right)^2 - B^2} \tag{5.16}$$

Figure 5.9 shows the dependence of the surface band bending of an n-type doped semiconductor as a function of $N_{st,d}$ for different values of $N_{st,a}$. If $N_{st,d} = N_{st,a} = N_{st}$, the surface band bending can be neglected for values less than $5 \cdot 10^{10}$ cm^{-2} where φ_0 increases for larger values of $N_{st,d}$ and saturates at large values of $N_{st,d}$ above 10^{13} cm^{-2}. The value of E_{FS} saturates

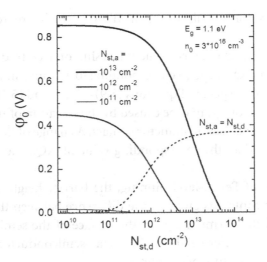

Figure 5.9. Dependence of the surface band bending at a free surface of an n-type doped semiconductor on the density of donor-like surface defects for constant values of the density of acceptor-like surface defects ($N_{st,a} = 10^{11}, 10^{12}$ and 10^{13} cm^{-2}, thin, medium and thick solid line, respectively) and for $N_{st,d} = N_{st,a}$ (dashed line). The Fermi-energy at the surface is pinned if $N_{st,a}$ and/or $N_{st,d}$ are larger than about 10^{13} cm^{-2}. The surface band bending should be as low as possible for ohmic contacts.

at the middle of E_g (the band gap) for N_{st} larger than 10^{13} cm^{-2}, i.e. the Fermi-energy is pinned at midgap. For comparison, density of surface states larger than 10^{13} cm^{-2} was obtained for metal contacts with c-Si, gallium phosphide (GaP) and GaAs (Cowley and Sze, 1965). The ratio between N_D (the density of donors) and N_{st}^2 is important for Fermi-level pinning (see also task T5.1 at the end of this chapter).

For an n-type doped semiconductor and for large values of $N_{st,a}$, the value of E_{FS} can be pinned at energies closer to E_{VS} ($N_{st,a} \gg N_{st,d}$, large surface band bending) or closer to E_{CS} ($N_{st,a} \ll N_{st,d}$, very low surface band bending). Similarly, E_{FS} can be pinned at energies closer to E_{CS} ($N_{st,d} \gg N_{st,a}$, large surface band bending) or closer to E_{VS} ($N_{st,d} \ll N_{st,a}$, very low surface band bending) for a p-type doped semiconductor.

The ratio between the densities of donor and acceptor surface states depends on chemical reactions at metal–semiconductor contacts. The barrier height, for example, can sensitively depend on the thermodynamic heat of the reaction (Brillson, 1978). The barrier height at metal–semiconductor contacts is less for metals or metal alloys, which are reactive at the semiconductor surface.

A shift of E_{FS} by ΔE_{FS} is equivalent to a shift of $q \cdot \varphi_0$ by $q \cdot \Delta \varphi$. Changes of Q_{st} (ΔQ_{st}) and Q_{sc} (ΔQ_{sc}) corresponding to a change in E_{FS} and φ_0 can be calculated with respect to Equations (5.9)–(5.11) and (5.16)–(5.18). For example, a change of E_{FS} may be caused by the variation of the metal work function at the metal–semiconductor contact. A change of E_{FS} is impossible if ΔQ_{st} is larger than the corresponding value of ΔQ_{sc}, i.e. E_{FS} is pinned (see also task T5.2).

In the case of Fermi-level pinning, the barrier height at the metal–semiconductor contact is given by the difference between the conduction band edge and the Fermi-energy at the surface of the semiconductor. In this instance, the barrier height at the metal–semiconductor contact does not depend on the nature of the metal.

The barrier at a metal–semiconductor contact can be described by the following empirical equation:

$$\Phi_B = \Phi_0 + S \cdot (\Phi_m - \chi_s) \tag{5.17}$$

The value of Φ_0 corresponds to the barrier height in the Bardeen limit and is equal to 0 at the Schottky limit. The empirical S-parameter in Equation (5.17) is equal to 0 for the Bardeen limit and equal to 1 for the Schottky limit. Photovoltaic absorbers with covalent bonds are close to the Bardeen limit, for example, the parameter S is about 0.05 for c-Si (Kurtin *et al.*, 1970). The value of S increases for compound semiconductors with increasing ionic character of bonding, for example, S is about 0.25 for cadmium telluride (CdTe), and S is practically equal to 1 for ionically bonded semiconductors or insulators such as zinc oxide (ZnO) (Kurtin *et al.*, 1970).

The Fermi-energy at a metal–semiconductor contact is often close to the middle of the band gap of the semiconductor in the Bardeen limit, i.e. the semiconductor is in depletion. A depletion region near metal–semiconductor contacts is a barrier layer (Schottky barrier (Schottky, 1938)) for majority charge carriers (Figures 5.6(c) and 5.6(d)). Schottky barriers are not ohmic but charge-selective. The analysis of charge transport across Schottky barriers (see, for example, Sze (1981)) is beyond the scope of this book since Schottky barriers are usually not applied for solar cells due to the limitation of the open-circuit voltage (V_{OC}) — which cannot exceed the surface band bending — and to a very high surface recombination velocity at metal–semiconductor contacts. Additional measures are required for the realization of ohmic metal–semiconductor contacts close to the Bardeen limit if depletion at the semiconductor surface cannot be avoided.

5.3 Tunneling Ohmic Contacts

5.3.1 *Tunneling at metal–semiconductor contacts*

Space charge regions at charge-selective contacts become very thin for very high densities of donor or acceptor atoms and low values of band bending. For example, the width of the space charge region is about 3 nm for a density of donors of about $5 \cdot 10^{19}$ cm^{-3} and for a band bending of 0.3 V (Figure 4.5). Tunneling of free charge carriers through very thin barrier layers is key for the formation of tunneling ohmic contacts (see, for example, Yu (1970)).

The density of free charge carriers is larger than the effective density of states of the respective band in very highly doped semiconductors. This means that E_{F0} is within the valence or conduction bands. Semiconductors for which $E_{F0} < E_V$ or $E_{F0} > E_C$ are called degenerated. The doping type of degenerated semiconductors is denoted by an additional plus (n$^+$-type or p$^+$-doping type).

For degenerated semiconductors, the thickness of the space charge region near a metal contact can be calculated by modifying Equations (4.5′) and (4.5″).

$$x_{SCR}^{n+} = \sqrt{\frac{2 \cdot \varepsilon_r \cdot \varepsilon_0}{q \cdot N_D^+} \cdot \Phi_B} \qquad (5.17')$$

$$x_{SCR}^{p+} = \sqrt{\frac{2 \cdot \varepsilon_r \cdot \varepsilon_0}{q \cdot N_A^-} \cdot (E_g - \Phi_B)} \qquad (5.17'')$$

Figure 5.10 shows the band diagrams for tunneling metal–semiconductor contacts with n-type or p-type doped semiconductors.

For the formation of an ohmic contact with majority charge carriers in an n-type doped semiconductor, a highly n-type doped (n$^+$-type) layer is prepared on top of the n-type doped layer before the metal contact layer is deposited (for more details see Sze (1981)). The potential energy

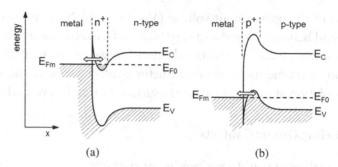

Figure 5.10. Band diagrams of tunneling metal–semiconductor contacts with n-type (a) and p-type (b) doped semiconductors. The arrows mark the tunneling transitions.

of electrons in the conduction band decreases from the n-type to the n^+-type doped region, and electrons can tunnel from the n^+-doped region into the metal contact. In a similar manner, an ohmic contact with holes in p-type doped semiconductors is formed by implementing a p^+-doped layer between the metal and the p-type semiconductor. The potential energy of holes in the valence band decreases towards the p^+-type doped region. Electrons can tunnel from the metal contact into unoccupied states in the valence band of the p^+-type doped region and recombine there with collected holes.

5.3.2 *Transmission probability of a tunneling barrier*

Quantum mechanical tunneling (see, for example, Griffiths (2004)) becomes possible due to the fact that an electron with certain total energy (E) can be considered as a wave with a certain wavelength, the de Broglie wavelength (λ_B):

$$\lambda_B = \frac{\hbar}{2\pi \cdot \sqrt{2 \cdot m_e^* \cdot E}} \tag{5.18}$$

The Planck constant ($\hbar$) is equal to $1.05 \cdot 10^{-34}$ Js. Incidentally, the Planck constants $\hbar$ and h differ by a factor of 2π, and $\hbar$ is traditionally used in equations describing tunneling.

The total energy of free electrons is of the order of hundreds of meV in photovoltaic absorbers. The de Broglie wavelength, for example, is about 1.2 or 3.8 nm for electrons with energies of 1 eV or 0.1 eV, respectively, i.e. the de Broglie wavelength of free electrons is of the same order as the thickness of a

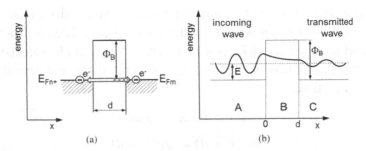

Figure 5.11. Particle (a) and wave (b) pictures of an electron at a rectangular barrier. Regions A, B and C denote the regions of propagation of the incoming wave, of the wave in the barrier and of the transmitted wave, respectively.

space charge region of a highly doped semiconductor. Therefore, an electron at the barrier can be treated as a wave. Figure 5.11 depicts the particle and wave pictures employed to describe charge transport and tunneling of a free electron at a rectangular barrier.

The wave function of the electron (ψ) with an effective mass (m_e^*) can be described by the one-dimensional stationary Schrödinger equation with a total energy (E) and potential energy (U_{pot}):

$$\frac{\partial^2 \psi(x)}{\partial x^2} + \frac{2 \cdot m_e^*}{\hbar^2} \cdot (E - \Phi_B) = 0 \tag{5.19}$$

A tunneling barrier is characterized by the transmission probability. For simplicity, electrons are considered as plane waves for calculating the transmission probability of electrons ($T_{tr,e}$). The product of the wave function and of the conjugate complex wave function describes the probability of finding an electron in a certain unit of space. The Schrödinger equation can be solved for the regions of the incoming wave (index A), of the wave in the barrier (index B) and of the transmitted wave (index C) with an approach for solving the Schrödinger equation, which takes into account reflected waves in the regions of the incoming wave and of the wave in the barrier.

$$\psi_A = A_{in} \cdot \exp(i \cdot k_A \cdot x) + A_{refl} \cdot \exp(-i \cdot k_A \cdot x) \tag{5.20'}$$

$$\psi_B = B_{in} \cdot \exp(i \cdot k_B \cdot x) + B_{refl} \cdot \exp(-i \cdot k_B \cdot x) \tag{5.20''}$$

$$\psi_C = C \cdot \exp(i \cdot k_C \cdot x) \tag{5.20'''}$$

The wave vectors are the same for regions A and C (k_A equal to k_C), and the wave vector is imaginary in the barrier region. The coefficient A_{in} can be set to 1, the coefficients A_{refl}, B_{in}, B_{refl} and C can be calculated by considering continuity of the wave function and of the first derivative of the wave function:

$$\psi_A(x = 0) = \psi_B(x = 0) \tag{5.21'}$$

$$\psi_B(x = d) = \psi_C(x = d) \tag{5.21''}$$

$$\frac{\partial \psi_A(x = 0)}{\partial x} = \frac{\partial \psi_B(x = 0)}{\partial x} \tag{5.21'''}$$

$$\frac{\partial \psi_B(x = d)}{\partial x} = \frac{\partial \psi_C(x = d)}{\partial x} \tag{5.21''''}$$

The transmission probability can be written as:

$$T_{tr} = C \cdot C^* \tag{5.22}$$

The solution of Equation (5.22) is given in textbooks (see, for example, (Griffith, 2004)). The commonly used approximation of the transmission probability for a rectangular barrier is given by the following equation:

$$T_{tr,e} = exp\left(-2 \cdot \sqrt{2 \cdot m_e^* \cdot (\Phi_B - E)} \cdot \frac{d}{\hbar}\right) \tag{5.23}$$

The barrier is usually distributed across a space charge region depending on doping, distribution of dopants near the metal contact, on metallurgical imperfections at the interface, on roughness at the interface and on other factors. The transmission probability can be calculated by integrating over the function of the barrier distribution in a space charge region ($U_{pot}(x)$):

$$T_{tr,e} = exp\left(-\frac{2}{\hbar} \cdot \int_0^{x_{SCR}} \sqrt{2 \cdot m_e^* \cdot (U_{pot}(x) - E)} \cdot dx\right) \tag{5.24}$$

The barrier of a space charge region at a metal contact with a highly doped n-type semiconductor can be approximated by a triangular shape as shown in Figure 5.12.

$$U_{pot}(x) - E = \Phi_B \cdot \left(1 - \frac{x}{x_{SCR}}\right) \tag{5.25}$$

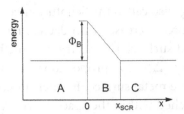

Figure 5.12. Triangular barrier for approximation of the space charge region at an ohmic tunneling metal–semiconductor contact.

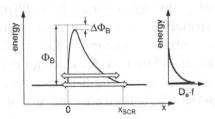

Figure 5.13. Barrier at a tunneling ohmic metal–semiconductor contact considering barrier lowering (caused by the image charge) and energy distribution of electrons crossing the barrier.

The electron transmission probability of a triangular barrier is obtained by considering Equations (5.24), (5.25) and (5.17′):

$$T_{tr,e} = exp\left(-\frac{8}{3} \cdot \sqrt{\frac{m_e^* \cdot \varepsilon_r \cdot \varepsilon_0}{N_D^+}} \cdot \frac{\Phi_B}{q \cdot \hbar} \right) \qquad (5.26)$$

To calculate the hole transmission probability of a triangular barrier at a metal contact with a highly p-type doped semiconductor, N_D^+, m_e^* and Φ_B in Equation (5.26) are replaced by N_A^-, m_h^* and $E_g-\Phi_B$.

Equation (5.26) is very useful for estimating the order of magnitude of the transmission probability at tunneling ohmic metal–semiconductor contacts. However, the values of T_{tr} are often underestimated due to barrier lowering and the energy distribution of free electrons (Figure 5.13). Furthermore, the real effective mass of the electron tunneling through the barrier and the exact shape of the barrier following from the solution of the Poisson equation have to be considered for a more precise analysis of T_{tr}.

A lowering of the barrier is caused by the image charge arising at the metal side of the contact when an electron is moving towards the barrier (image-force lowering, also called the Schottky effect; see, for example, Sze (1981)). The image force increases with decreasing distance between the electron and the metal surface. The decrease of the barrier caused by the image potential energy ($\Delta\Phi_B$) is proportional to the reciprocal distance of the electron from the metal surface. Furthermore, the maximum of the reduced barrier is reached within the space charge region. The additional reduction of the barrier, which is caused by the image force, increases with increasing density of dopants. The value of $\Delta\Phi_B$ can be about 10–20% of Φ_B for metal contacts with highly doped crystalline semiconductors.

Electrons tunneling through a barrier have different kinetic energy levels. The energy distribution of free electrons is given by the product of the density of states and the Fermi-function (see Chapter 2). The width of the barrier is reduced, and therefore T_{tr} is increased for electrons tunneling at higher energy. This so-called thermionic field emission enhanced tunneling becomes more important with decreasing density of dopants in highly doped semiconductors.

5.3.3 *Contact resistance of a triangular tunneling barrier*

The contact resistance (R_C) of a solar cell has to be compared with the tolerable series resistance (see Chapter 1), which is, for example, of the order of 10^{-1} or 10^{-4} Ωcm^2 for solar cells illuminated at air mass (AM)1.5 or at concentrated sunlight (concentration factor 1000), respectively. The reduction of the ohmic contact area at finger grids or at contact dots has to be taken into account as well. Therefore, the contact resistance of ohmic contacts should be of the order of 10^{-6} Ωcm^2 or even less for very efficient solar cells operated at concentrated sunlight.

The inverse contact resistance is defined as the derivative of the current density over the potential drop across the contact at 0 potential:

$$\frac{1}{R_C} = \frac{dI}{dU}\bigg|_{U \to 0} \tag{5.27}$$

The current across an ohmic contact, which can be supported by tunneling is proportional to the transmission probability of the barrier, to the density of free charge carriers at the side of the highly doped semiconductor, and to the thermal velocity of free charge carriers. The barrier is reduced under forward bias or under illumination by the product of the elementary charge and the potential drop across the barrier. The tunneling current of electrons can be expressed by the equation:

$$I_{tunn,e} = q \cdot N_D^+ \cdot v_{th} \cdot exp\left(-\frac{8}{3} \cdot \sqrt{\frac{m_e^* \cdot \varepsilon_r \cdot \varepsilon_0}{N_D^+}} \cdot \frac{\Phi_B - q \cdot U}{q \cdot \hbar}\right) \quad (5.28)$$

The resulting contact resistance of a tunneling ohmic contact is:

$$R_{c,tunn,e} = \frac{3 \cdot \hbar}{8 \cdot q \cdot v_{th}} \cdot \sqrt{\frac{1}{N_D^+ \cdot m_e^* \cdot \varepsilon_r \cdot \varepsilon_0}}$$

$$\cdot exp\left(\frac{8}{3} \cdot \sqrt{\frac{m_e^* \cdot \varepsilon_r \cdot \varepsilon_0}{N_D^+}} \cdot \frac{\Phi_B}{q \cdot \hbar}\right) \quad (5.29)$$

It is no mean feat to obtain correct values for the effective mass of the tunneling charge carrier and for the dielectric constant, taking into account the extremely short duration of the tunneling process. A good approximation is reached for metal contacts with c-Si by taking the static dielectric constant of c-Si. Equations (5.29) can be simplified if approximating the effective electron or hole masses by the mass of the free electron in vacuum ($m_e = 9.1 \cdot 10^{-31}$ kg) and the relative dielectric constant by the relative dielectric constant of c-Si.

$$R_{c,tunn,e} \approx \frac{\Omega cm^2}{1.26 \cdot 10^9} \cdot \sqrt{\frac{10^{19} cm^{-3}}{N_D^+}} \cdot exp\left(79 \cdot \sqrt{\frac{10^{19} cm^{-3}}{N_D^+}} \cdot \frac{\Phi_B}{eV}\right)$$
$$(5.30)$$

For obtaining $R_{c,tunn,h}$, the values of N_D^+, m_e^* and Φ_B in Equation (5.30) should be replaced by the values of N_A^-, m_h^* and $E_g - \Phi_B$. The resistance of the tunneling contact is plotted in Figure 5.14 as a function of the density of dopants.

Equation (5.30) is an approximation that does not take into account barrier lowering, the real barrier distribution in space, local morphology or the correct values of the effective mass and dielectric constant. The consideration of these factors demands sophisticated models and can lead to relatively large differences with the approximation given by Equation (5.30). Nevertheless, Equation (5.30) gives a clear idea about the orders

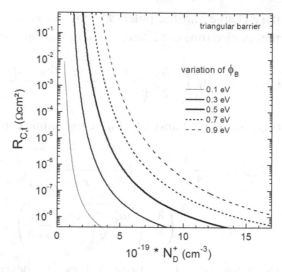

Figure 5.14. Resistance of an electron tunneling contact as a function of the density of donors for different values of Φ_B. The space charge region is approximated by a triangular barrier. The effective electron mass and the relative dielectric constant are set equal to the free electron mass and 12, respectively.

of magnitude of tunneling contact resistances and their dependence on barrier height and density of dopants.

Small changes in the density of dopants or in the barrier height can cause large changes in the contact resistance. For example, the resistance of the tunneling contact decreases from about 10^{-1} Ωcm^2 to 10^{-6} Ωcm^2 if increasing the density of dopants from $2 \cdot 10^{19}$ to $5 \cdot 10^{19}$ cm^{-3} for a barrier of 0.5 eV. Therefore, the precise control of doping as well as the conditioning of the metal–semiconductor interface is very important for solar cells.

The exponential dependence of the tunneling resistance on the barrier height shows the importance of the choice of the metal or metal alloy for the contact, as well as the importance of the control of interface states at ohmic contacts. Sophisticated technologies of contact formation including preparation of semiconductor surfaces, metal deposition and activation of chemical reactions at metal–semiconductor interfaces have been developed especially for c-Si and III–V semiconductors with respect to requirements for electronic and opto-electronic applications. For example, $Au_{0.88}Ge_{0.12}$

is used for the formation of ohmic contacts at n-type doped GaAs (Sze, 1981).

Ohmic contacts can be realized well with tunneling contacts between metals and highly doped semiconductors. The ohmic contact area can be strongly reduced for related tunneling contacts without remarkable power losses in the solar cell. In principle, there is no limitation of tunneling ohmic contacts for the implementation of (i) contact fingers at front contacts and (ii) contact dots at back contacts of solar cells (see Figure 5.3), and further, (iii) very low contact resistances allow operation of solar cells at highly concentrated sunlight.

5.3.4 *Tunneling resistance at degenerated pn-hetero-junctions*

The monolithic series connection of stacked solar cells plays an important role for the realization of very high solar energy conversion efficiencies with multi-junction solar cells (see Chapter 6). A monolithic series connection of stacked solar cells means the connection of the Fermi-energies of majority charge carriers in n-type and p-type doped semiconductors by an ohmic contact so that the photovoltages of both solar cells are added at the load resistance (see the equivalent circuit in Figure 5.15). The ohmic contact between the n-type and p-type doped semiconductors should not influence the function of charge-selectivity of both connected solar cells. In multi-junction solar cells, the n-type and p-type doped regions of recombination contacts often consist of semiconductors with different band gaps, E_{g1} and E_{g2} (i.e., a pn-hetero-junction; see also Chapter 8).

The contact between n-type and p-type doped semiconductors leads to the formation of space charge regions at both sides of the contact (see Chapter 4). The space charge regions are very thin for very high densities

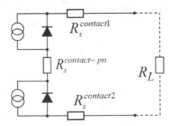

Figure 5.15. Equivalent circuit of two stacked solar cells with load, contact resistances between both solar cells and between metal leads and semiconductors.

of donors and acceptors at both the n-type and p-type doped sides of the pn-hetero-junction. The Fermi-energies in the bulk of highly doped n-type or p-type doped semiconductors are within the conduction or valence bands, respectively. Such semiconductors form degenerated pn-hetero-junctions. Degenerated pn-hetero-junctions are applied for integrated series connection in multi-junction solar cells with very high solar energy conversion efficiency (see Chapter 8).

The wave functions of electrons and holes at the n-type and p-type doped sides of a degenerated hetero-junction overlap within the space charge region. In this case, tunneling recombination of an electron from the highly doped n-type side with a hole from the highly doped p-type side becomes possible. Figure 5.16 shows a degenerated pn-hetero-junction. Hetero-junctions between two semiconductors are characterized by the offsets of the conduction and valence band edges (ΔE_C and ΔE_V, see also Chapter 8). The barrier heights for the tunneling of electrons ($\Phi_{B,e}$) and holes ($\Phi_{B,h}$) correspond to the respective potential drops across the space charge regions (see Chapter 4) at the n-type and p-type doped regions (see also task T5.5 at the end of this chapter).

For the example shown in Figure 5.16, the barrier heights for tunneling recombination of electrons and holes can be obtained from the following

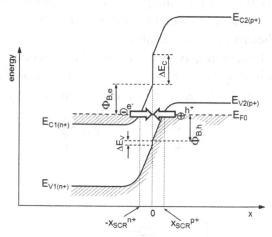

Figure 5.16. Band diagram of a degenerated pn-hetero-junction forming a tunneling recombination contact.

equations:

$$E_{g2} = \Phi_{B,e} + \Phi_{B,h} + \Delta E_C \qquad (5.31')$$

$$E_{g1} = \Phi_{B,e} + \Phi_{B,h} + \Delta E_V \qquad (5.31'')$$

The values of $\Phi_{B,e}$ and $\Phi_{B,h}$ depend on the densities of donors and acceptors and are connected via the condition of charge neutrality. The shapes of the barriers at degenerated pn-hetero-junctions can be well approximated by a triangular barrier shape, similar to tunneling metal–semiconductor contacts. Therefore, as for tunneling metal–semiconductor contacts, Equation (5.29) can be used for calculating the contact resistance of degenerated pn-hetero-junctions. The tunneling or recombination resistance can be calculated by adding the tunneling resistances of electrons and holes. For this purpose, Φ_B is replaced by $\Phi_{B,e}$ or $\Phi_{B,h}$ in Equation (5.29) and the corresponding values of m_e^*, m_h^*, N_A^-, N_D^+ and ε_r (the relative dielectric constant) are taken. The contact resistance of degenerated pn-hetero-junctions is limited by the space charge region at the side with the lower density of dopants due to the exponential dependence of the transmission probability on the thickness of the barrier layer.

The band diagram of an ohmic contact between moderately doped n-type and p-type semiconductors is shown in Figure 5.17. The structure of the complete ohmic contact consists of a stack of n/n$^+$/p$^+$/p-type doped

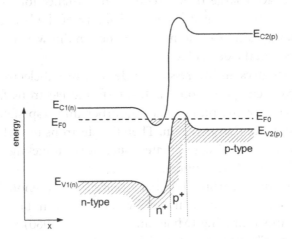

Figure 5.17. Band diagram of an ohmic contact between moderately doped n-type and p-type semiconductors.

semiconductors. It is interesting to remark that the potential energy of minority charge carriers increases towards the tunneling pn-contact so that both the density of minority charge carriers and the recombination rate are reduced (back surface field).

5.3.5 *Tunneling via defect states near metal–semiconductor contacts*

Sometimes, a photovoltaic absorber cannot be highly doped, and a suitable metal cannot be found for realizing an ohmic contact. In such a case, it may be useful to incorporate deep defect states in the absorber near the metal–semiconductor contact in order to obtain an ohmic character of the contact. The point is that deep defect states can be occupied by electrons or holes and that wave functions of occupied and unoccupied defect states can overlap in space. Charge carriers can be transferred between localized states if there is an overlap between the wave functions of the states in two neighboring defects. Therefore, mobile charge carriers can penetrate a barrier by tunneling via localized defect states if the distance between localized defect states is of the order of the tunneling lengths. The average distance between neighboring defects (Δx) is $(N_t)^{-3}$. For example, Δx is about 2.2 or 4.6 nm for defect densities of 10^{20} or 10^{19} cm^{-3}, respectively, which is of the order of tunneling lengths. The presence of deep defect states in the space charge region reduces the distance for tunneling and therefore increases the transmission probability of the barrier. However, very low contact resistances cannot be reached in this way due to multiple tunneling through the defect layer.

Figure 5.18 gives an impression of defect-assisted electron tunneling through a space charge region. First, transferred electrons fall from the conduction band of the photovoltaic absorber into trap states closest to the end of the space charge region. Then the electrons tunnel towards the metal surface via several defect states. Successive tunneling via localized defect states is also called hopping.

The tunneling rate from an initial localized state (denoted by index i) to a final localized state (denoted by index f) can be described by Miller–Abrahams tunneling (Miller and Abrahams, 1960) in the simplest approximation. The initial and final localized states are characterized by their distance (R_{if}) and by their energies (E_i and E_f).

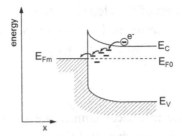

Figure 5.18. Band diagram of an ohmic metal–semiconductor contact with a large density of deep defect states in the n-type doped semiconductor near the contact.

The system has only two parameters for Miller–Abrahams tunneling: the maximum tunneling rate ($\omega_{0,tunn}$) and the reciprocal tunneling length (α_{tunn}). The maximum tunneling rate corresponds to a vibration frequency of a bond configuration at which charge carriers are localized and is of the order of $10^{12}\,s^{-1}$. The tunneling length is of the order of 0.5–1.5 nm. The tunneling step is thermally activated if the energy of the electron is larger in the final state than it is in the initial state.

$$\omega_{tunn} = \omega_{0,tunn} \cdot exp(-2 \cdot \alpha_{tunn} \cdot R_{if}) \cdot exp\left(-\frac{(E_i - E_f) - |E_i - E_f|}{2 \cdot k_B \cdot T}\right)$$
$$(5.32)$$

A rather crude approximation can be obtained for the current density supported by trap-assisted tunneling through a barrier. For this purpose, a narrow distribution of R_{if} and directed tunneling are assumed. The average displacement per hopping step then corresponds to a value close to Δx and can be written:

$$I_{tunn} \sim q \cdot N_D \cdot \sqrt[3]{N_t} \cdot \omega_{0,tunn} \cdot exp(-2 \cdot \alpha_{tunn} \cdot (N_t)^{-1/3}) \qquad (5.33)$$

Densities of donors (N_D) of $10^{19}\,cm^{-3}$ and of defects (N_t) of $10^{20}\,cm^{-3}$ are assumed for the estimation of I_{tunn}. An I_{tunn} of the order of 10 mA/cm² may be supported if considering an α_{tunn} of 1 nm and four tunnelings steps. This means that in principle, with respect to possible values of short-circuit current (I_{SC}), defect-assisted tunneling may be applied for contacting photovoltaic absorbers illuminated at light intensities of the order of AM1.5. However, one has to keep in mind that expression (5.33) does not consider an inhomogeneous distribution of defect states or a change of the distribution of the electric potential due to charging of trap states. Furthermore, the controlled incorporation of defects near metal–semiconductor interfaces may be challenging.

5.4 Summary

Ohmic contacts are the interfaces connecting the active part of a solar cell, i.e. the photovoltaic absorber and the charge-selective contact, with metal leads connected to the external load. Ohmic contacts are passive, i.e. they should not contribute to photo-generation and charge separation. Ideal ohmic contacts are also called recombination contacts. Fermi-level splitting is impossible at ideal ohmic contacts.

The surface recombination velocity is maximal at ohmic contacts in general. For reducing the surface recombination rate at ohmic contacts in solar cells, the density of photo-generated charge carriers has to be reduced for minority charges carriers. This can be done by (i) implementing barriers for minority charge carriers, for example, with back surface fields, (ii) reducing the area of ohmic contacts in addition to passivation of the surface area that is not covered by ohmic contacts or (iii) increasing the distance between the charge-selective contact and the ohmic contact to a value larger than the diffusion length of minority charge carriers. The implementation of a back surface field and the reduction of the ohmic contact area are decisive for reaching very high solar energy conversion efficiencies, for example, with c-Si (see Chapter 7) or GaAs (see Chapter 8) solar cells. The increase of the distance between the charge-selective and ohmic contacts to values larger than the diffusion length of minority charge carriers is important for thin-film solar cells based, for example, on chalcopyrite or CdTe thin-film absorber layers (see Chapter 9).

The electronic properties of an interface between a metal and a semiconductor depend on the metal work function, the electron affinity, doping of the semiconductor and on defect formation or chemical reactions at the metal–semiconductor interface. At ideal ohmic contacts, no barriers are formed for the transport of majority charge carriers from the semiconductor into the metal. This is the case if the metal work function ranges between the electron affinity and the Fermi-energy of an n-type doped semiconductor or between the ionization energy and the Fermi-energy of a p-type doped semiconductor. Therefore a metal or metal alloy with a reliable work function has to be chosen for forming an ohmic contact with a given doped semiconductor.

Barriers for the transport of majority charge carriers, the so-called Schottky barriers, are often formed at metal–semiconductor interfaces and the barrier heights are sometimes independent of the metal work function. In the extreme cases (i) the barrier height at the metal–semiconductor interface correlates exactly with the metal work function (Schottky limit) or (ii) the barrier height does not correlate at all with the metal work function (Bardeen limit). The Bardeen limit is caused by a very high density of electronic states in the forbidden band gap of the semiconductor at the metal–semiconductor interface. The Fermi-energy at a semiconductor surface is pinned at a certain value if the density of surface defects exceeds values of the order of 10^{13} cm^{-2}.

A transport barrier for majority charge carriers from a semiconductor towards a metal–semiconductor contact is characterized by the surface band bending at the semiconductor surface. A barrier layer against the transport of majority charge carriers acts as a charge-selective contact in a semiconductor by depleting majority charge carriers in the barrier region. The thickness of a surface space charge region must be drastically reduced for enabling tunneling transport of majority charge carriers from the semiconductor across the metal–semiconductor interface into the metal.

The thickness of a surface space charge region can be reduced to a few nm for highly doped semiconductors with densities of majority charge carriers larger than 10^{19} cm^{-3}. Therefore, very thin and highly doped semiconductor layers are deposited at semiconductor absorbers to form ohmic metal–semiconductor contacts.

The barrier layer at a contact between a metal and a highly doped semiconductor can be approximated by a triangular barrier, the width and the height of which are given by the extension of the space charge region and by the barrier height at the metal–semiconductor contact, respectively. The transmission probability of the barrier layer increases exponentially with the increasing square root of the density of majority charge carriers and with decreasing barrier height. The reduction of the effective barrier by image force, the precise shape of the barrier layer, the distribution of the kinetic energy of mobile charge carriers and local fluctuations of the surface band bending can be considered in more detailed models.

The tunneling or contact resistance of a space charge region at a metal–semiconductor contact can be as low as 10^{-6} Ωcm^2 for low barrier heights of about 0.3 eV and for very high densities of majority charge carriers of about $3 \cdot 10^{19}$ cm^{-3}. A contact resistance of about 10^{-6} Ωcm^2 is about five orders of magnitude lower than the tolerable series resistance of a solar cell illuminated at AM1.5. Therefore, the area of the ohmic contact can be strongly reduced in order to reduce surface recombination at the ohmic contact. Additionally, solar cells with related ohmic contacts can be illuminated at concentrated sunlight without remarkable resistive losses at the ohmic contacts.

Very low contact resistances are also reached for tunneling between highly n-type doped and highly p-type doped semiconductors. Tunneling or recombination contacts are very important for the realization of integrated series connection in multi-junction solar cells with the highest solar energy conversion efficiency at concentrated sunlight (see Chapter 8).

Sometimes defect-assisted tunneling through barrier layers may be useful in the reduction of the resistance at ohmic contacts, for example, in the case of a relatively large barrier height.

5.5 Tasks

T5.1: Surface band bending in the presence of surface defects

Ascertain the surface band bending at n-type doped semiconductors with equal densities of acceptor- and donor-like surface defects of 10^{13} and 10^{11} cm^{-2}. The densities of acceptor- and donor-like surface states are constant over the E_g. The densities of donors, the effective densities of states at the conduction band edge (N_C) and E_g of the semiconductors are (a) $N_D = 10^{16}$ cm^{-3}, $N_C = 3 \cdot 10^{19}$ cm^{-3} and $E_g = 1.1$ eV (c-Si) and (b) $N_D = 10^{17}$ cm^{-3}, $N_C = 5 \cdot 10^{17}$ cm^{-3} and $E_g = 1.42$ eV (GaAs). The relative dielectric constant is 12.

T5.2: Fermi-level pinning

Is it possible to shift the Fermi-energy at a semiconductor surface by 0.1 eV for a density of surface defects of (a) 10^{11} cm^{-2} and (b) 10^{13} cm^{-2}? The densities of acceptor- and donor-like surface states are equal and constant across the E_g. The values of N_C, N_D, ε_r and E_g are $3 \cdot 10^{19}$ cm^{-3}, 10^{16} cm^{-3}, 12 and 1.1 eV, respectively.

T5.3: Engineering of surface band bending with interface defects

Calculate the surface band bending for n-type ($N_D = 10^{18}$ cm^{-3}) and p-type ($N_A = 10^{19}$ cm^{-3}) doped semiconductors with different densities of acceptor- and donor-like surface defects of (a) $N_{st,a} = 3 \cdot 10^{13}$ and $N_{st,d} = 8 \cdot 10^{13}$ cm^{-2} and (b) $N_{st,a} = 10^{14}$ and $N_{st,d} = 2 \cdot 10^{13}$ cm^{-2}. The densities of acceptor- and donor-like surface states are assumed to be constant across the E_g. The values of N_C, N_V, ε_r and E_g are $1 \cdot 10^{18}$ cm^{-3}, $1 \cdot 10^{19}$ cm^{-3}, 12 and 1.42 eV, respectively. Discuss the relevance of the result for the formation of ohmic contacts.

T5.4: Transmission probability of tunneling contacts with inorganic and organic semiconductors

Inorganic and organic semiconductors can be highly doped while the effective electron mass can be very low for inorganic semiconductors (for example, $m_e^* = 0.066 \cdot m_e$) and the dielectric constant can be much lower for organic (for example, $\varepsilon_r = 3$) than for inorganic (for example, $\varepsilon_r = 12$) semiconductors. Compare the transmission probabilities of tunneling contacts with an inorganic and with an organic semiconductor for a barrier height of 0.4 eV and a density of donors of $5 \cdot 10^{19}$ cm^{-3}.

T5.5: Recombination contact between highly doped semiconductors

Estimate the tunneling resistance between two highly doped semiconductors with E_g of 1.32 and 1.84 eV. The semiconductor with $E_g = 1.32$ eV is n-type doped ($N_D^+ = 5 \cdot 10^{19}$ cm^{-3}) and the semiconductor with $E_g = 1.84$ eV is p-type doped ($N_A^- = 2 \cdot 10^{20}$ cm^{-3}). The effective electron and hole masses are 0.07 and 0.4 times the mass of the free electron, respectively. The dielectric constant and the conduction band offset are 12 and 0.3 eV, respectively. Discuss consequences for multi-junction solar cells under concentrated sunlight.

T5.6: Schottky barriers for charge separation and ohmic contacts

Schottky barriers can be used for charge separation. Discuss advantages and disadvantages in comparison with a pn-junction and separately formed ohmic contacts.

6

Maximum Energy Conversion Efficiency of Solar Cells

The diode equation of an ideal solar cell is obtained following the balance between thermal generation and radiative recombination (Shockley and Queisser, 1961; Würfel, 2005). The diode saturation current density of an ideal solar cell depends only on the band gap of the photovoltaic absorber and its temperature. The Shockley–Queisser limit describes the dependence of the solar energy conversion efficiency (η) of an ideal solar cell on the band gap (E_g) of its photovoltaic absorber illuminated at air mass (AM)1.5 and at 25°C. The maximum value of η is 32% for an E_g between 1.1 and 1.5 eV. The reduction of thermalization losses by spectral splitting is analyzed for multi-junction solar cells with several E_g. Optimum combinations of E_g are derived from the current-matching condition of multi-junction solar cells. Maximum efficiencies are calculated for tandem solar cells, triple- and quadruple-junction solar cells. The maximum efficiency is further increased under illumination with concentrated sunlight, for example, up to 66% for a quadruple-junction solar cell illuminated with sunlight concentrated 1000 times. The increase of the quantum efficiency by down-conversion and impact ionization and the influence of up-conversion on the maximum efficiency are discussed.

6.1 Diode Equation of the Ideal Single-Junction Solar Cell

6.1.1 *Thermal generation and radiative recombination rate constant*

Photons absorbed in a photovoltaic absorber are converted into free electrons and holes. A minimum of recombination of free electrons and holes is desired for a maximum energy conversion efficiency of solar cells.

One absorbed photon generates one electron-hole pair. The quantum efficiency (see Chapter 1) of an ideal photovoltaic absorber is equal to 0 for photons with energies ($h\nu$) less than the forbidden E_g (band gap) and equal to 1 for all incoming photons with energies equal to or larger than E_g.

$$QE(h\nu) = \begin{cases} 0 & (h \cdot \nu < E_g) \\ 1 & (h \cdot \nu \geq E_g) \end{cases} \tag{6.1}$$

Equation (6.1) describes the condition of an ideal photovoltaic absorber with a continuum of states in the valence and conduction bands at energies below the valence band edge and above the conduction band edge, respectively. Condition (6.1) requires that (i) losses caused by reflection are minimized to 0, (ii) optical transmission losses are reduced to 0, i.e. the product of the absorption coefficient and of the optical path is much larger than 1 for all photon energies equal to or larger than E_g (see Chapter 2) and (iii) there are no collection losses, i.e. diffusion lengths of minority charge carriers are much larger than the absorption length at any wavelength of absorbed light (see Chapter 4).

The thermal generation rate (G_0) describes the generation of free electrons and holes due to absorption of blackbody radiation emitted by the photovoltaic absorber (see Chapter 2). The thermal generation rate is only a function of E_g for ideal absorbers.

$$G_0(E_g) = \int_{E_g}^{\infty} \Phi_0(h\nu) \cdot d(h\nu) \tag{6.2}$$

The photon flux of blackbody irradiation ($\Phi_0(h\nu)$) follows directly from the Planck equation while T_0 is the temperature of the ideal photovoltaic absorber under operation of the solar cell.

$$\Phi_0(h\nu) = \frac{2 \cdot \Omega_s}{h^3 \cdot c^2} \cdot \frac{(h\nu)^2}{exp\left(\frac{h\nu}{k_B \cdot T_0}\right) - 1} \tag{6.3}$$

For an ideal photovoltaic absorber, the space angle (Ω_s) in Equation (6.3) is equal to the full space angle (4π).

The density of free electrons and holes is balanced by all possible recombination processes in thermal equilibrium. This principle is also called

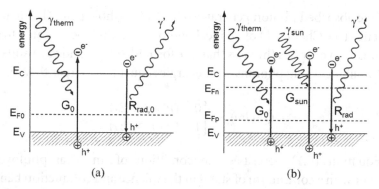

Figure 6.1. Energy diagram of an ideal photovoltaic absorber in thermal equilibrium (a) and under illumination (b). Photons emitted by blackbody radiation, photons emitted by the sun and photons emitted by radiative recombination of the photovoltaic absorber are denoted by γ_{therm}, γ_{sun} and γ', respectively.

the detailed-balance principle. Figure 6.1(a) shows thermal generation and radiative recombination.

The Shockley–Read–Hall and surface recombination rates (see Chapter 3) are minimized to 0 for ideal photovoltaic absorbers. Furthermore, the densities of free electrons and/or holes are considered to be low in thermal equilibrium and consequently Auger recombination can be neglected in ideal photovoltaic absorbers. Therefore, the thermal generation rate is balanced only by radiative recombination in an ideal photovoltaic absorber (R_{rad}):

$$G_0(E_g, T_0) = R_{rad,0} \qquad (6.4)$$

The radiative recombination rate constant of an ideal photovoltaic absorber (B_{ideal}) can be calculated using the definition of the thermal generation rate of an ideal photovoltaic absorber (Equation (6.2)) and the density of intrinsic charge carriers in thermal equilibrium (see Equation (2.36)):

$$B_{ideal} = \frac{G_0\left(E_g, T_0\right)}{n_i^2} \qquad (6.5)$$

For ideal photovoltaic absorbers, the radiative recombination rate constant depends only on the operation temperature and on the forbidden band gap, which is the only material parameter of an ideal photovoltaic absorber.

6.1.2 *Radiative recombination rate of an illuminated ideal absorber*

The densities of free electrons and holes of a photovoltaic absorber increase under illumination due to absorption of photons from sunlight (Figure 6.1(b)). The photo-generation rate of an ideal photovoltaic absorber (G_{sun}) is obtained by integrating over the known spectrum of the photon flux from the sun ($\Phi_{sun}(\hbar\omega)$). The value of G_{sun} depends only on the forbidden band gap of an ideal semiconductor.

$$G_{sun}(E_g) = \int_{E_g}^{\infty} \Phi_{sun}(h\upsilon) \cdot d(h\upsilon) \tag{6.6}$$

The radiative recombination rate of an illuminated semiconductor is given by Equation (3.11) where the radiative recombination rate constant is substituted by the expression in Equation (6.5) for ideal photovoltaic absorbers. Therefore it can be written:

$$R_{rad} = G_0(E_g, T_0) \cdot \frac{n \cdot p}{n_i^2} \tag{6.7}$$

The densities of free electrons and holes (n and p, respectively) are given by the Boltzmann equations (Equations (2.30′) and (2.30″)). Under illumination, the Fermi-energies split into separate Fermi-energies for free electrons (E_{Fn}) and free holes (E_{Fp}). Following the Boltzmann statistics of free electrons and holes in a semiconductor and following the definition of the density of intrinsic free charge carriers, the radiative recombination rate of an illuminated ideal photovoltaic absorber is given by the following expression:

$$R_{rad} = G_0(E_g, T_0) \cdot \exp\left(\frac{E_{Fn} - E_{Fp}}{k_B \cdot T_0}\right) \tag{6.8}$$

Equation (6.8) means that the radiative recombination rate of an illuminated ideal photovoltaic absorber depends only on the thermal generation rate, the operation temperature of the solar cell and the Fermi-level splitting ($E_{F,n}$–$E_{F,p}$).

For ideal solar cells, the Fermi-level splitting in the ideal photovoltaic absorber is conserved across the charge-selective contact (see Chapter 4) and between the two ohmic contacts (see Chapter 5). Therefore, the potential at the external leads is equal to the Fermi-level splitting in the ideal photovoltaic absorber divided by the elementary charge. The radiative recombination rate can be written (for more details see Würfel (2005)):

$$R_{rad} = G_0(E_g, T_0) \cdot \exp\left(\frac{q \cdot U}{k_B \cdot T_0}\right) \qquad (6.9)$$

The radiative recombination rate of an illuminated ideal solar cell depends only on the thermal generation rate of the ideal photovoltaic absorber and on the potential drop at the external contacts of the solar cell.

6.1.3 Diode saturation current of an ideal single-junction solar cell

The generation and recombination rates in Equations (6.2), (6.4), (6.6) and (6.9) correspond to particle fluxes through a unit area within a unit time. Therefore, thermal generation, photo-generation and recombination rates can be ascribed directly to the respective current densities.

The difference between the sum of the thermal and photo-generation rates and the recombination rate of an ideal photovoltaic absorber is related to the current density through a load connected with an ideal solar cell. The electric current has the opposite direction of the direction of the flowing electrons. Therefore, the following equation can be written for the current density of an ideal solar cell:

$$I = q \cdot R_{rad} - (q \cdot G_0 + q \cdot G_{sun}) \qquad (6.10)$$

Equation (6.10) can be transformed by taking into account Equation (6.9):

$$I = q \cdot G_0 \cdot \left[\exp\left(\frac{q \cdot U}{k_B \cdot T_0}\right) - 1\right] - q \cdot G_{sun} \qquad (6.11)$$

Equation (6.11) is the diode equation (see Chapter 1) of an ideal single-junction solar cell with an ideal photovoltaic absorber. The short-circuit current density is identical to the maximum short-circuit current density considered for the analysis of the ultimate solar energy conversion efficiency (see Chapter 2, Figure 2.3). The diode saturation current density of the ideal

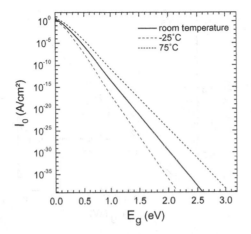

Figure 6.2. Dependence of the diode saturation current density of an ideal solar cell on the forbidden band gap of the ideal photovoltaic absorber at different temperatures.

solar cell is given by the thermal generation rate:

$$I_0 = \frac{2 \cdot q \cdot \Omega_S}{h^3 \cdot c^2} \cdot \int_{E_g}^{\infty} \frac{(h\upsilon)^2}{exp\left(\frac{h\upsilon}{k_B \cdot T_0}\right) - 1} \cdot d(h\upsilon) \qquad (6.12)$$

Equation (6.12) allows for the calculation of the diode saturation current density of ideal solar cells as a function of the forbidden band gap of the ideal photovoltaic absorber and of the operation temperature of the ideal solar cell. Figure 6.2 shows a plot of I_0 as a function of E_g.

For example, an ideal solar cell has a diode saturation current density of about 10^{-15} A/cm^2 at room temperature for an ideal photovoltaic absorber with a band gap equal to 1.1 eV.

For band gaps larger than about 0.3 eV, the dependence of I_0 on E_g fits excellently into the following equation:

$$I_0 = \frac{2 \cdot q \cdot \Omega_S \cdot k_B \cdot T_0}{h^3 \cdot c^2} \cdot (E_g)^2 \cdot exp\left(-\frac{E_g}{k_B \cdot T_0}\right) \qquad (6.13)$$

Equation (6.13) gives the opportunity to calculate with ease the diode saturation current density of ideal solar cells. Equation (6.13) also shows that the diode saturation current density of ideal solar cells is thermally activated with an activation energy equal to E_g.

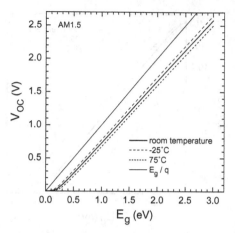

Figure 6.3. Dependence of the open-circuit voltage of an ideal solar cell on the forbidden band gap of the ideal photovoltaic absorber at different temperatures.

The open-circuit voltage (V_{OC}) is limited by I_0 (see Chapter 1, Equation (1.19)). Figure 6.3 shows the dependence of V_{OC} on the band gap of an ideal solar cell illuminated at AM1.5.

The values of V_{OC} are significantly below the band gaps divided by the elementary charge. Therefore, the solar energy conversion efficiency of ideal solar cells is significantly lower than the ultimate efficiency (see Chapter 2). Furthermore, V_{OC} becomes very low when I_0 increases up to the order of I_{SC} (short-circuit current), which is the case for band gaps less than about 0.2–0.3 eV. For this reason, band gaps of this range are not useful for photovoltaic solar energy conversion.

6.2 Maximum Efficiency of Ideal Single-Junction Solar Cells

6.2.1 *The Shockley–Queisser limit of solar energy conversion efficiency*

An ideal single-junction solar cell consists of an ideal photovoltaic absorber with one forbidden band gap and ideal charge-selective and ohmic contacts. The theoretical maximum energy conversion efficiency of ideal single-junction solar cells was calculated for the first time by Shockley and Queisser (1961) starting from the detailed-balance principle. The maximum efficiency of solar cells is generally obtained at a temperature of 25°C.

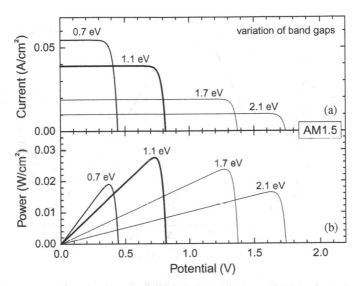

Figure 6.4. Current–voltage and power–voltage characteristics of ideal single-junction solar cells with different band gaps operated at room temperature and illuminated at AM1.5d.

The current–voltage characteristics of ideal single-junction solar cells can be calculated by using Equation (6.11) with the diode saturation current density and the maximum short-circuit current density given by Equations (6.12) and (2.2), respectively. Figure 6.4 shows current–voltage characteristics and the corresponding power–voltage characteristics of ideal single-junction solar cells with different band gaps illuminated at AM1.5d. The short-circuit current densities decrease and the open-circuit voltages increase with increasing band gap. A maximum power is reached at $0.029\,\text{W/cm}^2$ for a band gap of 1.1 eV, which corresponds to an energy conversion efficiency of roughly 32%.

The resulting energy conversion efficiencies of ideal single-junction solar cells are plotted in Figure 6.5 as a function of the forbidden band gap (25°C, AM1.5d). This dependence is also known as the Shockley–Queisser limit. The photovoltaic energy conversion efficiency is about 32–33% in the maximum and above 30% for ideal photovoltaic absorbers with band gaps between 1.0 and 1.6 eV.

The maximum solar energy conversion efficiency decreases towards lower band gaps and reaches values below 1% for band gaps less than 0.3 eV.

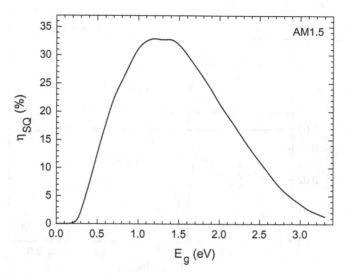

Figure 6.5. Dependence of the maximum solar energy conversion efficiency for ideal single-junction solar cells as a function of the band gap of ideal photovoltaic absorbers operated at 25°C and illuminated at AM1.5d (Shockley–Queisser limit).

Therefore, semiconductors with band gaps below 0.3 eV are not used for photovoltaic solar energy conversion.

The maximum energy conversion efficiency is still about 21% for a band gap of 2 eV and decreases to about 11% for a band gap of 2.5 eV. For band gaps above 3.2 eV the maximum solar energy conversion efficiency is less than 2%.

It is useful to compare the solar energy conversion efficiency of record solar cells (Green *et al.*, 2013) with the Shockley–Queisser limit of ideal solar cells with the same band gaps. Normalized differences between the efficiency in the Shockley–Queisser limit and the efficiency of record solar cells are plotted in Figure 6.6 for some solar cells based on single crystalline absorbers of crystalline silicon (c-Si) (wafer-based technology, see also Chapter 7) and of the III–V semiconductor family (epitaxy-based technology, GaAs, InP, see also Chapter 8), based on thin-film solar cells such as Cu(In,Ga)Se$_2$, CdTe, CuInS$_2$ and a-Si:H (absorber deposition on foreign substrates, see also Chapter 9) and based on nano-composite absorbers with quantum dots, dye molecules (DSSC — dye-sensitized solar cell), CH$_3$NH$_3$PbI$_3$ perovskite-sensitized or organic polymers (Chapter 10).

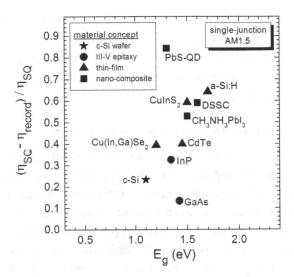

Figure 6.6. Deviation between the energy conversion efficiency in the Shockley–Queisser limit and record solar cells with the corresponding absorption gap for the different classes of materials (the energy conversion efficiencies of record solar cells are taken from Green *et al.* (2013) and, in the case of CH₃NH₃PbI₃, Burschka *et al.* (2013)).

The lowest deviations between the Shockley–Queisser limit and a world record solar cell are obtained for advanced single crystalline absorbers. The deviations are about 14, 23 and 32% gallium arsenide (GaAs) (E_g = 1.42 eV), c-Si (E_g = 1.1 eV) and indium phosphide (InP) (E_g = 1.34 eV) absorbers, respectively. The deviations increase for thin-film solar cells with an increasing band gap from 38% for copper indium gallium diselenide (Cu(In, Ga)Se₂, E_g = 1.2 eV), to about 40% for cadmium telluride (CdTe, E_g = 1.4 eV) and to about 56% for copper indium disulfide (CuInS₂, E_g = 1.5 eV) or 64% for hydrogenated amorphous silicon (a-Si:H, E_g = 1.7 eV).

The comparison between the Shockley–Queisser limit and the efficiency of record solar cells gives an impression of limitations imposed by materials and of technological achievements in semiconductor and solar cell technologies. For example, optical, resistive and recombination losses have been reduced to a minimum for solar cells with a GaAs absorber (see Chapter 8). Silicon crystals can be produced in large quantities and of the highest quality so that high solar energy conversion efficiencies are reached even in mass production (see Chapter 7). Fundamental materials limitations by disorder, for example, in amorphous silicon, can be reduced but not eliminated by technological optimization (see Chapter 9). Therefore, the Shockley–Queisser limit does not give a realistic limit for related materials. The second part of the book (Chapters 7–10) gives deeper insight into materials concepts for different classes of photovoltaic absorbers and solar cells.

6.2.2 Role of temperature and concentrated sunlight

The Shockley–Queisser limit is defined for thermal generation at 25°C and photo-generation at AM1.5. The energy conversion efficiency increases with decreasing operating temperature of the solar cell due to the thermal activation of the diode saturation current density and/or with increasing concentration of the sunlight. The upper limit of the solar energy conversion efficiency is given by the ultimate solar energy conversion efficiency (see Chapter 2). Figure 6.7 compares the Shockley–Queisser limit with the ultimate efficiency and the dependence of the solar energy conversion efficiency on the band gap of ideal solar cells operated at 75 and −25°C at AM1.5 and at concentrated sunlight with a concentration factor of 1000 by keeping the temperature at 25°C.

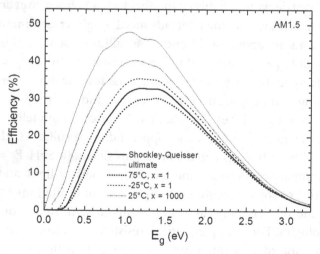

Figure 6.7. Comparison of the Shockley–Queisser limit (thick solid line) with the ultimate efficiency (thin solid line) and with the dependence of the maximum solar energy conversion efficiency on the band gap for ideal single-junction solar cells operated at AM1.5 at 75°C (thick dotted line) and −25°C (short dashed line) and operated at concentrated sunlight with a concentration factor of 1000 at 25°C (thin dotted line).

The ultimate efficiency has a maximum of about 48% at a band gap of about 1.0 eV. The efficiency in the Shockley–Queisser limit is 31% for an ideal solar cell with a band gap of 1.0 eV. The efficiency of the same ideal solar cell decreases to about 27% or it increases to about 34% if the temperature increases to 75°C or decreases to −25°C, respectively. On the other hand, the efficiency of an ideal solar cell with a band gap of 1.0 eV increases to about 40% if the sunlight is concentrated by 1000 times and if the operation temperature is kept at 25°C.

Relative changes of the efficiency are much more pronounced at lower band gaps. For example, the efficiency of an ideal solar cell with E_g equal to 0.3 eV operated at room temperature increases from about 1% at AM1.5 to about 10% at concentrated sunlight with a concentration factor of 1000. In contrast, the efficiency increases only from about 21% at AM1.5 to 24% at sunlight concentrated with a factor of 1000 for an ideal solar cell with a band gap of 2.0 eV. Therefore, the range of useful photovoltaic absorbers can be extended to lower band gaps under illumination with concentrated sunlight.

The maximum concentration factor of sunlight is limited due to the extension of the sun as a light source. An absorber cannot emit photons at a higher energy than it receives them, i.e. the maximum temperature of an absorber cannot be above the temperature at the surface of the sun (5800 K). The maximum concentration factor of sunlight can be calculated from the balance of energy fluxes that are received and emitted by an absorber. The maximum concentration factor of sunlight on earth is equal to 46200 (see also task T6.2 at the end of this chapter). However, there is no practical meaning of this value for photovoltaic solar energy conversion.

6.2.3 *Thermalization losses*

The maximum solar energy conversion efficiency is only 33% in the Shockley–Queisser limit and even the ultimate efficiency is less than 50% in the maximum. Thermalization is one of the most important fundamental loss mechanisms in photovoltaic solar energy conversion.

Photons of the sun spectrum have a broad energy distribution and can therefore excite free electrons and holes with broad distributions of kinetic energy as well (Figure 6.8). A distribution of kinetic energies is related to a certain temperature. The temperature corresponding to photo-excited free

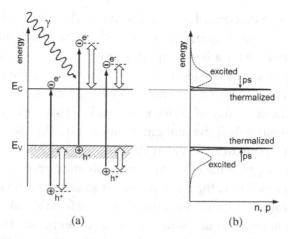

Figure 6.8. Photo-excitation of free electrons and holes at different kinetic energies denoted by the open double-ended arrows (a) and schematic energy distributions of excited and thermalized free electrons and holes (b).

electrons and holes is much higher than the temperature at which a solar cell is operated. Photo-excited free electrons and holes are therefore called hot electrons and hot holes. Hot electrons and hot holes lose their excess kinetic energy within a very short time due to interactions with atoms in the semiconductor. The cooling process of hot charge carriers is called thermalization and can last as long as several picoseconds. The excess kinetic energy of hot electrons and holes is transferred into vibrations of atoms in the photovoltaic absorber resulting in an increase of the temperature of the solar cell.

Free electrons and holes behave like an ideal gas. The kinetic energy of an ideal gas ($W_{ideal\ gas}$) is given by the following equation:

$$W_{ideal\ gas} = \frac{3}{2} \cdot k_B \cdot T \qquad (6.14)$$

Photons of the sun spectrum can heat photo-excited electrons and holes up to the temperature at the surface of the sun. In this case, photo-excited electrons and holes reach a maximum kinetic energy. For the estimation of thermalization losses in solar cells the value of $W_{ideal\ gas}$ has to be doubled since excited electrons and holes obtain high temperatures. Therefore, the maximum kinetic energy of excited electrons and holes is 1.5 eV ($k_B \cdot T_S$ is about 0.5 eV). This value has to be compared with the band gap of a given photovoltaic absorber. For example, the thermalization losses are about 50% for a solar cell with a semiconductor having a band gap equal to 1.5 eV. For comparison, the losses are about 58% or 68% for the ultimate efficiency or for the Shockley–Queisser limit, respectively, for a band gap of 1.5 eV. Thermalization losses increase with decreasing band gap. *Physics of Solar Cells. From Principles to New Concepts* by P. Würfel is recommended for further reading (Würfel, 2005).

6.3 Multi-Junction Solar Cells

6.3.1 *Spectral splitting for reduction of thermalization losses*

Thermalization losses are very low for a given photovoltaic absorber if the energies of absorbed photons have values between the band gap of the semiconductor and energies little larger than the band gap. The principle of multi-junction solar cells is based on the splitting of the broad energy spectrum of sunlight into several spectra with narrow distributions of photon energy (Figure 6.9). Each narrow spectrum of photons is assigned

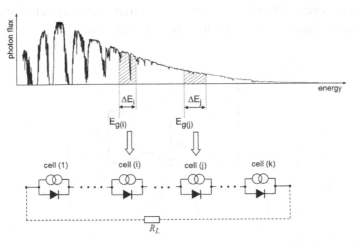

Figure 6.9. Principle of a multi-junction solar cell including spectral splitting of the sun spectrum into separate narrow spectra, assignment of an optimum band gap for each narrow spectrum and series connection of separate solar cells to one multi-junction solar cell. The photovoltaic absorber with the lowest band gap (bottom cell) has the index 1, the semiconductor with the highest band gap (top cell) the index k.

to one ideal photovoltaic absorber with a band gap equal to the lowest photon energy in the given narrow spectrum. The photons of each narrow spectrum are absorbed by the semiconductor with optimum band gap and converted into electric power at maximum efficiency with a separate solar cell. All separate solar cells with different absorber are connected in series so that a multi-junction solar cell can be treated like one solar cell connected to one load resistance.

Figure 6.9 depicts the principle of an ideal multi-junction solar cell with k band gaps while $E_{g(1)}$ and $E_{g(k)}$ denote the lowest and highest values, respectively. The solar cells with the band gaps $E_{g(1)}$ and $E_{g(k)}$ are called the bottom and top solar cells, respectively.

Figure 6.10 shows an example for the band diagram of a multi-junction solar cell with four different band gaps in thermal equilibrium. A multi-junction solar cell with four different band gaps contains four pn-homo-junctions as charge-selective contacts (see Chapter 4) — one for each separate solar cell — and three degenerated pn-hetero-junction as ohmic recombination contacts (see Chapter 5) for connecting the solar cells in series. The given example contains 16 layers of $n^+/n/p/p^+/n^+/n/p/p^+/n^+/n/p/p^+/n^+/n/p/p^+$. Real multi-junction solar cells (see Chapter 8) consist of even more layers. Obviously, such a layer system contains many junctions, which is characteristic of a multi-junction solar cell.

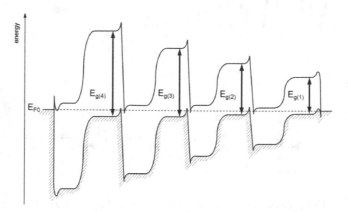

Figure 6.10. Schematic band diagram of a multi-junction solar cell with four different forbidden band gaps and its four corresponding pn-homo-junctions as charge-selective contacts and three degenerated pn-hetero-junctions as ohmic recombination contacts in thermal equilibrium.

The band gaps of multi-junction solar cells are numbered, starting with the lowest and finishing with the highest of the band gaps ($E_{g(1)}$, $E_{g(2)}$, $E_{g(3)}$, and $E_{g(4)}$ in Figure 6.10). First the sunlight passes into the ideal solar cell with the highest band gap so that all photons with energy equal to or larger than $E_{g(4)}$ are absorbed. All photons with energy less than $E_{g(4)}$ pass through to the ideal solar cell with $E_{g(3)}$. All photons with energies in the interval between $E_{g(4)}$ and $E_{g(3)}$ are absorbed in the solar cell with $E_{g(3)}$. Photons with energy less than $E_{g(3)}$ pass through to the ideal solar cell with $E_{g(2)}$ where all photons with energies between $E_{g(3)}$ and $E_{g(2)}$ are absorbed. Photons with energies below $E_{g(2)}$ continue through to the ideal solar cell with the lowest band gap where photons with energies in the interval between $E_{g(2)}$ and $E_{g(1)}$ are absorbed.

6.3.2 *Current-matching condition and optimum band gaps*

The spectrum of the photon flux is split into k intervals. The photons of each interval are absorbed by one separate solar cell. The lowest energy of each interval denoted by index j is assigned to a band gap $E_{g(j)}$. The short-circuit current density of the solar cell with the band gap $E_{g(j)}$ is equal to:

$$I_{SC(j)} = q \cdot \int_{E_{g(j)}}^{E_{g(j+1)}} \Phi_{sun}(h\upsilon) \cdot d(h\upsilon) \qquad (6.15)$$

The separate solar cells are connected in series in a multi-junction solar cell. Therefore the short-circuit current density of a multi-junction solar cell ($I_{SC(multi)}$) is limited by the lowest of the short-circuit currents of the separate solar cells ($I_{SC(i)}$).

$$I_{SC(multi)} = min(I_{SC(i)}) \qquad (6.16)$$

Equation (6.16) is a rigid condition since the separate solar cell with the lowest photocurrent limits the energy conversion efficiency of the complete multi-junction solar cell. Therefore, for maximum energy conversion efficiency, the current-matching condition of a multi-junction solar cell requires equal short-circuit current densities for all separate solar cells connected in series.

$$I_{SC(multi)} = I_{SC(i)}|_{i \in (1,k)} \qquad (6.17)$$

The energy conversion efficiency of a multi-junction solar cell can only be maximized if the current-matching condition is fulfilled. The current-matching condition is critical for attaining very high solar energy conversion efficiency in multi-junction solar cells. For ideal multi-junction solar cells the current-matching condition should be fulfilled by choosing the right values for the band gaps.

The highest short-circuit current density of a single-junction solar cell can be achieved with the lowest of the separate band gaps ($I_{SC(1)}$). For maximum $I_{SC(multi)}$, the short-circuit current density of a single-junction solar cell with the band gap of the bottom cell has to be divided into identical portions for each of the remaining solar cells:

$$I_{SC(multi)} = \frac{I_{SC}^{single}(E_{g(1)})}{k} = \frac{q}{k} \cdot \int_{E_{g(1)}}^{\infty} \Phi_{sun}(h\upsilon) \cdot d(h\upsilon) \qquad (6.18)$$

The values of the band gaps $E_{g(j)}$ are obtained by the following condition:

$$(k + 1 - j) \cdot I_{SC(multi)}|_{j \in (1,k)} = q \cdot \int_{E_{g(j)}}^{\infty} \Phi_{sun}(h\upsilon) \cdot d(h\upsilon) \qquad (6.19)$$

The values of the forbidden band gaps in a multi-junction solar cell can be obtained from the direct application of the dependence of I_{SC} on E_g for ideal single-junction solar cells following condition (6.19). Figure 6.11 shows an example for the determination of the optimum band gaps in a multi-junction solar cell containing four ideal photovoltaic absorbers with separate band gaps ($k = 4$). A minimum band gap of $E_{g(1)} = 0.74\,\text{eV}$ is chosen.

The short-circuit current density of an ideal single-junction solar cell with a band gap of 0.74 eV is 52 mA/cm^2. Therefore, with regard to Equation (6.18), $I_{SC(multi)}$ of the respective multi-junction solar cell is 13 mA/cm^2. The band gap of the solar cell with the index $j = 2$ is, with regard to Equation (6.19), equal to the band gap of a single-junction solar cell with a short-circuit current density of 39 mA/cm^2. Therefore, the value of $E_{g(2)}$ is equal to about 1.1 eV. Furthermore, the values of $E_{g(3)}$ and $E_{g(4)}$ are equal to 1.47 and 1.95 eV, respectively.

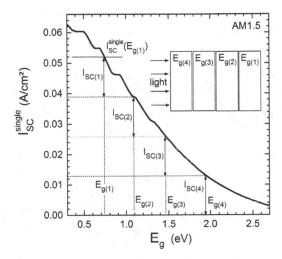

Figure 6.11. Determination of the optimum band gaps in a multi-junction solar cell with four separate solar cells and a band gap of the bottom cell of 0.74 eV.

Figure 6.12 shows the dependence of the band gaps $E_{g(i)}$ on $E_{g(1)}$ for tandem solar cells ($k = 2$), triple-junction solar cells ($k = 3$) and quadruple-junction solar cells ($k = 4$).

The dependencies do not follow smooth lines due to the dependence of the short-circuit current density of single-junction solar cells on the band gap. For example, the application of a band gap $E_{g(1)}$ of the bottom solar cell equal to 0.8 eV requires a band gap $E_{g(2)}$ of the top solar cell of about 1.45 eV for a tandem solar cell. The values of $E_{g(2)}$ decrease to about 1.3 eV and 1.05 eV for the respective triple- and quadruple-junction solar cells. The top solar cell of the triple-junction solar cell with $E_{g(1)}$ equal to 0.8 eV should have a band gap $E_{g(3)}$ equal to about 1.8 eV whereas the value of $E_{g(3)}$ is reduced to about 1.45 eV for the quadruple-junction solar cell. The band gap of the top solar cell of the quadruple-junction solar cell ($E_{g(4)}$) with $E_{g(1)}$ equal to 0.8 eV should be about 2.0 eV. The right combinations of optimum band gaps can be obtained in a similar way for any multi-junction solar cell.

6.3.3 *Dependence of the efficiency on the number of band gaps*

The open-circuit voltage of a multi-junction solar cell ($V_{OC(multi)}$) is equal to the sum of the open-circuit voltage of all separate solar cells ($V_{OC(i)}$),

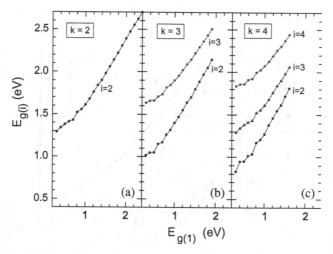

Figure 6.12. Dependence of the band gaps $E_{g(i)}$ ($i > 1$) on the band gap of the bottom solar cell ($E_{g(1)}$) for multi-junction solar cells with two (tandem solar cells (a)), three (triple-junction solar cells (b)) and four (quadruple-junction solar cells (c)) photovoltaic absorbers under illumination at AM1.5 and optimum current matching.

where k is the number of band gaps in the multi-junction solar cell (see also Figure 6.13).

$$V_{OC(multi)} = \sum_{i=1}^{k} V_{OC(i)} \qquad (6.20)$$

The ideal tunneling ohmic contacts for two neighboring separate solar cells connect the Fermi-energies of majority charge carriers with opposite signs at the same energy. The Fermi-level splitting of two neighboring separate solar cells is added in this way so that the potential drop across the external leads of a multi-junction solar cell corresponds to the sum of the Fermi-level splitting of all separate solar cells. This is shown in Figure 6.13 for the example of a multi-junction solar cell with four different band gaps under illumination at open-circuit condition.

The solar energy conversion efficiency of a multi-junction solar cell (η_{multi}) is obtained from the sum of the power of the separate solar cells.

$$\eta_{multi} = \frac{q}{k} \cdot \int_{E_{g(1)}}^{\infty} \Phi_{sun}(h\upsilon) \cdot d(h\upsilon) \cdot \frac{\sum_{i=1}^{k} (FF_i \cdot V_{OC(i)})}{P_{sun}} \qquad (6.21)$$

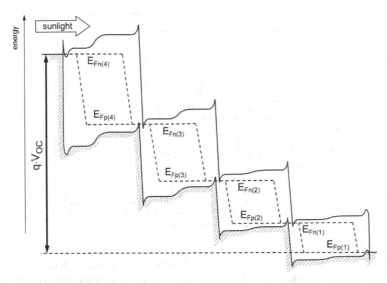

Figure 6.13. Schematic band diagram of a multi-junction solar cell with four different forbidden band gaps and corresponding four pn-homo-junctions as charge-selective contacts and three degenerated pn-hetero-junctions as ohmic recombination contacts under illumination at open-circuit condition.

Equation (6.21) shows that the energy conversion efficiency of multi-junction solar cells is larger than the efficiency of a single-junction solar cell due to the sum of the open-circuit voltages. In the following, η_{multi} will be analyzed as a function of the band gap of the bottom cell, of the number of band gaps, and of the concentration factor.

The diode saturation current densities ($I_{0(i)}$) are calculated using Equation (6.13) for each ideal solar cell with a band gap $E_{g(i)}$ within the multi-junction solar cell. The current–voltage and power–voltage characteristics are calculated for each separate ideal solar cell of the multi-junction solar cell by plugging $I_{SC(multi)}$ and $I_{0(i)}$ into the diode Equation (6.11).

Figure 6.14 shows an example for current–voltage and power–voltage characteristics of a quadruple-junction solar cell with $E_{g(1)}$, $E_{g(2)}$, $E_{g(3)}$ and $E_{g(4)}$ equal to 0.74, 1.1, 1.47 and 1.95 eV, respectively. As already mentioned, the short-circuit current density is 13 mA/cm^2 for the given quadruple-junction solar cell. The values of the open-circuit voltages are 0.47, 0.78, 1.14 and 1.63 V for $V_{OC(1)}$, $V_{OC(2)}$, $V_{OC(3)}$ and $V_{OC(4)}$, respectively. The corresponding open-circuit voltage of the quadruple-junction solar cell is therefore equal to 4.02 V. The values of the maximum power are 5, 9, 13 and

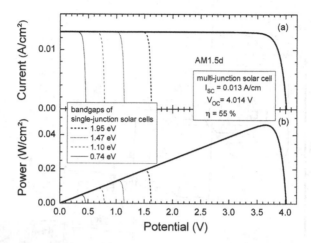

Figure 6.14. Current–voltage (a) and power–voltage (b) characteristics of an ideal quadruple-junction solar cell and the ideal single-junction solar cells with band gaps $E_{g(1)}$, $E_{g(2)}$, $E_{g(3)}$ and $E_{g(4)}$ equal to 0.74, 1.1, 1.47 and 1.95 eV (thin solid, thin dashed, dotted and thick dashed lines, respectively) within the quadruple-junction solar cell illuminated at AM1.5d.

19 mW/cm^2 for the single-junction solar cells with $E_{g(1)}$, $E_{g(2)}$, $E_{g(3)}$ and $E_{g(4)}$, respectively. The total power of the given ideal quadruple-junction solar cell is therefore equal to 46 mW/cm^2. The intensity of the direct sunlight is 84.4 mW/cm^2 at AM1.5d for the given sun spectrum. Therefore the solar energy conversion efficiency is about 55% for the given quadruple-junction solar cell. The efficiency of 55% is larger than the maximum solar energy conversion efficiency in the Shockley–Queisser limit (33%).

Figure 6.15 summarizes the dependencies on $E_{g(1)}$ of the solar energy conversion efficiencies of single-, triple- and quadruple-junction solar cells and tandem solar cells. A maximum efficiency of 44% is reached for tandem solar cells with $E_{g(1)} = 1.1$ eV and $E_{g(2)} = 1.7$ eV. The increase of the efficiency is very strong for low values of $E_{g(1)}$ due to the influence of the top solar cell with $E_{g(2)}$ around 1.3–1.5 eV. Combinations of $E_{g(1)}$ in the range between 0.4 and 1.6 eV with the respective values of $E_{g(2)}$ between 1.35 and 2.0 eV are suitable for increasing the solar energy conversion efficiency of tandem solar cells above the maximum efficiency in the Shockley–Queisser limit of single-junction solar cells.

Solar energy conversion efficiencies above 40% can be realized with ideal tandem solar cells for $E_{g(1)}$ between 0.6 and 1.3 eV ($E_{g(2)}$ between

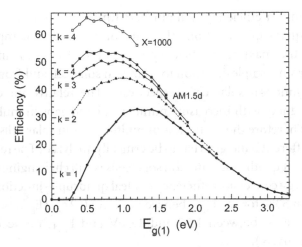

Figure 6.15. Dependence of the solar energy conversion efficiency on the band gap of the bottom solar cell ($E_{g(1)}$) of ideal single-junction solar cells (k = 1, circles), tandem solar cells (k = 2, triangles), triple-junction solar cells (k = 3, stars) and quadruple-junction solar cells (k = 4, squares) operated at room temperature and AM1.5d (filled symbols) and of quadruple-junction solar cells operated at room temperature and concentrated sunlight with concentration factor 1000 (open squares).

1.37 and 1.75 eV, respectively). This shows that semiconductors with a wide range of band gaps can be applied for the realization of high-efficiency tandem solar cells. The difference between the energy conversion efficiency of single-junction and tandem solar cells becomes small for values of $E_{g(1)}$ above 2 eV.

An efficiency slightly above 50% can be reached with triple-junction solar cells with the values of about 0.7, 1.15 and 1.75 eV for $E_{g(1)}$, $E_{g(2)}$ and $E_{g(3)}$, respectively. The highest efficiency of 36% at AM1.5 has been achieved for a triple-junction solar cell based on III–V semiconductors (see Chapter 8) with a similar combination of band gaps.

Solar energy conversion efficiencies above 48% can be obtained with ideal triple-junction solar cells for $E_{g(1)}$ between 0.6 and 1.1 eV corresponding to $E_{g(2)}$ between 1.15 and 1.45 eV and $E_{g(3)}$ between 1.65 and 1.95 eV, respectively.

The application of quadruple-junction solar cells presents the opportunity of increasing the solar energy conversion efficiency to values of about 54–55% for a combination of band gaps with energies of 0.7, 1.02, 1.40 and 1.92 eV for $E_{g(1)}$, $E_{g(2)}$, $E_{g(3)}$ and $E_{g(4)}$, respectively. Incidentally, the increase of the maximum efficiency is about 1.08 times for the transition from a triple- to a quadruple-junction solar cell. For purposes of comparison, the increase of the maximum efficiency is about 1.33 or 1.14 times for the transition from a single-junction to a tandem solar cell and from a tandem to a triple-junction solar cell, respectively. The relative increase of the efficiency reduces with increasing number of band gaps in multi-junction solar cells. Therefore the usefulness of multi-junction solar cells with more than four different band gaps has to be critically analyzed. The realization of highly efficient quadruple-junction solar cells is still challenging in practice.

The energy conversion efficiency of ideal quadruple-junction solar cells is larger than 53% for for $E_{g(1)}$ between 0.5 and 0.9 eV ($E_{g(2)}$ between 0.95 and 1.17 eV, $E_{g(3)}$ between 1.35 and 1.6 eV and $E_{g(4)}$ between 1.85 and 2.0 eV, respectively).

The solar energy conversion efficiency can be further increased by concentrating the sunlight. An example is given for the dependence of the solar energy conversion efficiency on $E_{g(1)}$ of a quadruple-junction solar cell illuminated with sunlight concentrated 1000 times. The solar energy conversion efficiency reaches 66%. This demonstrates the high potential of quadruple-junction solar cells operated at highly concentrated sunlight. The maximum efficiency is obtained for band gaps with energies of 0.5, 0.95, 1.35 and 1.84 eV for $E_{g(1)}$, $E_{g(2)}$, $E_{g(3)}$ and $E_{g(4)}$, respectively. The values of these band gaps are red shifted in comparison to the band gaps of the quadruple-junction solar cell with maximum efficiency at AM1.5. The shift to lower band gaps of a quadruple-junction solar cell with maximum efficiency at concentrated sunlight is caused by the fact that the solar energy conversion efficiency increase is much stronger for lower band gaps, as mentioned before.

So far, the highest solar energy conversion efficiency of 44.7% has been reached with a quadruple-junction solar cell operated at concentrated sunlight with a concentration factor of 297 (Fraunhofer, 2013).

6.4 Alternative Concepts for Increasing the Efficiency

6.4.1 Increased quantum efficiency with higher energetic photons

A part of thermalization losses may be avoided if the energy of one photon exceeding the energy of the doubled band gap can be converted into the

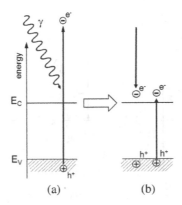

Figure 6.16. Schematic of impact ionization including the absorption of one photon with high energy (a) and the excitation of two electron-hole pairs (b).

excitation of two free electron-hole pairs. This is possible due to impact ionization. A free charge carrier with high kinetic energy collides with an electron in the valence band and excites it into the conduction band. The quantum efficiency of an absorption step followed by impact ionization is equal to 2 (Figure 6.16).

Quantum efficiencies can exceed 1 (Kolodinski *et al.*, 1993) and can become even larger than 2 as reported for optical experiments on quantum dots (multi-exciton generation, see also Chapter 10) (Schaller and Klimov, 2004).

Another way to increase the quantum efficiency to values above 1 makes use of so-called down-conversion. The energy of one photon with high energy can be split into the energy of several photons with lower energy in down-conversion processes (Figure 6.17). In down-conversion, an electron is excited from a ground state with energy E_1 to an excited state E_3. The excited electron relaxes to an intermediate excited state with energy E_2 between E_1 and E_3 by emitting a photon. Furthermore, the electron relaxes from the state at E_2 to the ground state by emitting a second photon. The emitted photons can be absorbed by a photovoltaic absorber if the energy of the emitted photons is equal to or larger than the band gap of the photovoltaic absorber. Therefore, the quantum efficiency of a solar cell can be increased by covering the surface of a solar cell with a down-converter for photons exceeding the doubled band gap. Impact ionization and down-conversion are equivalent in this sense.

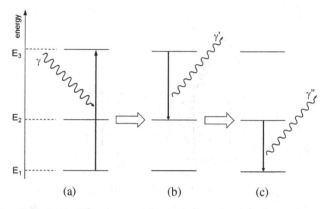

Figure 6.17. Schematic of down-conversion including the absorption of one photon with high energy (a) and the emission of two photons with lower energies ((b) and (c)).

The forbidden band gap of an ideal photovoltaic absorber remains unchanged if processes of impact ionization or down-conversion are incorporated into the analysis of the maximum solar energy conversion efficiency. Therefore, only I_{SC} is changed in the diode equation of an illuminated solar cell. The short-circuit current of a solar cell can be expressed by the following equation:

$$I_{SC} = q \cdot \sum_{i=1}^{n} i \cdot \int_{i \cdot E_g}^{(i+1) \cdot E_g} \Phi_{sun}(h\upsilon) \cdot d(h\upsilon) \qquad (6.22)$$

Equation (6.22) describes the maximum effect of impact ionization or down-conversion. However, electrons can also be excited from states deeper in the valence band, which results in an excited electron with less energy. This behavior leads to a reduction of the influence of impact ionization on I_{SC}. Therefore, photon energies much larger than the doubled band gap are required for a significant influence of impact ionization on I_{SC}. As a more realistic assumption, the influence of impact ionization on I_{SC} will start with a doubled quantum efficiency at photon energies larger than the tripled band gap.

$$I_{SC} = q \cdot \left(\int_{E_g}^{2 \cdot E_g} \Phi_{sun}(h\upsilon) \cdot d(h\upsilon) + \sum_{i=1}^{n} i \cdot \int_{(i+1) \cdot E_g}^{(i+2) \cdot E_g} \Phi_{sun}(h\upsilon) \cdot d(h\upsilon) \right)$$

$$(6.22')$$

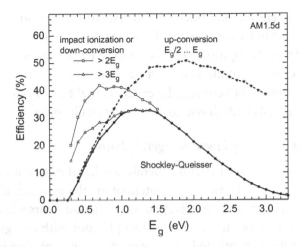

Figure 6.18. Comparison of the Shockley–Queisser limit (thick solid line) with the dependence of the efficiency on the band gap of an ideal solar cell taking into account impact ionization or down-conversion beginning at $2 \cdot E_g$ (squares) or $3 \cdot E_g$ (triangles) and taking into account up-conversion for photons with energy $E_g/2$ and E_g (stars).

Equation (6.22) can be rewritten in such a way that the values of the short-circuit current densities of ideal solar cells related to the Shockley–Queisser limit can be used to calculate I_{SC}:

$$I_{SC} = q \cdot \sum_{i=1}^{n} \int_{i \cdot E_g}^{\infty} \Phi_{sun}(h\upsilon) \cdot d(h\upsilon) = \sum_{i=1}^{n} I_{SC,i \cdot E_g}^{single} \qquad (6.23)$$

Figure 6.18 compares the dependencies of the energy conversion efficiency on the band gap of an ideal single-junction solar cell for the Shockley–Queisser limit and for considering impact ionization.

Maximum impact ionization and optimum down-conversion can play a significant role for band gaps below 1.5 eV since the photon flux is already very low at photon energies above 3 eV. The energy conversion efficiency reaches a value of about 42% at a band gap of 0.7 eV and values of the energy conversion efficiency above 39% are obtained for band gaps between 0.6 and 1.2 eV with maximum impact ionization or optimum down-conversion. These values of the energy conversion efficiency are higher than the maximum in the Shockley–Queisser limit and demonstrate the potential for impact ionization and/or down-conversion. However, one has to take into account that the energy conversion efficiency will not exceed the maximum efficiency in the Shockley–Queisser limit if considering the more realistic assumption expressed by Equation (6.22′). Therefore, it seems difficult to improve significantly the solar energy conversion efficiency if assuming even moderate losses on the maximum influence of impact ionization or down-conversion. The experimental task of improving high-efficiency solar cells with down-conversion is challenging.

6.4.2 *Up-conversion of lower energetic photons*

The use of photons with energies below the band gap of a photovoltaic absorber in a solar cell has a large potential for the increase of the energy conversion efficiency, and especially so for semiconductors with wide band gaps. For this purpose, the energy of two photons with energy below the band gap should be converted into the energy of one photon with energy equal to or larger than the band gap of the semiconductor. Processes in which the energy of two photons is converted into a higher energy of one photon are called up-conversion. The quantum efficiency is 0.5 or less for up-conversion.

Figure 6.19 shows an example of a transition scheme for up-conversion in a system with six energy levels. An electron is excited from its ground

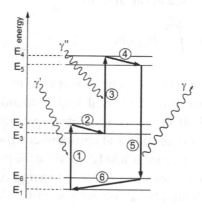

Figure 6.19. Transition scheme for up-conversion of photon energies including elementary steps of excitation of an electron from a ground state E_1 to a state E_2 (1), fast relaxation of the electron from state E_2 to state E_3 from where recombination to states at even lower energies is strongly reduced (2), subsequent excitation from state E_3 to state E_4 (3), fast relaxation of the electron from state E_4 to state E_5 from where recombination to states E_2 and E_3 is strongly reduced (4), radiative relaxation from state E_5 to a state E_6 including photon emission (5) and relaxation from state E_6 back to the ground state E_1 (6).

state E_1 to an excited state E_2 by a photon with low energy. The excited electron relaxes very rapidly into a state E_3 at a slightly reduced energy. The excited electron is metastable in state E_3, i.e. further relaxation to lower energies takes a much longer time compared to the time it takes for the second excitation step up to the energy E_4, again caused by another photon with low energy. The electron relaxes from state E_4 to state E_5 at a slightly reduced energy. Relaxation of the excited electron from state E_5 to the states at energies E_2 and E_3 is much slower than relaxation to the state E_6. Therefore, the electron relaxes from state E_5 at high energy to state E_6 at low energy under the emission of a photon with high energy. In the last step, the electron relaxes from E_6 to the ground state at energy E_1 from which a next up-conversion cycle can start.

Up-conversion demands sophisticated optical or molecular systems with reliable transitions. It seems that up-conversion is practically impossible for a broad spectrum of incoming photons. The increase of the waiting time of an excited electron before the second excitation occurs is challenging since relaxation back to the ground state must be efficiently suppressed during this waiting time. The waiting time can be reduced under concentrated sunlight.

Up-converting optical elements can be incorporated into solar cells, for example, between the back surface of the solar cell and a back reflector without changing the photovoltaic absorber. An optimum up-converter will be able to convert all photons with energy between the half of the band gap of the ideal photovoltaic absorber and its band gap to photons with energies above the band gap. Therefore the short-circuit current density can be calculated in the following way for a solar cell with an optimum up-converter:

$$I_{SC} = \frac{q}{2} \cdot \int_{E_g/2}^{E_g} \Phi_{sun}(h\upsilon) \cdot d(h\upsilon) + q \cdot \int_{E_g}^{\infty} \Phi_{sun}(h\upsilon) \cdot d(h\upsilon) \quad (6.24)$$

The dependence of the energy conversion efficiency on the band gap of a single-junction solar cell optimal up-conversion is plotted in Figure 6.18. The relative increase of the energy conversion efficiency is tremendous for solar cells with large band gaps. For example, the efficiency increases from about 3.5% for a solar cell with a band gap of 3 eV in the Shockley–Queisser limit to about 39% for a solar cell with the same band gap but with additional optimum up-conversion.

The maximum energy conversion efficiency of about 50% may be reached for a single-junction solar cell with up-conversion and a band gap of 1.9 eV. The energy conversion efficiency is above 48% for solar cells with band gaps between 1.4 and 2.2 eV and optimum up-converters. However, the search for reliable up-converters is still ongoing.

Electrolytic water splitting requires potentials of more than 1.23 eV (Walter *et al.*, 2010). Therefore, the successful development of suitable up-converters may become very important, especially for the production of solar fuels by photocatalytic water splitting.

6.5 Summary

All losses that can be avoided in ideal solar cells have to be minimized to zero for achieving the maximum solar energy conversion efficiency. These losses include (i) optical losses caused by reflection and transmission, (ii) resistive losses caused by series and shunt resistances and (iii) recombination losses caused by defects in the bulk and at the surfaces of the photovoltaic absorber. Furthermore it is assumed that radiative recombination is much stronger than Auger recombination so that Auger recombination can be neglected for the analysis of the maximum efficiency. As a consequence, the maximum efficiency depends only on the band gap (E_g) of a photovoltaic absorber, on its temperature and on the spectrum and intensity of sunlight.

A photovoltaic absorber absorbs the part of the blackbody radiation with photon energies equal or larger than E_g resulting in thermal generation of free electrons and holes. The thermal generation rate ($G_0(E_g,T_0)$) is balanced by the radiative recombination rate so that the rate constant can be derived as a function of the temperature and E_g.

Under illumination, the radiative recombination rate is proportional to the product of $G_0(E_g,T_0)$ and the exponential of the Fermi-level splitting ($E_{Fn}-E_{Fp}$) divided by $k_B \cdot T_0$. In an ideal solar cell, the Fermi-level splitting is equal to the potential difference at the external leads multiplied by the elementary charge ($q \cdot U$) due to ideal charge-selective and ohmic contacts. The current density of an illuminated solar cell follows from the radiative recombination, thermal generation and photo-generation rates.

The resulting diode saturation current density is given by $q \cdot G_0(E_g, T_0)$. The calculation of the diode saturation current density from the thermal generation rate is the key for the analysis of the maximum solar energy conversion efficiency of solar cells.

At a given temperature and for a given sun spectrum, the maximum efficiency depends only on E_g. The Shockley–Queisser limit describes the dependence of the maximum efficiency on E_g under illumination at AM1.5 and at a temperature of 25°C. The maximum efficiency in the Shockley–Queisser limit is about 32% for solar cells with E_g between 1.1 and 1.5 eV. A comparison of the efficiency of real solar cells with the efficiency in the Shockley–Queisser limit shows the potential for solar energy conversion of a particular photovoltaic absorber and of advanced technologies for producing solar cells with minimum losses. The lowest deviation between the Shockley–Queisser limit and the efficiency of record solar cells is obtained for a GaAs absorber.

Losses are dominated by thermalization in the Shockley–Queisser limit. A further increase of the maximum efficiency of solar cells is only possible by reducing thermalization losses. For this purpose (i) spectral splitting of the sun spectrum and (ii) separate conversion with an optimum E_g for each part of the sun spectrum are required.

Solar energy conversion efficiencies above the Shockley–Queisser limit can be realized with stacked multi-junction solar cells where the separate solar cells are connected in series. The bottom cell has the lowest band gap ($E_{g(1)}$). The top cell has the highest band gap ($E_{g(k)}$, k is the number of band gaps in the multi-junction solar cell). Each subsequent solar cell has a band gap ($E_{g(i)}$) lower than the band gap of the solar cell in front of it $E_{g(i+1)}$ but larger than the band gap of the solar cell that follows ($E_{g(i-1)}$). Therefore, each solar cell with $E_{g(i)}$ absorbs photons in the interval between $E_{g(i+1)}$ and $E_{g(i)}$.

In a multi-junction solar cell, the short-circuit current density (I_{SC}) has to be the same in each separate solar cell (current-matching condition) since I_{SC} of the multi-junction solar cell is limited by the solar cell with the lowest I_{SC}. The optimum combinations of band gaps in multi-junction solar cells, respectively, can be easily derived from the current-matching condition. The optimum band gaps of the bottom and top cells are 1.1 and 1.7 eV, respectively, in a tandem solar cell.

The maximum solar energy conversion efficiency is about 44, 50 and 55% for tandem, triple-junction and quadruple-junction solar cells, respectively, operated at room temperature and illuminated at AM1.5. The maximum efficiency can be further increased under illumination with concentrated sunlight, for example, up to 66% for a quadruple-junction solar cell illuminated with sunlight concentrated 1000 times. A solar energy conversion efficiency as high as 43% has been achieved with a triple-junction solar cell illuminated with concentrated sunlight (Green *et al.*, 2013).

The maximum efficiency of solar cells may also be increased by increasing the quantum efficiency of photo-generation. For this purpose, impact ionization can be applied. The number of photons absorbed by the solar cell can be increased by introduction of optical down-converters for photons with energies above the doubled band gap of the photovoltaic absorber. Impact ionization and down-conversion are equivalent. The implementation of optical up-converters at the back side of solar cells provides a further alternative with which to overcome the Shockley–Queisser limit. However, in reality increasing the solar energy conversion efficiency with concepts of impact ionization, down-conversion and up-conversion remains challenging.

6.6 Tasks

The tasks T6.1–T6.4 are well-suited for working in groups. Figures T6.1 and 2.4 are used as nomograms for ascertaining values of I_{SC} and fill factors (FF) by interpolation without the need for integration over the sun spectrum or for simulation of current–voltage characteristics.

T6.1: Shockley–Queisser limit

Ascertain the Shockley–Queisser limit for solar cells with ideal absorbers with band gaps of (a) 0.3, (b) 1.3, (c) 2.0 and (d) 2.7 eV by using Figures T6.1 and 2.4.

T6.2: Maximum efficiency of tandem solar cells

Ascertain the efficiency of ideal tandem solar cells with a band gap of the bottom cell of (a) 0.5, (b) 1.0 and (c) 1.5 eV operated at 50°C by using Figures T6.1 and 2.4.

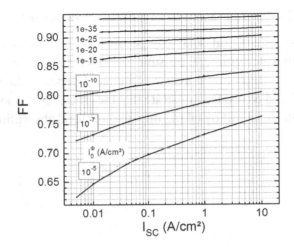

Figure T6.1.   Fill factor as a function of I_{SC} and I_0.

T6.3: Band gaps of absorbers in multi-junction solar cells

Ascertain the optimum band gaps of the absorbers in a multi-junction solar cell with five different absorbers if the band gap of the bottom cell is (a) 0.5 or (b) 1.0 eV.

T6.4: Concentrator multi-junction solar cells

Calculate the maximum solar energy conversion efficiency of an ideal triple-junction solar cell with a band gap of the bottom cell of 0.7 eV under illumination with concentrated sunlight for concentration factors of 20, 200, 500 and 1000 by using Figures T6.1 and 2.4.

T6.5: Maximum concentration factor of sunlight

Ascertain the maximum concentration factor of sunlight by using the balance between the power of sunlight and the power of thermal radiation.

T6.6: Thermodynamic limit of solar energy conversion

Ascertain the maximum solar energy conversion efficiency in the thermo-dynamic limit. Heating of an absorber or medium from which work is obtained via a Carnot process is taken into account.

T6.7: Critical remark on solar cells with metal intermediate bands

Some people believe that solar energy conversion efficiencies above the Shockley–Queisser limit are possible for solar cells with an absorber which has a so-called metal intermediate band between the valence and conduction band edges. Demonstrate that the efficiency cannot exceed the Shockley–Queisser limit by taking into account ideal absorption.

Part II

Materials Specific Concepts

7

Solar Cells Based on Crystalline Silicon

Crystalline silicon (c-Si) solar cells are the backbone of large-scale photovoltaic energy production. Over more than half a century of technological development, the energy conversion efficiency of c-Si solar cells have increased from 6% (Chapin *et al.*, 1954) to more than 24% (Zhao *et al.*, 1998) and the production costs have diminished drastically. This chapter starts with principle design rules of c-Si solar cells based on 200–400-μm thick c-Si wafers consisting of a thick p-type doped base, a thin n-type doped emitter, a front-contact finger grid, a back contact and an antireflection coating. Architectures for reducing shading losses by the front contact and for reducing resistive losses by front and back contacts are discussed briefly. The role of segregation at the liquid–solid interface is illuminated for the fast growth of homogeneously doped and extremely pure c-Si crystals. Wafering by means of wire sawing and by kerfless transfer is mentioned. Processes employed for the emitter formation, such as diffusion of phosphorus in c-Si, laser-assisted doping, ion implantation and deposition of doped amorphous silicon are described. Recombination losses at c-Si surfaces are minimized by cleaning, structuring and passivation treatments including etching, oxidation, hydrogen passivation, deposition of amorphous silicon nitride or aluminum oxide and formation of a back surface field. Concepts of high-efficiency c-Si solar cells are compared at the end of this chapter.

7.1 Principle Architecture of c-Si Solar Cells and Resistive Losses

7.1.1 *Base and emitter of conventional c-Si solar cells*

The thickness of a photovoltaic absorber should be about three times larger than its absorption length (see Chapter 2). Without sophisticated measures for light trapping, the thickness of a c-Si absorber has to be of the order of 200–400 μm with regard to the absorption spectrum of c-Si (see Figures 2.8 and 2.11). A flat silicon crystal with a thickness of the order of 200–400 μm is stiff enough to be mechanically stable. Such flat self-supporting silicon crystals are known as wafers. The stiffness of c-Si wafers defines their handling during the production of c-Si solar cells (wafer-based technology). Wafer-based technologies have strong advantages such as high throughput, lightweight handling and excellent binning into narrow intervals for the separation of different power classes into modules.

The thick light-absorbing layer of a c-Si solar cell is called the base. To achieve a high energy conversion efficiency, the lifetime of minority charge carriers in the base of a c-Si solar cell has to be significantly larger than the minimum lifetime (about 100 μs for c-Si; see Chapter 3) so that the diffusion length of minority charge carriers exceeds the absorption length by about ten times (see also Figure 4.11).

The minority carrier lifetime is limited by Auger recombination in c-Si absorbers (see Figure 3.14). With regard to the definition of the diffusion length (Equation (3.5)), of the lifetime (Equation (3.2)) and with reference to the dependence of the Auger recombination lifetime on the density of free charge carriers of doped c-Si (Equation (3.15)), the conditions for the densities of majority free electrons or holes is given by the following equations:

$$n_0 \leq \sqrt{\frac{D_p}{100 \cdot C_A \cdot d_{abs}^2}} \tag{7.1'}$$

$$p_0 \leq \sqrt{\frac{D_n}{100 \cdot C_A \cdot d_{abs}^2}} \tag{7.1''}$$

The diffusion coefficient of electrons (D_n) is 26 cm^2/s in moderately p-type doped c-Si, whereas the diffusion coefficient of holes (D_p) in n-type doped c-Si is about 3.5 times less than D_n (Jacoboni *et al.*, 1977).

The densities of donors and acceptors of the n-type and p-type doped regions, respectively, should be as high as possible in order to maximize the diffusion potential of a pn-junction (Equation (4.12)) since the diffusion potential is the upper limit of the open-circuit voltage (V_{OC}; see Chapter 4).

The c-Si wafer defines the doping of the base of c-Si solar cells. A p-type doped c-Si base has two significant advantages in comparison to an n-type doped c-Si base. First, a p-type doped c-Si absorber can be higher doped than an n-type doped c-Si absorber with the same d_{abs} (thickness of the absorber layer) due to the higher diffusion coefficient of electrons, i.e. a higher diffusion potential can be reached with a p-type doped base. Second, and this is a technological reason, silicon crystals can be very homogeneously p-type doped because the segregation coefficient of boron is close to 1 over a wide range of growth parameters (as discussed in this chapter). Therefore, the base of conventional c-Si solar cells is p-type doped with a density of acceptors of $2-4 \cdot 10^{16}\,\mathrm{cm}^{-3}$.

The resistance of the base can be calculated if the density and the mobility of free holes (μ_p) are known. The conductivity of p-type doped c-Si is:

$$\sigma_p = q \cdot p_0 \cdot \mu_p \tag{7.2}$$

With a typical mobility of holes of about $270\,\mathrm{cm}^2/(\mathrm{Vs})$ for moderately doped p-type c-Si (Jacoboni *et al.*, 1977), the resistance of the base of a c-Si solar cell (R_{base}^{c-Si}) is:

$$R_{base}^{c-Si} = \frac{d_{abs}}{q \cdot p_0 \cdot \mu_p} \approx 0.02\ \Omega\mathrm{cm}^2 \tag{7.3}$$

For purposes of comparison, the tolerable series resistance (Equation (1.29′)) is about $0.17\ \Omega\mathrm{cm}^2$ for a record c-Si solar cell with V_{OC} and I_{SC} (short-circuit current) equal to $0.709\,\mathrm{V}$ and $40.7\,\mathrm{mA/cm}^2$, respectively, at standard test conditions (Zhao *et al.*, 1995). Therefore, the resistance of the base does not limit the energy conversion efficiency of a c-Si solar cell illuminated at AM1.5. However, limitation of the efficiency of c-Si solar cells by the resistance of the base becomes important under illumination at concentrated sunlight when R_{base}^{c-Si} becomes larger than the tolerable series resistance.

The n-type doped layer of a conventional c-Si solar cell is called the emitter. The density of donors in the emitter should be as high as possible for maximizing the diffusion potential of the pn-junction. It is useful to increase the density of donors in the emitter to a value close to the effective density of states at the conduction band edge (N_C about $3 \cdot 10^{19}$ cm^{-3} for c-Si (Sze, 1981)). For a maximum density of donors the thickness of the emitter (d_{em}) can be obtained.

$$d_{em} \leq \sqrt{\frac{D_p}{100 \cdot C_A \cdot n_0^2}} \tag{7.4}$$

The Auger coefficient is about $2 \cdot 10^{-30}$ cm^6/s for c-Si (Yablonovitch and Gmitter, 1986, (a)). Therefore, the thickness of the n-type doped emitter should be of the order of 100 nm in c-Si solar cells. Usually d_{em} is a factor of two to four times larger than 100 nm for technological and/or design reasons of the c-Si solar cell. An n-type doped emitter is doped with phosphorus in conventional c-Si solar cells. An n-type doped base and a p-type doped emitter can be considered in a similar way in general (see task T7.1 at the end of this chapter).

The diffusion potential of conventional c-Si solar cells ranges between 0.90 and 0.94 V. An additional improvement of the diffusion potential is possible by increasing the density of acceptors in the base. For this purpose, the thickness of the base has to be reduced (Equation (7.1″)). This is only possible if the optical path of light is increased by introducing light trapping (see Chapter 2).

7.1.2 *Front-contact finger grid and emitter resistance*

A c-Si solar cell is illuminated from the emitter side for absorbing photons with a low absorption length as close as possible to the pn-junction. The metal front-contact area should be minimized so that sunlight can penetrate to the emitter and base. The front contact consists of thin grid-like metallic fingers collecting electrons from the emitter. The single grid fingers are connected via bus bars with an external lead.

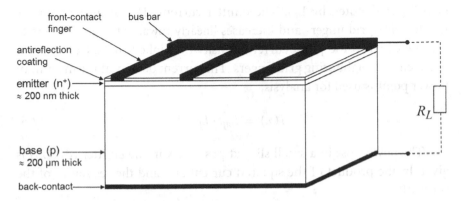

Figure 7.1. Schematic of the simplest c-Si solar cell.

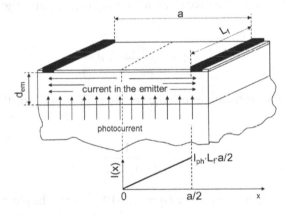

Figure 7.2. Photocurrent and current in the emitter between two grid fingers.

The high reflectivity of c-Si is reduced with an antireflection coating layer (see Chapter 2). Figure 7.1 depicts the schematic of the simplest c-Si solar cell.

The series resistance of the emitter (R_{em}^{c-Si}) is defined by d_{em}, the distance between grid fingers (a) and by the specific resistance of the n-type doped emitter (ρ_n). Figure 7.2 shows the geometry of the emitter and two neighboring grid fingers under photocurrent generation.

The photocurrent flows homogeneously into the emitter and is collected by equal parts at the neighboring grid fingers (length of a grid finger denoted by L_f). The emitter current (I) is 0 in between the neighboring grid fingers and increases linearly towards the grid fingers to a value equal to the photocurrent density multiplied by half of the area between the neighboring grid fingers. The current density at the maximum power point is used for analysis.

$$I(x) = I_{mp} \cdot L_f \cdot x \qquad (7.5)$$

The power loss in a small slice at position x in the emitter ($dP_s(x)$) is given by the product of the squared current at x and the resistance of the slice (dR).

$$dP_s(x) = I(x)^2 \cdot dR \qquad (7.6)$$

The resistance of the slice depends on ρ_n, d_{em}, L_f and the thickness of the slice (dx).

$$dR = \frac{\rho_n}{d_{em} \cdot L_f} \cdot dx \qquad (7.7)$$

The specific resistance of the emitter is equal to the following expression:

$$\rho_n = \frac{1}{q \cdot n_0 \cdot \mu_{n+}} \qquad (7.8)$$

where q is the elementary charge ($q = 1.6 \cdot 10^{-19}$ As). The electron mobility in highly n-type doped c-Si (μ_{n+}) is about 100 cm^2/(Vs) (Jacoboni *et al.*, 1977). Therefore, the specific resistance of the emitter in a c-Si solar cell is about 5 mΩcm.

Taking into account Equations (7.5)–(7.8), the power loss related to half the area between two grid fingers can be obtained with the following equation:

$$P_{em} = I_{mp}^2 \cdot L_f^2 \cdot \frac{\rho_n}{d_{em} \cdot L_f} \cdot \int_0^{a/2} x^2 \, dx \qquad (7.9)$$

The power loss in the emitter is given by the squared current collected over half the distance between two grid fingers and the specific emitter resistance ($R_{em}^{c\text{-}Si}$) normalized to the considered generation area.

$$P_{em} = \left(I_{mp} \cdot L_f \cdot \frac{a}{2}\right)^2 \cdot \frac{R_{em}^{c\text{-}Si}}{\frac{a}{2} \cdot L_f} \tag{7.10}$$

The combination of Equations (7.9) and (7.10) leads to the following expression for the resistance of the emitter of a c-Si solar cell:

$$R_{em}^{c\text{-}Si} = \frac{\rho_n \cdot a^2}{12 \cdot d_{em}} \tag{7.11}$$

The distance between grid fingers is usually of the order of 0.2 cm and the thickness of the emitter is about 200 nm for c-Si solar cells. This results in an emitter resistance of about 0.05 Ωcm^2. Therefore, the resistance of the emitter is significantly smaller than the tolerable series resistance of a c-Si solar cell illuminated at air mass (AM)1.5.

The sum of the base and emitter resistances is about 0.07 Ωcm^2. Therefore the sum of the resistances of the ohmic emitter and base contacts, the grid fingers and the bus bar should be equal to or less than 0.1 Ωcm^2 for c-Si solar cells with high energy conversion efficiency.

7.1.3 *Minimization of shading caused by the front-contact finger grid*

The metal finger grid shades a part of the absorber layer from the irradiation of sunlight. To minimize shading losses, the area of the front-contact finger grid should be as small as possible. In large-scale production of solar cells, the front-contact finger grid is usually screen-printed from a paste containing silver particles. The width of grid fingers and the distance between the grid fingers are about 100 μm and 2 mm, respectively. The bus bars occupy a similar size area as the grid fingers. The corresponding shading losses reduce the active area of a solar cell by about 8–9%.

Shading losses can be only reduced if the area occupied by the front-contact finger grid is decreased (Figure 7.3). This can be done, for example, by increasing the distance between grid fingers and increasing the thickness of the emitter in order to avoid increasing resistive losses. However, the quantum efficiency decreases for photons absorbed in the emitter if d_{em} increases since d_{em} becomes larger than the diffusion length in the emitter.

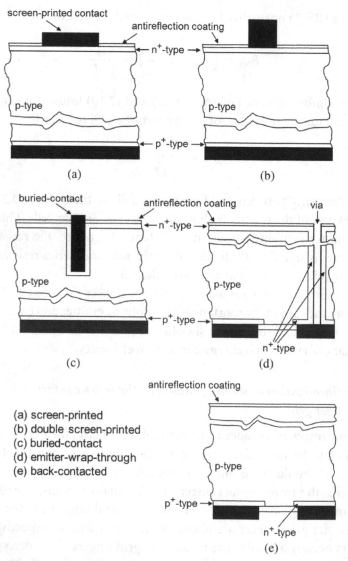

Figure 7.3. Local cross sections of c-Si solar cells with a screen-printed grid finger (a), a double screen-printed grid finger (b), a buried front-contact grid finger (c), a back-contacted wrap-through-emitter (d) and with charge-selective and electron- and hole-collecting contacts at the back surface (e). The solar cells are illuminated from the top. Light scattering surface structures such as texturing (see Chapter 2) are not considered.

Therefore, a significant reduction of shading losses is practically impossible without changing the architecture of front contacts.

The area occupied by the front-contact finger grid can be reduced by increasing the thickness of the grid fingers. The ratio between the height and the width is also called the aspect ratio. The resistance of a grid finger and hence its cross section should be kept constant when increasing the aspect ratio. The aspect ratio is limited for screen-printing. The aspect ratio can be increased by screen-printing the finger grid twice, one on top of the other (see, for example, Dullweber *et al.* (2012)), as shown in Figure 7.3(b).

The aspect ratio can be strongly increased if filling trenches or grooves in the emitter with a metal (buried-contact approach (Wenham *et al.*, 1994), see Figure 7.3(c)). Trenches that are about ten times finer than grid fingers produced by screen-printing can be formed, for example, by laser scribing and subsequent etching. Shading losses can be reduced to 3% or less by the buried-contact approach.

In addition to the reduction of shading losses, the buried-contact approach has additional advantages. First, the emitter resistance can be significantly reduced for buried contacts due to the finer grid in comparison to printed contacts. For a finer grid, d_{em} can be reduced. The reduction of d_{em} leads to an increase of the quantum efficiency for photons absorbed in the emitter. Second, buried contacts can be combined with selective doping along the groove walls of buried front contacts so that recombination losses in the emitter are further reduced. Third, grooves can be coated with diffusion barriers for copper so that silver, which is usually used for screen-printed contacts, can be replaced by copper. The replacement of silver by copper is important since the limited availability of silver in the earth's crust limits the worldwide implementation of photovoltaic energy production.

The deposition of ohmic front contacts at the back side of the solar cell is the most radical step for reduction of shading losses. This is demonstrated for the so-called emitter-wrap-through concept and for the back-contacted architectures in Figures 7.3(d) and 7.3(e), respectively.

In the emitter-wrap-through concept, the emitter is extended from the front side to the back side of the solar cell by using via holes. The base and emitter contacts have to be structured such that the emitter and the base can be contacted separately with an interdigitated finger grid at the back side of the c-Si solar cell (see, for example, Ulzhöfer *et al.* (2008)). Via holes are drilled with strong laser pulses into silicon wafers. The area occupied by via holes is much smaller than the area occupied by buried contacts. In addition, charge separation occurs on the front and back sides of a solar cell with the emitter-wrap-through concept, i.e. the diffusion length of the c-Si wafer can be reduced so that the base can be higher doped and therefore the diffusion potential increased.

In the concept of back-contacted solar cells, the emitter is eliminated from the front side of the solar cell and a pn-junction contacted with an interdigitated finger grid is deposited at the back side of the solar cell. The elimination of the emitter from the front side has the advantage that the emitter resistance can be strongly reduced and that the quantum efficiency can be increased for photons with high energy and a short absorption length. Furthermore, the drilling of via holes is avoided in back-contacted solar cells. However, the diffusion length of the base has to be increased by a factor of two in comparison to a conventional solar cell and excellent surface passivation of the front side of the back-contacted solar cell is required.

The concept of electron and hole collecting contacts at the back side of the solar cell can be realized well by laser-assisted technologies (see, for example, Huljic *et al.* (2006)) and has a strong potential for reaching very high solar energy conversion efficiency. The structuring of back contacts on large areas demands the application of powerful lasers being used for laser ablation of material, i.e. for precise removal of silicon and of protective coatings (see, for example, Engelhardt *et al.* (2007)).

7.1.4 *Resistances at emitter and base contacts*

The sum of the contact resistances of the emitter and of the base should be of the order of $0.05\ \Omega\text{cm}^2$ or less. Resistive losses through contacts increase with decreasing contact area of the front contact (A_{fc}) if the current remains nearly unchanged. In a first approximation, the series resistance of a c-Si solar cell caused by the front contact (R_{fc}^{c-Si}) is related to the tunneling resistance (see Equation (5.30)) of electrons and scales with the ratio between A_{fc} and the area of the solar cell (A_{cell}).

$$R_{fc}^{c-Si} \approx R_{c,tunn,e} \cdot \frac{A_{cell}}{A_{fc}} \qquad (7.12)$$

Front and back contacts are screen-printed on c-Si solar cells in conventional technology. The screen-printing paste contains metallic particles and a binder giving the viscosity of the paste needed for printing. The contacts are fired after the screen-printing step so that a conducting metal layer is formed and the organic binder is removed. The formation of the tunneling junction at ohmic contacts is a rather critical technological step in which additional barriers arising between the grid finger and the emitter need to be avoided. For this reason, the screen-printing paste contains additional chemical components such as a frit, which enables the penetration of silver onto the c-Si surface by adding a small amount of 2–4% of lead (Schubert *et al.*, 2006). Resistances of tunneling contacts can be of the order of $10^{-3} \ldots 10^{-7}$ Ωcm^2 depending on doping of c-Si and the barrier height between the metal and c-Si (see Chapter 5).

Figure 7.4 shows a typical idealized profile of phosphorus atoms across an emitter between the ohmic front contact and the base. The density of phosphorus atoms is above 10^{20} cm^{-3} at the ohmic metal contact and close to the solubility limit. The tunneling resistance of the ohmic contact is very low for a very high density of phosphorus atoms (see Chapter 5). The density of phosphorus atoms in the emitter underneath the highly doped

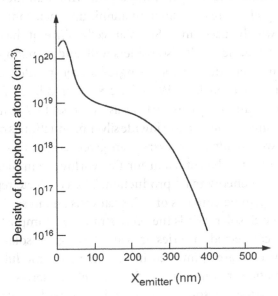

Figure 7.4. Typical doping profile of phosphorus atoms across the n-type doped emitter in a c-Si solar cell. Depth 0 corresponds to the ohmic contact; 10% of compensation with boron in the base is reached at about 350 nm for the given example.

contact layer is about 10^{19} cm^{-3}. Full compensation with boron atoms in the p-type doped base is reached at a depth of about 400 nm. The area of the ohmic contact at the emitter is about one to two orders of magnitude lower than the area for screen-printed or buried-contact solar cells, respectively. The reduced emitter contact area is not critical for the series resistance of c-Si solar cells.

The ohmic back contact with the p-type doped base can cover the whole contact area. Aluminum atoms are acceptors in c-Si. Therefore, aluminum is used for the back contact since it easily forms a metal/p$^+$/p ohmic contact (see Chapter 5) due to the diffusion of aluminum into the near contact region of the p-type doped base of the c-Si solar cell.

7.1.5 Connection of c-Si solar cells in photovoltaic modules and role of silver

Separate c-Si solar cells are connected in series by contacting the bus bars of the front contact of a solar cell with the bus bars of the back contact of the following solar cell. For this purpose, the bus bars contain silver which can be easily soldered (in comparison with aluminum). Soldering bars containing silver are deposited on aluminum back contacts.

Silver is widely used in c-Si solar cells since it has the highest conductivity of the metals. It also solders well, and its nanoparticles form well-connected layers during the firing of screen-printed contacts. The amount of silver needed for $1\,W_p$ of a c-Si solar cell is of the order of $0.1\,g/W_p$ (Herron and Podewils, 2011) (also see task T7.2 at the end of this chapter). For comparison, the worldwide silver production is of the order of $2 \cdot 10^4$ t/a and worldwide silver reserves are given as about $5 \cdot 10^5$ t (Brooks, 2011). This would not be sufficient for the worldwide implementation of photovoltaic (PV) energy mass production. The complete replacement of silver in screen-printed contacts of c-Si solar cells and in soldering systems for connecting c-Si solar cells is therefore strategically important.

Solar cells connected in series are called strings of solar cells. Several strings of solar cells are connected in parallel in PV modules. Figure 7.5 shows a schematic for connecting c-Si solar cells in series and parallel in a PV module. Strings of solar cells are produced with so-called stringers using inductive energy transfer for heating and soldering. The relatively high number of mechanical handling steps needed for the conventional

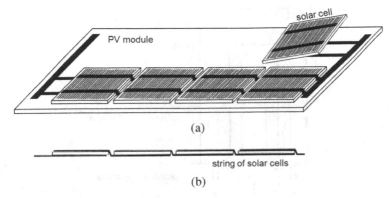

Figure 7.5. Schematic of series and parallel connection of c-Si solar cells in a PV module (a) and schematic cross section of a string of solar cells (b).

soldering of each solar cell is a disadvantage for the conventional production of PV modules.

Materials can be locally heated with precise laser pulses. Metal layers absorbing laser light with photons of a certain wavelength can be combined with substrates and laminates, which are transparent to those wavelengths. Laser pulses penetrate the laminate and are absorbed by tinned copper ribbons (Gast *et al.*, 2008). Soldering with laser pulses reduces the number of handling steps in the production of PV modules.

7.2 Homogeneously Doped c-Si Wafers with a Low Density of Defects

A c-Si wafer is a slice of a crystal with properties which decisively influence the performance of the solar cell that will be built on the wafer. Homogeneously doped wafers with a very low density of defects in the bulk are required for a maximum of photocurrent generation and a minimum diode saturation current density. Only crystallization via solid–liquid interfaces enables the growth of large silicon crystals within a reasonable time frame. Wafers are usually cut by wire sawing from large c-Si crystals.

7.2.1 *Siemens process of very pure polycrystalline silicon*

Silicon crystals are grown from very pure so-called electronic grade or solar grade polycrystalline silicon. Solar grade means that the density of

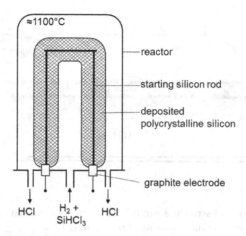

Figure 7.6. Schematic of the Siemens process of silicon deposition from trichlorsilane and hydrogen on hot silicon rods. The vast majority of pure silicon for c-Si solar cells is produced using the Siemens process.

impurity atoms remaining in silicon crystals is reduced to a value at which a defined lifetime of minority charge carriers is guaranteed. A certain energy conversion efficiency can be reached for solar cells produced from a given solar grade crystal.

Distillation of trichlorsilane ($SiHCl_3$) is key for the production of polycrystalline silicon with the highest purity. Figure 7.6 shows a schematic of the very robust Siemens process (developed in the 1950s; see, for example, Gutsche (1958)) consisting of chemical vapor deposition (CVD) of silicon from H_2 and $SiHCl_3$ on hot silicon rods (Bischoff, 1954; Schweickert *et al.*, 1956).

The silicon rod is heated resistively to a temperature of 1100°C. Trichlorsilane is reduced under a hydrogen atmosphere (the simplified overall reaction is given by Equation (7.13)) at the surface of a silicon rod.

$$SiHCl_3 + H_2 \rightarrow Si + 3HCl \qquad (7.13)$$

A disadvantage of the Siemens process is the high amount of energy needed for production of very pure CVD polysilicon.

7.2.2 *Methods of c-Si crystal growth with liquid–solid interfaces*

The very pure polycrystalline silicon is then further processed into large crystals for c-Si solar cells. Three types of crystallization are used most

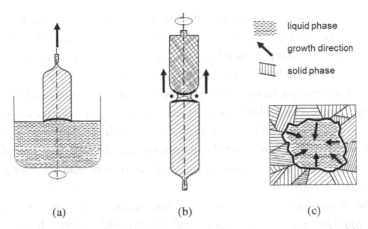

Figure 7.7. Cross sections and directions of crystallization during Czochralski (a) and float zone (b) growth of c-Si single crystals and during block casting of multi-crystalline c-Si blocks (c). The arrows mark the pulling ((a) and (b)) and the direction of growth (c).

often in c-Si photovoltaics. In the first method, a growing silicon crystal is pulled out from a silicon melt. This method is also known as Czochralski growth (Czochralski, 1918; see Figure 7.7(a)).

In the second method, a silicon rod is fixed and a narrow molten zone is floated along the silicon rod from one end to the other end during controlled crystallization. This method is called float zone (FZ) growth (see Figure 7.7(b)).

Czochralski growth and FZ growth of c-Si crystals start at a seed crystal defining the orientation of the growing crystal. Silicon single crystals grown by the Czochralski and FZ techniques have a cylindrical shape and can be as large as 40 cm in diameter and 2 m in length (Zulehner, 2000).

In the third method, troughs are filled with molten silicon and crystallization starts under well-controlled cooling (this method is known as block casting; see Figure 7.7(c)). Block casting results in large prismatic crystals of polycrystalline or multi-crystalline silicon (mc-Si).

Each of these three methods has advantages and disadvantages. From the point of view of production costs, block casting is the cheapest method and FZ growth of c-Si crystals is the most expensive. However, in contrast to Czochralski and FZ growth, the growth direction and velocity cannot be well controlled in a block-casting process. Grain boundaries between crystals of different sizes and orientations contain a large number of unsaturated silicon bonds (so-called dangling bonds) which are recombination active defects. Therefore, the solar energy conversion efficiency that can be reached with mc-Si solar cells produced by block casting is lower than that for c-Si solar cells produced by Czochralski or FZ growth.

As an alternative to conventional block casting, directed solidification on large mono-crystalline silicon seed plates is possible under precise temperature control and large quasi-mono-crystalline silicon (qc-Si) ingots can be obtained (see, for example, Toor (2012)). The performance of solar cells produced on qc-Si is close to the performance of solar cells produced on Czochralski-grown c-Si.

The permanent contact between the silicon melt and the internal silica crucible during the process of Czochralski growth leads to the incorporation of oxygen into the c-Si crystal (Zulehner, 2000). Precipitates of silicon oxide are incorporated over the whole c-Si crystal grown by the Czochralski method. In comparison to the pure bulk of c-Si, the generation of recombination active defects is increased at c-Si/SiO_2 interfaces.

The silicon rod does not have contact with the moving heating coil during FZ growth of c-Si crystals. Therefore practically no additional impurity atoms are incorporated into FZ-grown c-Si crystals. In addition, a FZ process can be repeated several times for the same crystal, which improves the quality even further.

The highest quality is therefore achieved for c-Si crystals grown by the FZ method. Float zone-grown c-Si crystals are necessary for getting very high energy conversion efficiencies of c-Si solar cells. Incidentally, the present c-Si world record solar cell was prepared on FZ-grown silicon (Zhao *et al.*, 1998).

Large silicon crystals are grown at liquid–solid interfaces. New crystal unit cells are formed at the crystal surface during the crystal growth. Impurity atoms present during the formation of new crystal unit cells are able to escape back into the liquid by diffusion from the region where new crystal unit cells are formed (see Figure 7.8) so that silicon atoms can occupy vacant lattice positions.

The escape distance of impurity atoms from the growing crystal back into the liquid is of the order of 1–2 lattice constants (a_{lc}). The time that is needed for the escape of impurity atoms from the growing crystal surface back into the liquid is given by the squared escape distance divided by the

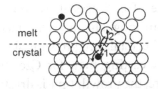

Figure 7.8. Schematic of an interface between a crystal and a melt with an escape sequence of an impurity atom from the crystal surface into the liquid.

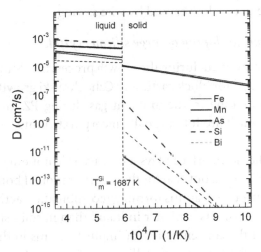

Figure 7.9. Dependence of the diffusion coefficients of impurity atoms in liquid and solid silicon and of the self-diffusion coefficient of silicon on the reverse temperature (data obtained by using equations given by Tang *et al.* (2009)).

diffusion coefficient of the impurity atom in the crystal at a temperature close to the melting point (D_{ic}). Then, the upper limit of an optimum velocity of crystal growth (v_{cg}) can be estimated:

$$v_{cg} = \frac{D_{ic}}{a_{lc}} \tag{7.14}$$

The lattice constant of c-Si is about 0.543102 nm (Massa *et al.*, 2009). The D_{ic} of impurity atoms in the crystal near the melting point vary over a wide range (Figure 7.9). The lower values of D_{ic} are important for the crystal growth and are of the order of 10^{-10} cm^2/s. Therefore the growth rate of c-Si crystals should be of the order of, or below, 0.2 mm/s.

A usual growth rate of c-Si crystals is about 0.1 mm/s. This means that the growth of a 2-m long silicon crystal takes about six hours. The distance over which c-Si crystals are grown is shorter by about one order of magnitude for block casting than for Czochralski or FZ growth of silicon crystals. Therefore, the growth time is drastically reduced for block-casted c-Si. The drastically shorter growth time is one reason for reduced production costs of block-casted c-Si solar cells. Incidentally, about half of the large silicon crystals are lost in the process of cutting c-Si wafers from the crystals due to kerf loss caused by wire sawing.

7.2.3 *Homogeneous doping of large silicon crystals*

Silicon crystals are doped during the growth process by adding highly doped silicon into the melt for block casting or Czhochralski growth, or by adding gases such as diborane to the ambient gas during FZ growth. Dopants are impurities which are intentionally incorporated into the growing c-Si crystals.

The solid–liquid interface plays a decisive role not only for the quality of a growing c-Si crystal but also for the incorporation of both dopants, and unwanted impurities, i.e. atoms forming recombination active defects. The solubility of the elements is higher in liquid than in solid silicon resulting in an increase of the concentration of impurity atoms in the liquid at the solid–liquid interface during crystallization. The enrichment of impurity atoms in the liquid at a freezing solid–liquid interface is called segregation. The incorporation of impurity atoms into a growing crystal is controlled by segregation.

In the following, the dependence of the concentration of a given impurity will be obtained as a function of the volume of the growing crystal. The initial volume of the melt (V_0) and an initial number of impurity atoms in the melt (N_{i0}) are given. The initial concentration of impurity atoms in the melt is denoted by C_{i0}. The volume of the crystal (V_s) increases and the volume of the melt (V_m) decreases during crystal growth. The concentration of impurity atoms in the melt (C_m) is given by the number of impurity atoms in the melt (N_{im}) and the difference between the initial volume and the volume of the growing crystal.

$$C_m = \frac{N_{im}}{V_0 - V_s} \qquad (7.15)$$

The decrease of the number of impurity atoms in the melt (dN_{im}) is proportional to the concentration of impurity atoms in the melt and the increase of the volume of the growing crystal (dV_s). The proportionality factor is the segregation coefficient (k_s).

$$dN_{im} = -k_s \cdot C_m \cdot dV_s \qquad (7.16)$$

The segregation coefficient is specific for different impurities and depends on diffusion of the impurity in the melt, on temperature, on the lattice side where a given impurity can be incorporated into a growing crystal, on the velocity of crystal growth and other factors such as crystal orientation and morphology of the liquid–solid interface.

The concentration of impurity atoms in the melt can be replaced by Equation (7.15), and the resulting equation can then be transformed:

$$\int_{N_{i0}}^{N_{im}} \frac{dN_{im}}{N_{im}} = -k_s \cdot \int_0^{V_s} \frac{dV_s}{V_0 - V_s} \qquad (7.17)$$

Equation (7.17) has the following solution:

$$N_{im} = N_{i0} \cdot \left(1 - \frac{V_s}{V_0}\right)^{k_s} \qquad (7.18)$$

The number of impurity atoms in the melt remains unchanged if the segregation coefficient is equal to 0, i.e. impurity atoms are not incorporated into the growing crystal. This is desired for unwanted impurity atoms, which would form recombination active defects in the c-Si crystal.

The concentration of impurity atoms in the growing crystal (C_{is}) corresponds to the decrease of the number of impurity atoms in the melt divided by the change of the volume of the growing crystal.

$$C_{is} = -\frac{dN_{im}}{dV_s} \qquad (7.19)$$

Equation (7.19) can be solved with ease by using Equation (7.18). The following dependence of the concentration of impurity atoms in a growing crystal on the segregation coefficient is obtained:

$$\frac{C_{is}}{C_{i0}} = k_s \cdot \left(1 - \frac{V_s}{V_0}\right)^{k_s - 1} \qquad (7.20)$$

If the segregation coefficient is equal to 1 then the concentrations of the impurity are equal in the melt and in the growing crystal. A segregation coefficient close to 1 is desired for dopants in silicon crystals. The advantage of a dopant atom is that it replaces a silicon atom at the correct position in the c-Si lattice. Therefore, the segregation coefficient of dopant atoms can be close to 1. This is the case for boron atoms (Hall, 1953).

The segregation coefficient of boron is close to 1 over a wide range of growth rates of silicon crystals (k_s = 0.8 near the melting point (Trumbore, 1960)). For purposes of comparison, the segregation coefficients of phosphorus and aluminum are 0.35 and 0.002, respectively, near the melting point of silicon (Trumbore, (1960). Therefore, silicon crystals can be doped with boron much more homogeneously than with the other dopants. This is one important reason why boron-doped silicon crystals are preferred for solar cells.

A homogeneous doping of silicon crystals is important for a narrow distribution of the lifetime and of the diffusion potential for solar cells produced on wafers cut from the grown crystal. The concentrations of impurity atoms in a silicon crystal normalized to C_{i0} is shown in Figure 7.10 as a function of the volume fraction of the growing crystal for segregation coefficients of 0.99, 0.8 (B), 0.35 (P) and 0.002 (Al).

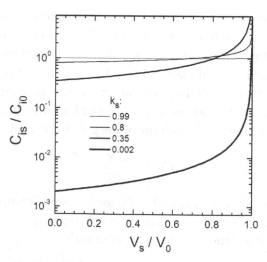

Figure 7.10. Density of impurity atoms in the crystal normalized to the density of impurity atoms in the liquid before starting crystal growth as a function of the volume fraction of the growing crystal.

The concentration of impurity atoms generally increases with increasing volume fraction of the growing crystal due to the growing concentration of impurity atoms in the liquid. For a segregation coefficient of 0.99, the concentration ratio increases from 0.99 at the beginning of crystallization to 0.997 after crystallization of half of the volume and to 1.014 after crystallization of 90% of the initial liquid volume.

For segregation coefficients of 0.8, 0.35 and 0.002 the concentration ratios increase by 15%, 57% and 100%, respectively, after crystallization of only half of the volume. The concentration ratios increase much quicker after 90% crystallization of the initial liquid volume and, consequently, these parts of the silicon crystal are not used for producing solar cells.

During crystal growth, a surface atom layer is covered by a new one. The composition of the new layer is in equilibrium with the composition in the liquid for very slow crystal growth. The segregation coefficient becomes larger than the segregation coefficient in equilibrium (k_0) if the time interval for deposition of one mono-layer (Δt_{ML}) becomes shorter than the time needed for reaching equilibrium in one mono-layer (Δt_{equ}). If the growth rate is too fast then the growing surface layer does not reach equilibrium, and so impurity atoms adsorbed at the liquid–solid interface cannot escape into the liquid. As a consequence the segregation coefficient increases with increasing growth rate (Hall, 1953).

$$k_s = k_0 + (k_{s'} - k_0) \cdot exp(-v_i/v_{cg}) \qquad (7.21)$$

The coefficient $k_{s'}$ corresponds to the concentration of impurity atoms in the layer at the liquid–solid interface. As a definition, the values of Δt_{ML} and Δt_{equ} are equal at the growth rate v_i.

The dependence of the segregation coefficient on the growth rate is key for very homogeneous doping of large silicon crystals within the PV industry. For this purpose, the growth rate and the temperature are controlled very precisely during the entire growth process of large silicon crystals.

7.2.4 Refinement of silicon crystals by segregation

The values of segregation coefficients usually range between 0 and 1. In some cases, segregation coefficients practically equal to 0 or even a little larger than 1 are possible, for example, for oxygen incorporation into silicon crystals during FZ growth in oxygen atmosphere (k_s slightly lower than 1 (Carlberg, 1986), $k_s = 1.25 \pm 0.17$ (Yatsurugi *et al.*, 1973)).

Unwanted impurities are atoms which cannot usually replace silicon atoms in their host lattice positions. For this reason, the segregation coefficients of atoms forming recombination active defects are much lower than the segregation coefficient of dopants (refining effect, see Zulehner (2000)). For example, the segregation coefficient of iron is only $8 \cdot 10^{-6}$, i.e. less than one iron atom from 10^5 iron atoms in the melt is incorporated into the growing crystal (Trumbore, 1960). This fact makes it possible to grow doped silicon crystals with Shockley–Read–Hall recombination lifetimes longer than the Auger recombination lifetime (see Chapter 3).

The segregation coefficient of impurity atoms decreases with decreasing growth rate. This effect gives the opportunity of further reducing the concentration of unwanted impurities at very low growth rates.

The concentration of unwanted impurities with a low segregation coefficient can be reduced drastically by repetitive melting and crystallization during FZ processing of the same crystal. Unwanted impurities are collected in the melt and transported in the floating zone to the end of the long c-Si crystal. Silicon crystals of the absolute highest purity are produced by repetitive FZ processing (Pfann, 1957).

In silicon crystals, segregation of impurity atoms, which cannot replace silicon atoms from their host lattice positions, is strongly enhanced in regions with imperfections such as precipitates or grain boundaries. The directed segregation of unwanted impurities at precipitates is also known as gettering (see Chapter 3). One has to take into account that defects at the interface between the c-Si host crystal and a precipitate can cause states in the band gap of c-Si, i.e. they are recombination active. The additional incorporation of so-called getterers into c-Si crystals is useful if the number of gettered atoms forming recombination active defects is significantly larger than the number of defects introduced by the getterer.

Segregation can be applied for alternative production technologies of solar grade polycrystalline silicon instead of the rather expensive Siemens process. The melting temperature of silicon is higher than the melting temperature of Al–Si alloys. Metallurgical silicon with a purity of 98% can be cleaned by crystallization from an Al–Si melt to a purity better than 99.9999% (Sollmann, 2009). In this process, metallurgical silicon is dissolved in an aluminum melt. Silicon crystallizes in the form of flakes during cooling of the aluminum melt. This is the basic refinement step since most of the impurities remain in the melt due to low segregation coefficients. After draining the liquid aluminum, the silicon flakes are washed in an acid to get rid of the thin aluminum layer coating the flakes. The temperature of the silicon flakes is increased to the melting temperature of silicon and the rest of the aluminum segregates at the surface. The concentration of unwanted impurities can be further reduced during subsequent Czochralski growth due to gettering at oxide precipitates. However, the concentration of unwanted impurities remains relatively high, and so the energy conversion efficiency of c-Si solar cells produced with solar grade silicon from an Al–Si process is significantly lower than the efficiency of c-Si solar cells fabricated with silicon from the Siemens process. Incidentally, aluminum drained after crystallization of silicon flakes contains about 10–12% of silicon, which makes the aluminum hard. Hard aluminum is used for lightweight construction in the automobile industry, i.e. solar grade silicon can be obtained as a by-product of production of hard aluminum.

7.2.5 Wafering of c-Si crystals

Practically all c-Si wafers are cut by wire sawing from large silicon crystals or ingots (see Figure 7.11(a)). A long steel wire moves rapidly through the ingot as a slurry is continuously added to the kerf. Diamond crystallites in the slurry cause the abrasion between the wire and the silicon. Many wafers can be cut from a silicon crystal by wire sawing within one run. The wafers

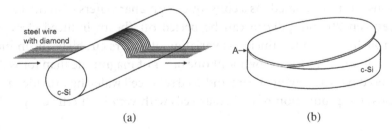

| (a) | (b) |

Figure 7.11. Schematic of mass cutting of c-Si wafers from large crystals by wire sawing (a) and of kerfless cutting of a transferrable c-Si wafer along a layer enriched with structural defects (denoted by A in (b)). Well-defined defect layers can be introduced, for example, by high-dose ion implantation of hydrogen or by porous silicon formation.

are planar after wire sawing. A defect layer with a thickness of the order of 10 μm at the silicon wafer has to be removed by polishing and wet chemical etching after wire sawing.

About half of a silicon crystal is lost during wire sawing due to abrasion. This is a major disadvantage of wire sawing. A further disadvantage of wire sawing is the limitation to a certain thickness of c-Si wafers since the wafers should remain stiff for further handling during the production of solar cells. For example, silicon wafers with a thickness of about 50 μm are not stiff enough to be self-supporting.

The reduction of the wafer thickness combined with light trapping and advanced surface passivation is crucial for a further increase of the energy conversion efficiency of c-Si solar cells (see Chapter 4). For the even greater increase of the energy conversion efficiency, a transfer of flexible c-Si wafers onto rigid substrates is required during subsequent processing. This can be achieved by cutting wafers from c-Si crystals along a buried layer enriched with structural defects (Figure 7.11(b)). The structural defects serve as breaking points for the lift-off of a flexible c-Si solar cell and its transfer onto a rigid substrate.

A buried layer enriched with structural defects can be created by implantation of protons at high dose (beam-induced solar cell wafering) combined with a thermal treatment (Henley *et al.*, 2011). Protons implanted with an energy of 2–4 MeV accumulate in a layer at a distance of the order of 50–100 μm from the surface of the silicon crystal. The solubility of hydrogen in c-Si is only of the order of 10^{15} cm^{-3} (see Van de Walle (1994) and references therein). Therefore, structural defects such as dislocation loops and voids are formed during thermal annealing in the regions where protons are accumulated. As a consequence, planar wafers with thicknesses of between 20 and 150 μm can be peeled off (beam-induced cleaving). The roughness and the thickness variation of wafers cut by beam-induced solar cell wafering are less by about one order of magnitude than for wafers produced by wire sawing. Beam-induced solar cell wafering provides a tool for possible production of c-Si solar cells with very high efficiency at low cost.

Peeling of 25-μm thin c-Si layers is possible on large areas by exfoliation, which makes use of stress induced between a c-Si crystal and a metal layer during cooling (Rao *et al.*, 2012).

Epitaxial silicon wafers can be transferred by implementing a porous silicon layer (Haase *et al.*, 2012). In this process, a porous silicon double layer is formed by electrochemical etching at different current densities. The porous silicon top layer is reorganized during sintering in a hydrogen atmosphere and the epitaxial c-Si layer can be deposited at 1100°C in trichlorsilane and hydrogen atmosphere (Haase *et al.*, 2012). The deposition rate of silicon epitaxial layers from trichlorsilane and hydrogen is between 0.4 and 3 μm/min (Hammond, 2001). This process can be combined well with back-contact solar cells (Haase *et al.*, 2012).

7.3 Formation of the Emitter

7.3.1 *Diffusion from inexhaustible and exhaustible sources*

The n-type doped emitter is usually formed by diffusion of phosphorus in c-Si from an inexhaustible and/or from an exhaustible source into the p-type doped base while the density of donors in the emitter is much higher than the density of acceptors in the base (overcompensation of acceptors). Quasi-unlimited sources of phosphorus such as phosphorus oxy chloride ($POCl_3$) or phosphine (PH_3) are used for doping of the emitter in conventional technology of c-Si solar cells.

The propagation in time and space of the density of a diffusing impurity (denoted by index N_i) with a diffusion coefficient (denoted by D_i) can be calculated by solving the one-dimensional diffusion equation:

$$\frac{\partial N_i(x.t)}{\partial t} = D_i \cdot \frac{\partial^2 N_i(x, t)}{\partial x^2} \tag{7.22}$$

An inexhaustible phosphorus source permanently provides phosphorus atoms at the solubility limit in c-Si. Therefore the density of impurity atoms at the surface is constant during the whole diffusion process:

$$N_i(0, t) = N_{is} \tag{7.23}$$

The value of N_{is} corresponds to the saturation density of a diffusing species. N_{is} increases with increasing temperature. For the diffusion of phosphorus the values of N_{is}, for example, are equal to $3 \cdot 10^{20}$ and $5.5 \cdot 10^{20}$ cm^{-3} at 800 and 900°C, respectively (Solmi *et al.*, 1996). Incidentally, at temperatures above 750°C the saturation density of mobile

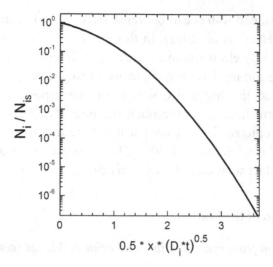

Figure 7.12. Dependence of the normalized density of impurity atoms diffusing from an inexhaustible source as a function of the complementary error function of the quotient of the distance and the doubled square root of the product of the diffusion coefficient of the impurity and time.

phosphorus atoms becomes larger than the density of activated phosphorus atoms (Solmi *et al.*, 1996).

The density of impurities beyond the emitter surface is 0 at the start of diffusion. Then the solution of Equation (7.22) is equal to the product of N_{is} and the complementary error function, the values of which are tabulated.

$$\frac{N_i(x, t)}{N_{is}} = erfc\left(\frac{x}{2 \cdot \sqrt{D_i \cdot t}}\right) \tag{7.24}$$

Figure 7.12 shows the dependence of the normalized density of diffusing impurity atoms as a function of the complementary error function. The depth at which the density of diffusing phosphorus reaches the density of acceptors in the base can be considered to be the penetration depth of phosphorus or thickness of the emitter (H_n). Therefore, the solution of the complementary error function can be found from the ratio between the density of acceptors in the base and N_{is}. The penetration depth of phosphorus is reached for a solution of the complementary error function between 2.6 and 3.0.

The thickness of the emitter depends on the diffusion coefficient of phosphorus and time:

$$H_n \approx 5.6 \cdot \sqrt{D_i \cdot t} \qquad (7.25)$$

Diffusion with an inexhaustible source of phosphorus is usually performed at a temperature of between 800 and 900°C. The inexhaustible source is removed after the first diffusion step, defining the total amount of phosphorus in the emitter. The optimal density of free electrons in the emitter is much lower than the saturation density of phosphorus in c-Si. Therefore, a second diffusion step is applied for forming the optimum doping profile of the emitter. The second diffusion step corresponds to diffusion from an exhaustible source with a total amount of diffusing species of Q_{is}.

In the simplest case, a delta function can be assumed as the initial distribution function before starting diffusion. The solution of the diffusion equation with an exhaustible source is then given by an exponential function. Equation (7.26) gives an idea of the development of a doping profile during diffusion:

$$N_i(x, t) = \frac{Q_{is}}{\sqrt{\pi \cdot D_i \cdot t}} \cdot exp\left(-\frac{x^2}{4 \cdot D_i \cdot t}\right) \qquad (7.26)$$

The second diffusion step is usually performed at temperatures equal to or larger than 1000°C. Diffusion times of thousands of seconds are needed to form the right doping profile in the emitter. As an example, Figure 7.13 shows diffusion profiles for an initial density of dopants of 10^{14} cm^{-2} and a diffusion coefficient of $3 \cdot 10^{-15}$ cm^2/s (Fahey *et al.*, 1989) after diffusion times of 20, 200 and 2000 s.

The diffusion coefficient of phosphorus in c-Si must be known for the precise control and calculation of diffused doping profiles across the emitter. The diffusion coefficient of impurity atoms is thermally activated and depends on the diffusion mechanism.

$$D_i = D_{i0} \cdot exp\left(-\frac{E_{Ai}}{k_B \cdot T}\right) \qquad (7.27)$$

Dopants change their lattice position during diffusion. The change of the lattice position of an impurity atom corresponds to a local chemical

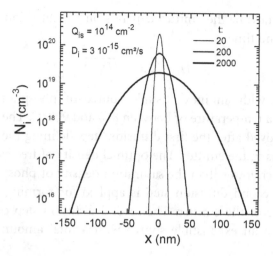

Figure 7.13. Profile of the density of impurity atoms diffusing from an exhaustible source with an areal density of 10^{14} cm^{-2} at position 0 after different diffusion times for a diffusion coefficient of $3 \cdot 10^{-15}$ cm^2/s.

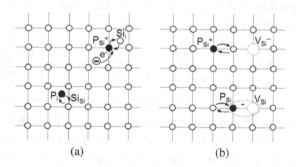

Figure 7.14. Illustration of diffusion of phosphorus in c-Si via interstitials (a) and via double negatively charged vacancies (b).

reaction, which can involve different reaction partners such as point defects. Vacancies, i.e. an atom missing in its lattice position, or interstitials, i.e. atoms in a place between lattice positions, are very prominent point defects.

Diffusion of phosphorus in c-Si can be provided by different elementary steps such as kick-out reactions with self-interstitials, via vacancies or other point defects where the charge state of the defect has to be considered as well (see, for example, Fahey *et al.* (1989)). Figure 7.14 illustrates diffusion via interstitials and double negatively charged vacancies.

Diffusion of phosphorus is dominated at low densities of phosphorus atoms by kick-out reactions of interstitial phosphorus atoms (P_i) with silicon atoms at lattice positions (Si_{Si}). This process is in equilibrium with kick-out reactions of self-interstitials (Si_i) with ionized phosphorus atoms at lattice positions (P_{Si}^+) and free electrons (Bentzen *et al.*, 2006):

$$Si_{Si} + P_i \leftrightarrow Si_i + P_{Si}^+ + e^- \tag{7.28}$$

The density of double negatively charged vacancies in silicon crystals increases with increasing density of phosphorus atoms so that diffusion via vacancies becomes dominating at very high densities of phosphorus atoms (Bentzen *et al.*, 2006):

$$V_{Si}^{--} + P_{Si}^+ \leftrightarrow (V_{Si} - P_{Si})^- \tag{7.29}$$

The dominating diffusion mechanism can change depending on the density of diffusing species, on the charge of diffusing species, on the presence of other charged or compensating species in silicon, and on temperature. Furthermore, the superposition of different diffusion processes has to be taken into account.

The experimental values of the diffusion coefficient of phosphorus in c-Si are about 10^{-16}, 10^{-15} and 10^{-12} cm^2/s at 800, 900 and 1200°C, respectively (Fahey *et al.*, 1989). The diffusion coefficient of copper in c-Si is about ten orders of magnitude larger than the diffusion coefficient of phosphorus in c-Si ($5 \cdot 10^{-5}$ cm^2/s at 900°C, see, for example, Istratov *et al.* (1998)). Therefore, copper contamination must be avoided during thermal processing of c-Si solar cells.

Diffusion of the phosphorus emitter occurs around the whole c-Si wafer. Therefore, the pn-junction has to be disconnected between the front and back contacts of c-Si solar cells. This is realized by the edge isolation which can be performed, for example, by a strong laser damaging the pn-junction at the edges of the wafer.

7.3.2 *Laser-assisted doping and ion implantation of the emitter*

With a thickness of 200–400 nm, the emitter is thinner than the silicon wafer by 500–1000 times. It saves a lot of processing energy if only the region of the emitter is heated during the formation of the emitter instead of heating the whole silicon wafer as in a conventional diffusion process.

This is possible by inducing melting with laser pulses with a duration of nanoseconds.

Heat can be locally transferred to c-Si wafers during pulsed laser irradiation. Heat transfer depends on heat conductivity (k(T)), heat capacitance (C(T)) and density (ρ) of a given medium as well as on phase transitions. For a given energy source (Q_{source}), the following equation of heat transfer can be written:

$$\frac{\partial}{\partial t}[C(T) \cdot \rho(T) \cdot T(x,t)] = \frac{\partial}{\partial x}\left[k(T) \cdot \frac{\partial T(x,t)}{\partial x}\right] + Q_{source}(x,t)$$

(7.30)

For a given initial temperature distribution, an equation for the temperature similar to the diffusion equation can be obtained, and a thermal diffusivity corresponding to a thermal diffusion coefficient (D_{th}) can be defined:

$$D_{th}(T) = \frac{k(T)}{C(T) \cdot \rho(T)}$$

(7.31)

The values of the heat conductivity, heat capacitance and density depend on temperature. Values of the order of $1\,W/(cm{\cdot}K)$, $2.33\,g/cm^3$ and $0.8\,W{\cdot}s/(g{\cdot}K)$ can be assumed for the heat conductivity, density and heat capacitance of silicon, respectively (Shanks et al., 1963). The respective thermal diffusivity is of the order of $0.5\,cm^2/s$, which corresponds to the thermal diffusivity of c-Si measured at 400 K (Shanks et al., 1963). The thermal diffusivity decreases to a value of about $0.1\,cm^2/s$ near to the melting temperature (Shanks et al., 1963).

A thermal diffusion length can be estimated for the time of interaction between the laser pulse and the silicon wafer (Δt):

$$L_{th}(T) = \sqrt{D_{th}(T) \cdot \Delta t}$$

(7.32)

The dependence of L_{th} on Δt is plotted in Figure 7.15 for a constant D_{th} of $0.5\,cm^2/s$. For example, at moderate heating, heat is transferred over about 200 nm within one nanosecond or over about $10\,\mu m$ within one microsecond. The thermal diffusion length decreases during heating due to the decrease of D_{th}, and so laser-induced melting requires times periods longer than Δt (a factor of 2–5). Nevertheless, Figure 7.15 gives an idea of the duration of laser pulses necessary for different processing steps. For

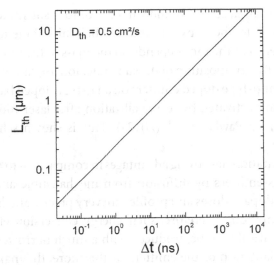

Figure 7.15. Estimated dependence of the thermal diffusion length on the interaction time between a laser pulse and silicon for a constant thermal diffusion coefficient of 0.5 cm^2/s.

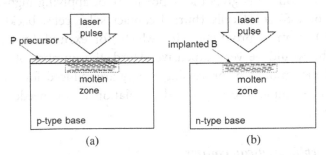

Figure 7.16. Laser doping of an n^+-doped emitter on a p-type doped base (a) and laser-induced activation of an implanted p-type doped emitter on an n-type doped base (b).

example, local melting of a c-Si surface layer with a thickness of the order of 100 nm is possible with nanosecond laser pulses strongly absorbed in c-Si with energies of the order of 0.5 J/cm^2.

The required density of phosphorus atoms for the emitter on p-type can be adjusted by diffusion of phosphorus from a precursor layer into a molten surface layer within a time of about 100 ns (laser-assisted doping) (Eisele *et al.*, 2009). This is schematically shown in Figure 7.16(a).

Dopants for the emitter can also be provided by ion implantation. Ion implantation means that dopants are ionized and accelerated in the high electric field of an accelerator and directed onto a substrate. Accelerated ions penetrate into the substrate due to their high kinetic energy. The penetration depth of the ions depends on the mass of the ions, their kinetic energy and on the composition of the substrate. Ion implantation is applied, for example, for p-type doped emitters on an n-type doped base. Implanted boron atoms are activated by recrystallization after laser-induced melting (see, for example, Pawlak *et al.* (2012)). This is shown schematically in Figure 7.16(b).

Ion implantation has strong advantages in comparison to the formation of phosphorus emitters by diffusion from inexhaustible and exhaustible sources. First, dopant doses and profiles are very precise and homogeneous for ion implantation. Therefore, solar energy conversion efficiencies for c-Si solar cells are able to be produced with a much narrower distribution using ion implantation of the emitter. Furthermore, the maximum of the distribution is shifted to higher efficiencies (Hieslmair *et al.*, 2012). Second, local and patterned doping with donors and acceptors is possible with ion implantation. This gives the opportunity of applying high-efficiency concepts of c-Si solar cells (buried-contact solar cells, back-contacted solar cells) to mass production for which, for example, patterning by lithography would be very expensive. Third, the number of processing steps and the amount of processing energy are reduced for ion implantation. For example, the step of edge isolation is not needed after ion implantation.

7.3.3 *Amorphous silicon emitter*

Hydrogenated amorphous silicon (a-Si:H; see also Chapter 9) is a disordered semiconductor that can be doped and which has a band gap of about 1.7 eV (Cody *et al.*, 1981). At a-Si:H/c-Si hetero-junctions, the valence band offset (ΔE_V), i.e. the difference between the valence band edges of a-Si:H and c-Si, is about 0.46 eV (Schmidt *et al.*, 2007). The conduction band offset is about 0.14 eV, with respect to ΔE_V and the band gaps of c-Si and a-Si:H.

The density of deep defect states has to be as low as possible at a charge-selective contact (Chapter 4). The density of deep defect states in un-doped a-Si:H can be as low as about 10^{15} cm^{-3} (Jackson and Amer, 1982) and dangling bonds at the c-Si surface are passivated efficiently with hydrogen, which is present in a-Si:H (see also Section 7.4.4.). For this reason a very

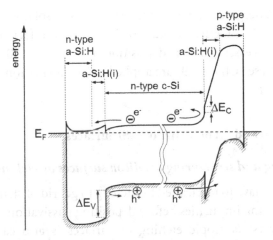

Figure 7.17. Schematic band diagram across a HIT solar cell with n-type a-Si:H/a-Si:H(i) n-type c-Si/a-Si:H(i)/p-type a-Si:H in thermal equilibrium. Defect states in the band gap of a-Si:H are omitted for clarity. One of the contact layers is transparent.

thin layer of un-doped a-Si:H (a-Si:H(i), with a thickness several nm) is placed between the doped c-Si and doped a-Si:H layers.

With regard to the decisive charge-selective and passivating component, c-Si solar cells with a-Si:H contacts are known as HIT (*h*etero-junction with *i*ntrinsic *t*hin layer) solar cells (Sawada *et al.*, 1994). The highest solar energy conversion efficiencies have been obtained for HIT solar cells with an n-type doped c-Si base.

Figure 7.17 shows a schematic band diagram across a HIT solar cell with a layer structure of n-type a-Si:H/a-Si:H(i)/n-type c-Si/a-Si:H(i)/p-type a-Si:H. The c-Si base and the a-Si:H layers are about 200 μm and 20–50 nm thick, respectively. Photo-generated electrons are reflected at the barrier of the charge-selective n-type doped c-Si/a-Si:H(i)/p-type doped a-Si:H contact. Photo-generated holes can penetrate the thin barrier by tunneling via defect states (see Section 5.3.5.). A strong advantage of HIT solar cells is that photo-generated holes are reflected at the electron extracting contact due to the large valence band offset. Therefore, HIT solar cells have a c-Si absorber embedded between two charge-selective contacts, one for electrons and the other one for holes.

The front contact in HIT solar cells consists of a transparent conducting oxide (TCO, see Chapter 9) layer covering the complete solar cell. For this reason the a-Si:H emitter layer can be very thin.

The solar energy conversion efficiency of HIT solar cells can be very high (above 20%) (Taguchi *et al.*, 2005) despite the fact that only low-temperature processes (such as deposition of a-Si:H at 200°C by plasma-enhanced CVD, see Chapter 9) are applied for the formation of the charge-selective contact.

7.4 Passivation and Structuring of c-Si Surfaces

7.4.1 *Cleaning and structuring of silicon surfaces by etching*

Silicon surfaces have to be cleaned by etching to get rid of structural defects and contaminating impurities before depositing passivation, antireflection or contact layers. Isotropic etching of surfaces is also called chemical polishing. Many cleaning procedures of c-Si surfaces are based on oxidation of surface atoms or molecules followed by dissolution of oxidized species. Volatile oxidized species can leave the solution and enter the surrounding atmosphere. An example of an oxidizing species is hydrogen peroxide (H_2O_2).

Structural defects introduced during wafering of c-Si crystals are removed by mechanical and chemical polishing. Chemical polishing of c-Si wafers is performed in solutions containing hydrofluoric acid (HF), nitric acid (HNO_3), H_2O and, for example, acetic acid ($HC_2H_3O_2$) (Robbins and Schwartz, 1959, 1960). SiO_2 is etched in aqueous HF solutions (see, for example, Knotter (2000)). Etch rates are controlled by temperature and concentrations of components in etch solutions.

The so-called RCA (Radio Corporation of America) process (RCA clean, Kern and Puotinen, 1970; Kern, 1984) is an industrial standard for cleaning of c-Si surfaces. In a pre-cleaning step, the silicon surface is oxidized in HNO_3 at 80°C and etched in buffered HF. Mainly organic rests are oxidized during the cleaning step at 80°C in a solution of H_2O_2 and NH_4OH. Residuals of metal contaminations form volatile chlorides during the second cleaning step at 80°C in a solution of HCl and H_2O_2. Finally, cleaned silicon wafers have to be rinsed in deionized water.

The purity of deionized water, of all the chemicals and of the surrounding ambience is very important in order to avoid recontamination of cleaned c-Si surfaces. For this reason, c-Si solar cells with very high energy conversion efficiency have to be processed in specially designed clean rooms, none of the handling tools are allowed to contain metal and only ultra-clean chemical solutions may be used.

Silicon surfaces have to be structured or textured for light trapping (see Chapter 2). Processes of anisotropic etching in alkaline solutions such as potassium hydroxide (KOH) (Bean, 1978) are applied for texturing c-Si surfaces of c-Si solar cells. As an alternative, the starting porous silicon formation in HF-containing solutions can also be used in order to roughen a c-Si surface (see, for example, Smith and Collin (1992)).

The etch rate of c-Si in alkaline solutions has a minimum in the ⟨111⟩ direction since there is only one free bond per silicon surface atom available (Figure 7.18(b)). In contrast, the etch rate of c-Si is much higher in the ⟨100⟩ direction since there are two free bonds per silicon surface atom available (Figure 7.18(a)). As a consequence, facets of the most stable c-Si(111) surface remain after etching of c-Si in alkaline solutions (Figure 7.19(a)).

On c-Si wafers oriented in the ⟨100⟩ direction, facets of c-Si(111) surfaces form pyramids, which are excellent for light trapping in c-Si solar cells. Silicon dioxide is stable in alkaline solutions. Therefore SiO_2 masks can be applied for etching ordered arrays of inverted pyramids on c-Si wafers oriented in the ⟨100⟩ direction (Bean, 1978) (Figure 7.19(b)). Incidentally,

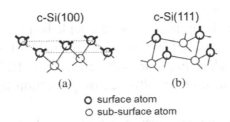

c-Si(100)   c-Si(111)

(a)   (b)

O surface atom
O sub-surface atom

Figure 7.18. Bond configurations at the c-Si(100) (a) and c-Si(111) (b) surfaces.

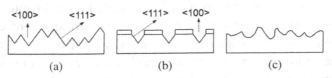

⟨100⟩   ⟨111⟩   ⟨111⟩   ⟨100⟩

(a)   (b)   (c)

Figure 7.19. Schematic cross sections of c-Si(100) surfaces structured with random pyramids (a), with inverted pyramids etched with a SiO_2 mask (b) and of a roughened c-Si surface (c).

light trapping is introduced by scattering at inverted pyramids in the present world record c-Si solar cell (Zhao *et al.*, 1998).

Anisotropic etching in alkaline solutions is not very useful for mc-Si solar cells. Alternatives are based on isotropic roughening of silicon surfaces (Figure 7.18(c)) by stain etching in solutions containing HNO_3, sodium nitrite and HF (Archer, 1960) or by the formation of macropores during electrochemical treatment in fluoride-containing organic electrolytes (Ponomarev and Levy-Clement, 1998).

7.4.2 The c-Si/SiO$_2$ interface

Silicon dioxide layers are grown on c-Si wafers by thermal oxidation of c-Si at high temperatures of the order of 1000°C. The growth rates of SiO_2 are, for example, about 1.7 nm/min and 20 nm/min at 1100°C in dry and wet oxygen, respectively (Deal and Grove, 1965). The reactions can be described by the following equations for dry and wet oxidation:

$$Si + O_2 \rightarrow SiO_2 \tag{7.33}$$

$$Si + 2H_2O \rightarrow SiO_2 + 2H_2 \tag{7.34}$$

The oxidation process is limited by diffusion of oxidizing species such as O_2^- through the growing SiO_2 layer for layer thicknesses above 20–30 nm (Deal and Grove, 1965).

Silicon dioxide is an insulator with a wide band gap of about 9.3 eV (Weinberg *et al.*, 1979), and SiO_2 is chemically stable at room temperature in most acidic, alkaline and other solutions except HF. Therefore, silicon dioxide is an excellent electrically inactive protection layer for c-Si solar cells.

The c-Si/SiO$_2$ interface played a decisive role for the invention of the MOSFETs (metal oxide semiconductor field-effect transistors), the heart of modern electronics (Kahng, 1960). The reason for this is that electronic devices can be excellently passivated (see Chapter 3) with c-Si/SiO$_2$ interfaces. The surface of the present world record c-Si solar cell was also passivated with a c-Si/SiO$_2$ interface (Zhao *et al.*, 1998).

In the following, the passivation mechanisms at c-Si/SiO$_2$ interfaces will be described. Figure 7.20 shows a schematic bond configuration at a c-Si/SiO$_2$ interface. At the c-Si/SiO$_2$ interface Si–Si bonds are replaced by Si–O bonds. There is a very thin SiO_x transition layer at the c-Si/SiO$_2$

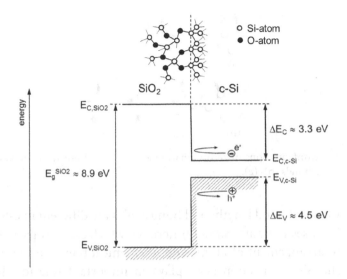

Figure 7.20. Schematic bond configuration and band diagram at a c-Si/SiO$_2$ interface.

interface where the oxidation state of silicon atoms varies gradually from Si$^{(0)}$ (oxidation state of silicon atoms in c-Si) and Si$^{(1+)}$ via Si$^{(2+)}$ and Si$^{(3+)}$ to Si$^{(4+)}$ (oxidation state of silicon atoms in SiO$_2$) (Himpsel *et al.*, 1988). The energies of bonding and anti-bonding states of oxidized silicon atoms are deep in the valence and conduction bands of c-Si, respectively. This means that oxidized silicon surface atoms do not have surface states in the band gap of c-Si.

The energies of the valence or conduction band edges of SiO$_2$ are much lower or higher than for c-Si, respectively. The related band offsets of the conduction (ΔE_C) and valence (ΔE_V) bands at the c-Si/SiO$_2$ interface are defined as the differences between the conduction band edges of c-Si ($E_{C,c-Si}$) and SiO$_2$ ($E_{C,SiO2}$) and between the valence band edges of c-Si ($E_{V,c-Si}$) and SiO$_2$ ($E_{V,SiO2}$), respectively. The values of ΔE_C and ΔE_V are very large for c-Si/SiO$_2$ interfaces and amount to -3.3 and 4.5 eV, respectively. The values of the band offsets can vary over about 0.2 eV depending on oxidation technology and on orientation of the silicon wafer (see, for example, Keister *et al.* (1999) and Ribeiro *et al.* (2009)). Therefore the c-Si/SiO$_2$ interface provides huge barriers for free electrons and holes in the conduction and valence bands of c-Si.

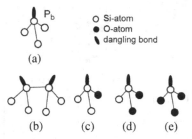

Figure 7.21. Configurations of recombination active (a) and not recombination active silicon dangling bonds (b)–(e).

The values of bond lengths and bond angles are different in c-Si and in SiO_2. This causes stress at c-Si/SiO_2 interfaces, which can only partially relax by local re-arrangement of atoms and of chemical bonds in the interface region. The SiO_x transition layer plays an important role for the local relaxation of stress at c-Si/SiO_2 interfaces. The relaxation of remaining stress at c-Si/SiO_2 interfaces is realized by omitting some of the chemical $\equiv$Si–O bonds at the c-Si/SiO_2 interface. This leads to the formation of unsaturated or silicon dangling bonds, the most prominent defects at c-Si/SiO_2 interfaces (Figure 7.21). In principle, thermally oxidized c-Si/SiO_2 interfaces cannot be free from dangling bonds.

Silicon dangling bonds at silicon atoms with three silicon back bonds, i.e. at non-oxidized silicon atoms, can be recombination active. Not all of such dangling bonds form recombination centers. For example, dangling bonds at silicon dimers are not recombination active (Dittrich *et al.*, 2002), in contrast to silicon dangling bonds pointing in the $\langle 111 \rangle$ direction from so-called P_b recombination centers (Poindexter *et al.*, 1981) as shown by directly correlating the density of surface states in the middle of the band gap (D_{it}) of c-Si and the density of P_b defect centers (Caplan *et al.*, 1979). Dangling bonds at silicon atoms with oxygen back bonds can form trap states at the c-Si/SiO_2 interface (Figure 7.21(c,d)) or trap states for holes in SiO_2 (Figure 17.21(e)) (Caplan *et al.*, 1979).

The density of P_b defect centers at c-Si/SiO_2 interfaces can be minimized by controlling the orientation, surface roughness and oxidation regime. In general, the lower the surface roughness and the lower the oxidation rate, the lower the density of P_b centers at the c-Si/SiO_2 interface. The lowest density of P_b defect centers has been reached for thermal oxidation of c-Si(100) wafers in dry oxygen (Poindexter *et al.*, 1981). In contrast, much higher densities of P_b defect centers were observed after oxidation of c-Si(111) wafers (Caplan *et al.*, 1979). This is another reason why c-Si(100) wafers are preferred for c-Si solar cells.

The surface recombination velocity increases with increasing density of surface states and is proportional to the capture cross sections of electrons and holes (see Chapter 3). Values of D_{it} of the order of 10^{11}–10^{12} cm^{-2}eV^{-1} can be reached by thermal oxidation of c-Si (Poindexter *et al.*, 1981). As a good approximation, the capture cross sections of P_b defect centers are about 10^{-16} and 10^{-14} cm^2 for holes and electrons, respectively (see Aberle *et al.* (1992) and references therein). Defect states with electron capture cross sections of 10^{-14} and 10^{-16} cm^2 have been ascribed to P_b defect centers and to silicon dangling bonds at silicon atoms with one oxygen back bond, respectively (Albohn *et al.*, 2000). The surface recombination velocity still remains as high as 10^3–10^4 cm/s after thermal oxidation of c-Si surfaces. However, surface recombination velocities below 10–100 cm/s are required for c-Si solar cells with very high energy conversion efficiencies (see Chapter 3).

7.4.3 *Hydrogen passivation of silicon dangling bonds*

Hydrogen passivation is a general concept that is applied whenever the density of silicon dangling bonds should be reduced or kept as low as possible. The density of dangling bonds at c-Si/SiO$_2$ interfaces can be further reduced by so-called hydrogen passivation, during which silicon dangling bonds react with hydrogen atoms and form Si–H bonds. Usually hydrogen passivation of c-Si/SiO$_2$ interfaces is performed during post-annealing in so-called forming gas, which contains hydrogen and NO$_2$. Figure 7.22 shows the principle of hydrogen passivation of a silicon dangling bond at a c-Si/SiO$_2$ interface. The density of dangling bonds at c-Si/SiO$_2$ interfaces can be further reduced by hydrogen passivation to values even below 10^{10} cm^{-2}, i.e. surface recombination velocities between 10 and 100 cm/s and even less can be realized (Stephens *et al.*, 1994).

Hydrogen passivation takes also place at c-Si surfaces coated with SiN$_x$:H or a-Si:H layers where hydrogen is introduced together with the reactant gas during layer deposition. Silicon dangling bonds are also passivated in amorphous silicon solar cells with hydrogen (see Chapter 9).

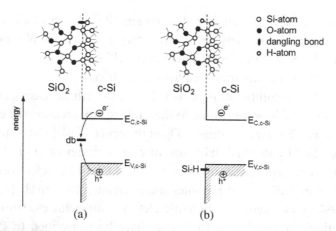

Figure 7.22. Schematic bond configuration and band diagram of the c-Si surface before (a) and after (b) hydrogen passivation of a silicon dangling bond (db).

Furthermore, the preparation of ultra-clean hydrogen-passivated c-Si surfaces by wet chemical treatments in acidic HF-containing solutions is an important pretreatment for subsequent thermal oxidation or plasma-enhanced deposition steps of passivation layers on c-Si surfaces at low temperatures. The density of surface states at hydrogenated c-Si(111) surfaces can be reduced to very low values up to 10^{10} cm^{-2}. Extremely low surface recombination velocities of about 0.7 cm/s have been demonstrated for c-Si surfaces immersed in HF (Yablonovitch *et al.*, 1986b).

The oxidation state of silicon atoms at the c-Si/SiO$_2$ interface does not change during hydrogen passivation (Himpsel *et al.*, 1988). Annealing in vacuum at temperatures above 400–500°C leads to an increase of the density of dangling bonds at the c-Si/SiO$_2$ interface, i.e. $\equiv$Si–H interface bonds can dissociate into a silicon dangling bond and a hydrogen atom (Stathis, 1995).

$$\equiv Si - H \rightarrow \equiv Si \cdot + H^0 \qquad (7.35)$$

Hydrogen-passivated c-Si surfaces are also unstable at high negative electric fields (negative-bias-temperature instability, see, for example, Helms and Poindexter (1994)). The dissociation of hydrogen from passivated silicon dangling bonds is enhanced in the presence of holes at the c-Si/SiO$_2$ interface leading to the following electrochemical reaction:

$$\equiv Si - H + A + h^+ \rightarrow \equiv Si \cdot + AH^+ \qquad (7.36)$$

The component A in Equation (7.36) is usually related to water molecules, which can occur at c-Si/SiO$_2$ interfaces (Helms and Poindexter, 1994). The correlation between the fixed positive charge and the density of surface states has also been demonstrated by electrochemical passivation and de-passivation cycling under electron injection (Dittrich *et al.*, 2001).

The negative-bias-temperature instability (Equation (7.36)) has consequences for the limitation of the minimum surface recombination velocity at c-Si/SiO$_2$ interfaces. The lowest surface recombination velocities have been achieved for c-Si with low density of free holes, i.e. for n-type doped c-Si (Aberle, 2000).

In HIT solar cells (Figure 7.17), hydrogen passivation of c-Si surfaces is performed at temperatures below 200°C (Taguchi *et al.*, 2005; Tsunomura *et al.*, 2009) and hydrogen passivation is realized at c-Si/a-Si:H(i) interfaces instead of c-Si/SiO$_2$ interfaces. Therefore, the negative-bias-temperature instability does not limit surface passivation in HIT solar cells.

Photo-generated charge carriers can recombine at defect states at the hydrogenated c-Si surface and at defect states in a-Si:H(i) available by tunneling. Consequently, only an interface region with a thickness of a couple of nanometers at the a-Si:H(i) side is relevant for recombination in HIT solar cells. If taking into account a density of dangling bonds in a-Si:H(i) of $3 \cdot 10^{15}$ cm^{-3} (Jackson and Amer, 1982), electronic states with an areal density of less than 10^9 cm^{-2} are available for recombination at the a-Si:H(i) side of the charge-selective contacts of HIT solar cells. The density of interface states at hydrogenated c-Si surfaces can be reduced to values of the order of and below 10^{10} cm^{-2} eV^{-1} (see, for example, Rauscher *et al.* (1995)). Since the complete area of the c-Si absorber is passivated with a-Si:H(i) and since the density of free charge carriers is very low in a-Si:H(i), the surface recombination rate (see Chapter 3) can be strongly reduced in HIT solar cells due to hydrogen passivation.

7.4.4 *Passivation with the c-Si/SiN$_x$:H interface*

The band gap of silicon nitride is about 5.4 eV (Miyazaki *et al.*, 2003). Silicon nitride is chemically very stable. The conduction and valence band offsets are about 2.8–2.4 and 1.5–1.9 eV, respectively, at c-Si/Si$_3$N$_4$ interfaces, see, for example, Miyazaki *et al.* (2003) and Higuchi *et al.* (2007), i.e. barriers

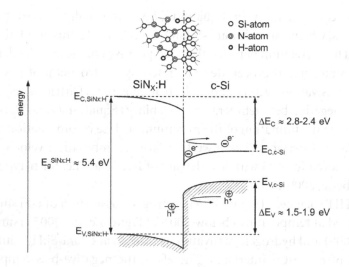

Figure 7.23.   Bond configuration and band diagram at the c-Si/SiN$_x$:H interface.

for electrons and holes can be used for passivation of c-Si surfaces with silicon nitride.

Silicon nitride can be deposited on c-Si surfaces by plasma-enhanced deposition processes containing hydrogen (also denoted as SiN$_x$:H), which enables hydrogen passivation of remaining silicon dangling bonds. Furthermore, a fixed positive charge is formed in silicon nitride near the c-Si/SiN$_x$:H interface after plasma-enhanced deposition at temperatures below 500°C (Leguijt *et al.*, 1996). Figure 7.23 shows a schematic bond configuration and the band offsets at a c-Si/SiN$_x$:H interface.

The fixed positive charge introduces an electric field in c-Si, thereby attracting free electrons and repelling free holes. Very low surface recombination velocities can be reached at c-Si/SiN$_x$:H interfaces due to the low density of surface states and due to the built-in electric field caused by the fixed positive charge. A surface recombination velocity as low as 4 cm/s has been demonstrated even for standard p-type doped c-Si wafers covered with SiN$_x$:H (Lauinger *et al.*, 1996).

The passivation of c-Si solar cells with multi-functional silicon nitride layers has important technological advantages. First, three measures of surface passivation are realized within one deposition step, namely the (i) creation of barriers for free electrons and holes due to large conduction and valence band offsets, (ii) passivation of remaining silicon dangling bonds by hydrogen which is part of the reaction gas and (iii) creation of an electric field by positive fixed charge in the SiN_x:H layer near the c-Si/SiN_x:H interface (field-effect passivation). Second, SiN_x:H has a refractive index less than that of c-Si and, consequently, the SiN_x:H layer serves as an antireflection coating. Third, the low deposition temperature of SiN_x:H is much less critical for diffusion of unwanted impurities in c-Si than thermal oxidation. This allows additional cost reduction regarding the clean room atmosphere and handling tools needed for passivation technologies with thermal oxidation.

7.4.5 *The c-Si/Al$_2$O$_3$ interface*

Crystalline silicon surfaces can be excellently passivated with Al_2O_3 layers synthesized by atomic layer deposition (ALD) (Hoex *et al.*, 2008). Besides the reduced density of surface states, a large fixed negative charge of the order of 10^{12}–10^{13} cm^{-2} near the c-Si/Al_2O_3 interface provides additional passivation by the field effect (Dingemans *et al.*, 2011). Surface recombination velocities at the p-type doped base side for c-Si solar cells with a base passivated by Al_2O_3 are as low as for c-Si solar cells with a base passivated by SiO_2 (Schmidt *et al.*, 2008).

Figure 7.24 shows a schematic of the bond configuration and of the band diagram at the c-Si/Al_2O_3 interface. The band gap of very thin Al_2O_3 layers grown by ALD depends sensitively on the deposition parameters and can range between 6.2 and 7 eV (Bersch *et al.*, 2008). The oxidation state of silicon atoms changes at the c-Si/Al_2O_3 interface, i.e. there is an ultra-thin SiO_x layer at the c-Si/Al_2O_3 interface. The extension and the variation of the oxidation state of silicon atoms at and near the c-Si/Al_2O_3 interface are important for the offsets of the conduction and valence band edges, which can vary between 2.1 and 2.8 eV and between 2.9 and 3.7 eV, respectively (Bersch *et al.*, 2008).

The surface recombination velocity is determined by the capture cross sections of electrons (σ_{se}) and holes (σ_{sh}) and by the density of interface states (see Chapter 3). The values of σ_{se} and σ_{sh} are about $7 \cdot 10^{-15}$ and $4 \cdot 10^{-16}$ cm^2, respectively, and the density of interface states is about $1 \cdot 10^{11}$ eV^{-1} cm^{-2} near midgap at the c-Si/Al_2O_3 interface (Werner *et al.*, 2012).

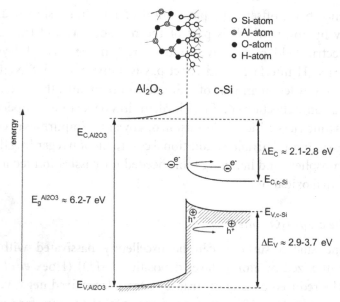

Figure 7.24. Bond configuration and schematic band diagram at the c-Si/Al$_2$O$_3$ interface.

The effective surface recombination velocity at highly p-type doped c-Si/Al$_2$O$_3$ interfaces is even less by about one or three orders of magnitude in comparison with the passivated highly p-type doped c-Si/SiO$_2$ or c-Si/a-SiN$_x$:H interfaces, respectively (Hoex *et al.*, 2008). Even surface recombination velocities of highly p-type doped emitters can be as low as 100–300 cm/s in cases of passivation with Al$_2$O$_3$. An energy conversion efficiency of 23.2% has been demonstrated for c-Si solar cells with an n-type base passivated with SiO$_2$ and a highly p-type doped emitter passivated with Al$_2$O$_3$ (Benick *et al.*, 2008).

The surface of n-type doped silicon can also be well passivated with Al$_2$O$_3$ layers (Hoex *et al.*, 2008). For example, effective surface recombination velocities below 0.8 cm/s have been achieved with plasma ALD of Al$_2$O$_3$ on n-type doped silicon surfaces (Dingemans *et al.*, 2010).

The high thermal stability of passivated c-Si/Al$_2$O$_3$ interfaces enables treatments of c-Si solar cells at high temperatures even after the deposition of the passivation layer. For example, Al$_2$O$_3$ passivation layers can be locally opened by laser ablation (laser ablation means the removal of surface atoms from a substrate by strong short laser pulses). Additionally, opened areas can be further doped and contacted without increasing the effective surface recombination velocity. This opens technological opportunities for mass production of c-Si solar cells with very high energy conversion efficiency at low costs.

7.4.6 Back surface field and local ohmic contacts

The surface recombination velocity at ohmic contacts should be very large since Fermi-level splitting is impossible by definition (see Chapter 5). Therefore, the surface recombination rate can only be reduced by reducing the density of photo-generated minority charge carriers at the ohmic contact of the base. A barrier repelling photo-generated minority electrons and attracting majority holes is necessary to reduce the density of photo-generated electrons at the ohmic contact of the p-type doped base of a c-Si solar cell. The region of a related barrier should be wide enough to avoid recombination due to tunneling of electrons into the highly p-type doped ohmic contact region. The introduced controlled barrier region is called the back surface field since it is related to the back contact of a c-Si solar cell (Figure 7.25).

The principle realization of a back surface field is possible for the ohmic contact of the base because the density of free holes is about $2 \ldots 5 \cdot 10^{16}$ cm^{-3}. The effective density of states at the valence band edge is about two to three orders of magnitude larger than p_0. Therefore, the

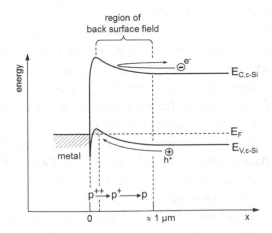

Figure 7.25. Band diagram across the ohmic contact at the p-type doped base of a c-Si solar cell with back surface field.

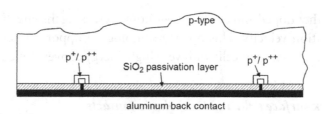

Figure 7.26. Schematic cross section of the back contact of a c-Si solar cell with a local back surface field.

potential energy of free electrons increases by about 0.12 . . . 0.18 eV towards the highly p-type doped region near the ohmic contact. The barrier height is further increased due to extremely high doping (denoted by p^{++}) near the ohmic contact of the order of 10^{20} cm^{-3}. The region of the back surface field is adjusted to an extension of about 1 μm by a concentration gradient of acceptors.

The ohmic contact area has to be strongly reduced (also see Chapter 5) in c-Si solar cells with very high energy conversion efficiency (local back surface field). The area between local ohmic contacts is passivated, for example, with a c-Si/SiO$_2$ interface (Figure 7.26). A p^{++} doped region forms at the c-Si/Al contact during firing since Al atoms are acceptors in c-Si.

The concept of the present world record c-Si solar cell makes use of locally diffused ohmic back contacts (PERL — *p*assivated *e*mitter *r*ear *l*ocally diffused; Wang *et al.* (1990)).

Conventional fabrication of local ohmic back contacts with a back surface field requires expensive photolithography technologies for placing well-defined windows in the SiO$_2$ passivation layer. The back surface field and the highly doped p^+-type doped region can be formed by boron diffusion and the contacting aluminum can be deposited by evaporation through the windows in the passivation layer.

For mass production of c-Si solar cells with local ohmic back contacts, the aluminum back-contact layer is deposited directly onto the passivation layer (Figure 7.27(a)). A strong short laser pulse induces local melting of the aluminum surface layer, subsequent local damage of the passivation layer, local alloying between silicon and aluminum and some diffusion of aluminum into c-Si (Figure 7.27(b)). As a result, a laser fired local ohmic back contact is formed (Schneiderlöchner *et al.*, 2002).

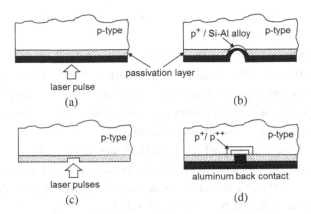

Figure 7.27. Schematic cross sections of p-type doped c-Si coated with a passivation and aluminum layers before (a) and after formation of laser-fired local p^+-Si–Al alloy/Al contact (b), of p-type doped c-Si coated with a passivation during opening of a window by laser ablation (c) and after diffusion of a local back surface field and deposition of the aluminum back-contact layer (d).

Laser ablation can be used for opening well-defined windows in a passivation layer deposited on the p-type doped base (Figure 7.27(c)). The local back surface field and the highly doped regions can be formed by conventional and/or laser-assisted diffusion. Finally, the aluminum back-contact layer is deposited (Figure 7.27(d)). The high thermal stability of c-Si/Al$_2$O$_3$ passivation layers is an advantage of this kind of technology.

7.5 Summary

The absorption length is of the order of 100 μm for a large part of photons in the sun spectrum with energies above the band gap of c-Si. For this reason thick c-Si wafers (i.e., with a thickness 200–400 μm) are demanded for c-Si solar cells with high energy conversion efficiency.

Recombination of photo-generated charge carriers in the thick c-Si absorber should be as low as possible. The diffusion length of photo-generated minority charge carriers has to be much longer than the thickness of the c-Si absorber. This requirement limits the maximum density of majority charge carriers (usually free holes introduced by boron doping) in the base to the order of 10^{16} cm^{-3} due to Auger recombination as well as the maximum density of defects in bulk c-Si crystals to the order of 10^{11} cm^{-3} due to Shockley–Read–Hall recombination. As a consequence, homogeneously doped c-Si crystals with very low densities of defects have to be grown for c-Si solar cells.

Purification by distillation of trichlorsilane $SiHCl_3$ precursor for c-Si crystal growth and refinement by segregation at the liquid–solid interface are applied for getting very pure silicon crystals. Silicon crystals grown by FZ melting have the highest purity and are suitable absorbers for c-Si solar cells with solar energy conversion efficiencies above 24%.

Etching and cleaning of c-Si surfaces are crucial for reaching high energy conversion efficiencies since defects caused by wafering have to be removed, c-Si surfaces have to be structured for light trapping and surface contaminations must be removed or avoided.

Photo-generated charge carriers have to be collected at the highest possible potential difference between charge-collecting electron and hole contacts. For this demand the emitter is doped with a dopant (phosphorus for n-type doping) density as high as 10^{19} cm^{-3}. Technologies of thermal diffusion, ion implantation, thermal annealing or laser processing can be applied for the preparation of the emitter. As an alternative, doped a-Si:H /c-Si hetero-junctions can be applied as charge-selective contacts as well in c-Si solar cells (HIT concept).

Sophisticated measures of c-Si surface passivation are required for reducing the density of surface states to values below 10^{10} cm^{-2} for c-Si solar cells with very high energy conversion efficiency.

Silicon surfaces can be passivated with c-Si/SiO_2, c-Si/SiN_x:H, c-Si/ a-Si:H(i) or c-Si/Al_2O_3 interfaces. The surface recombination rate is additionally reduced by a fixed positive or negative charge near c-Si/SiN_x:H and c-Si/Al_2O_3 interfaces, respectively, as well as by the back surface field at ohmic contacts. In c-Si solar cells with a pn-homo-junction, the passivation layer has to be opened locally for contacting the base and the emitter with ohmic contacts. For this purpose, etching with masks or local thermal processing with laser pulses is applied.

There are numerous types of c-Si solar cells, which are mainly distinguished by their architecture (for example, buried-contact solar cells, back-contacted solar cells, local-contact solar cells, emitter-wrap-through solar cells …) and/or by their technology or passivation concepts (for example, screen-printed c-Si solar cell, passivated emitter and rear locally diffused, passivated emitter and rear contact, hetero-junction with intrinsic thin layer, etc.). Each concept has individual physical and technological advantages and disadvantages, important for the production costs per W_p of c-Si solar cells and modules, and which are especially important for the further reduction of the energy needed for the production of c-Si solar cells with high energy conversion efficiency.

The basic experience obtained during the optimization of c-Si solar cells with solar energy conversion efficiencies above 24% is summarized in Figure 7.28 for the PERL (a) and HIT (b) concepts.

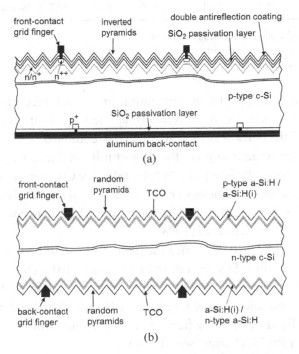

Figure 7.28. Schematic cross sections of c-Si solar cells with highest solar energy conversion efficiency based on a pn-homo-junction (a) and on a-Si:H/c-Si hetero-junctions (b) as charge-selective contacts. In the PERL concept, most of the c-Si surface area is passivated with c-Si/SiO$_2$ interfaces opened locally for ohmic contacts (a). In the HIT concept, the complete c-Si surface area is passivated with a c-Si/a-Si:H(i) interface and the n-type doped c-Si absorber is sandwiched between two charge-selective contacts (b). Inverted (a) and random (b) pyramids are etched onto the c-Si(100) absorbers for light trapping. A double antireflection coating and an aluminum rear contact as back reflector are applied for minimization of optical losses in the PERL concept (a). For minimization of optical losses in the HIT concept a TCO is optimized additionally as an antireflection coating and as a layer for total internal reflection (b).

Table 7.1. Values of the thickness of the c-Si absorber (d_{c-Si}), short-circuit current density (I_{SC}), open-circuit voltage (V_{OC}), fill factor (FF) and solar energy conversion efficiency (η) for different c-Si solar cell concepts. References for the data are given.

	d_{c-Si} (μm)	I_{SC} (mA/cm^2)	V_{OC} (V)	FF (%)	η (%)	Reference
PERL	260	42.7	0.706	82.8	25.0	Zhao *et al.* (1998)
HIT	98	39.6	0.750	83.2	24.7	Panasonic (2013)
Kerfless epitaxy	43	38.14	0.682	77.4	20.1	Moslehi *et al.* (2012)
Exfoliation	25	33.6	0.580	76.7	14.9	Rao *et al.* (2012)
LPE on p$^+$	16.8	27.2	0.660	82.2	14.7	Werner *et al.* (1993)

PERL, passivated emitter rear locally diffused
HIT, hetero-junction intrinsic thin film
LPE, liquid phase epitaxy

Different measures for minimization of optical, recombination and resistive losses have been successfully applied for both concepts. The reduction of the thickness of c-Si solar cells combined with a reduction of surface recombination gives the opportunity of further increasing the energy conversion efficiency of c-Si solar cells (see Chapter 4). Record solar cells with different thicknesses are compared in Table 7.1. Until now the thickest c-Si solar cell (260 μm) made by the PERL concept has had the highest efficiency (Zhao *et al.*, 1998).

The highest V_{OC} (0.75 V) and the highest fill factor have been reached with a 98-μm thin c-Si solar cell with an area of 101.8 cm^2 made by the HIT concept (Panasonic, 2013). This demonstrates the importance of the reduction of the thickness of c-Si solar cells. Additional optical losses in HIT solar cells are caused by free carrier absorption and by fundamental absorption in the TCO layer (see Chapter 9).

The high potential of rather thin c-Si solar cells has been shown by the model experiment of Werner where an energy efficiency of 14.7% and V_{OC} of 0.66 V were demonstrated for a 16.8-μm thin c-Si epitaxial layer without any measures of antireflection coatings or light trapping (Werner *et al.*, 1993). Kerfless exfoliation (Rao *et al.*, 2012) and kerfless transfer of epitaxial c-Si layers (Moslehi *et al.*, 2012) can be realized on large areas and solar energy conversion efficiencies of 14.9% and 20.1% were obtained on 25- and 43-μm thin c-Si solar cells, respectively. However, high energy conversion efficiencies can hardly be expected for the exfoliation technology due to the intimate c-Si/metal contact during generation of stress at high temperature.

The energy conversion efficiency of conventional c-Si PV modules decreases with increasing temperature by a factor of about −0.45%/K. Operation temperatures of 40–60°C are easily reached under full illumination of PV modules in summer so that the solar energy conversion efficiency is reduced by about 10% in comparison to operation at 25°C. A further advantage of HIT solar cells is that the temperature dependence of the solar energy conversion efficiency is reduced to −0.25%/K (Taguchi *et al.*, 2005), which is caused by the lower thermal generation rate at the a-Si:H side of the charge-selective contact.

The energy payback time and the energy payback factor are important parameters of c-Si solar cells in the context of sustainable renewable energy production. The energy payback time of conventional c-Si solar cells is about 1.5 years in the south of Italy (Wild-Scholten *et al.*, 2010). The degradation rates of c-Si PV modules are about 0.5% per year on average (Jordan and Kurtz, 2013) and c-Si PV systems working well for longer than 30 years are known. Therefore the energy payback factor of c-Si solar cells and modules is larger than 10–15 with respect to warranty for PV modules given by producers. The energy payback factor can be even higher than 50 depending on the operation regime and on the total operation time. The reduction of the thickness of c-Si wafers and the consequent penetration of low-temperature and/or local laser heating processes in mass production of c-Si solar cells will further reduce the energy payback time of c-Si solar cells so that overall energy payback factors of the order of 100 can be expected for c-Si-based photovoltaics in the future.

7.6 Tasks

T7.1: n-type doped base and p-type doped emitter

Calculate the density of free electrons in an n-type doped base and ascertain the thickness of a p-type doped emitter of a c-Si solar cell.

T7.2: Silver required for c-Si-based photovoltaics

Bus bars at front and back contacts and grid fingers at front contacts of c-Si solar cells are screen-printed from a silver-containing paste in industrial processes. The specific conductivity and the density of silver are about $6 \cdot 10^7$ S/m and $10.5\,g/cm^3$, respectively. Estimate the mass of silver which would be required for a complete support of electricity for mankind produced only with c-Si solar cells and discuss the importance of replacement of silver in c-Si solar cells. About $2 \cdot 10^4$ tons of silver are produced globally in one year (Brooks, 2011).

T7.3. Boron-doped c-Si(100) wafers

Why are boron-doped c-Si(100) wafers used for the production of conventional c-Si solar cells?

T7.4. Diffusion of phosphorus in c-Si

Calculate the thickness of a phosphorus emitter after diffusion from an inexhaustible source for 1000 s at 900°C.

T7.5. Diffusion of copper in c-Si

Calculate the spread of copper after diffusion from a localized copper contamination of one mono-layer at 900°C for 10, 100 and 1000 s.

T7.6. V_{OC} of PERL and HIT c-Si solar cells

Explain why a higher V_{OC} can be reached for c-Si solar cells made by the HIT concept than those made by the PERL concept.

T7.7. Potential of kerfless wafer transfer technologies

Discuss the potential of kerfless wafer transfer technologies for realizing high energy payback factors with c-Si solar cells under the conditions of low-cost mass production.

8

Solar Cells Based on III–V Semiconductors

The absolute highest solar energy conversion efficiencies (η) are reached for single- and multi-junction solar cells based on III–V semiconductors illuminated at both non-concentrated and highly concentrated sunlight. This chapter illuminates properties of III–V semiconductors and their interfaces which are decisive for very high values of η. Binary, ternary and quaternary III–V semiconductors are introduced and variations of band gaps with stoichiometry are described. High diffusion length of minority charge carriers in comparison with the absorption length as well as n-type and p-type doping over a wide variable range can be achieved with III–V semiconductors. Barrier layers, passivation layers and tunneling junctions can be realized at hetero-junctions between III–V semiconductors. The principle of epitaxy of III–V semiconductor layers is explained for the growth of sophisticated layer systems realizing functions of optical absorption, charge separation, charge transport, surface passivation and ohmic contacts for monolithically stacked multi-junction solar cells based on III–V semiconductors. Formation principles of ohmic metal contacts with III–V semiconductors are given. Architectures are derived for tandem, triple-junction and quadruple-junction solar cells based on III–V semiconductors.

8.1 III–V Semiconductor Family

8.1.1 *Binary III–V semiconductors*

Semiconductors of the III–V family are built from elements of the third (aluminum — Al, gallium — Ga and indium — In) and fifth (nitrogen — N, phosphorus — P, arsenic — As and antimony — Sb) groups of the

III	IV	V
B 5 *boron*	C 6 *carbon*	**N** 7 *nitrogen*
Al 13 *aluminum*	Si 14 *silicon*	**P** 15 *phosphorus*
Ga 31 *gallium*	Ge 32 *germanium*	**As** 33 *arsenic*
In 49 *indium*	Sn 50 *tin*	**Sb** 51 *antimony*

Figure 8.1. Part of the periodic table with Al, Ga and In of the third group and N, P, As and Sb of the fifth group, which form the family of III–V semiconductors.

periodic table of elements (Figure 8.1). Atoms are bonded covalently in III–V semiconductors.

The twelve binary III–V semiconductors contain only one kind of atoms of the third and fifth groups of the periodic table ($A^{III}B^{V}$, A or B corresponds to Al, Ga, In or to N, P, As, Sb, respectively).

The band gap of III–V semiconductors decreases with increasing atom numbers of the components. For example, the band gap of the III–V semiconductors with the lowest (AlN) and highest (InSb) atom numbers are 6.2 and 0.18 eV, respectively. The band gaps of the binary III–V semiconductors (GaN, InN, AlP, GaP, InP, AlAs, GaAs, InAs, AlSb, GaSb and InSb) are visualized in Figure 8.2 in relation to the metal atom. The most prominent III–V semiconductors are GaAs (1.42 eV) and InP (1.34 eV) with band gaps in the optimum range for photovoltaic solar energy conversion (see Chapter 6). The band gaps of GaP, AlAs and InAs are 2.26, 2.16 and 0.36 eV, respectively (see, for example, Vurgaftman *et al.* (2001) and references therein).

The temperature dependence of band gaps of semiconductors has been described by the following empirical equation (Varshni, 1967b):

$$E_g = E_{g0} - \frac{\alpha \cdot T^2}{T + \beta} \tag{8.1}$$

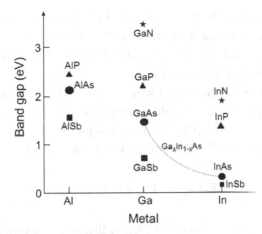

Figure 8.2. Visualization of the band gaps of binary III–V semiconductors at room temperature. The dashed line schematically shows the behavior of the band gap for alloying between GaAs and InAs.

The band gaps at 0 K (E_{g0}) are equal to 1.519, 2.24, 0.417, 2.35 and 1.424 eV for GaAs, AlAs, InAs, GaP and InP, respectively. The constants α and β are equal to 0.5405 meV/K and 204 K for GaAs, to 0.7 meV/K and 530 K for AlAs, 0.276 meV/K and 93 K for InAs, to 0.577 meV/K and 372 K for GaP and to 0.363 meV/K and 162 K for InP (Vurgaftman *et al.*, 2001).

8.1.2 *Ternary and quaternary III–V semiconductors*

Atoms of the third or fifth groups of the periodic table can be partially replaced by other atoms of the third or fifth groups of the periodic table. The replacement of given atoms in a compound by atoms with similar chemical properties is also called alloying.

In ternary III–V semiconductors atoms of the third or fifth groups of the periodic table are partially replaced by other atoms of the third or fifth groups of the periodic table ($A_x^{III}B_{1-x}^{III}C^V$ or $A^{III}B_x^VC_{1-x}^V$, where the composition parameters x and y range between 0 and 1).

In quaternary III–V semiconductors, both atoms of the third and fifth groups of the periodic table are partially replaced by other atoms of the third and fifth groups of the periodic table ($A_x^{III}B_{1-x}^{III}C_y^VD_{1-y}^V$, where the composition parameters x and y range between 0 and 1).

The band gap of ternary III–V semiconductors can be varied smoothly between the band gaps of the corresponding binary III–V semiconductors by alloying. For example, the band gap of InAs can be extended to higher values by partially replacing indium atoms with gallium atoms. On the other hand, the band gap of GaAs can be reduced by partially replacing gallium atoms with indium atoms.

The band gap of a compound semiconductor $A_x B_{1-x}$ is a superposition of the band gaps of the separate semiconductors $E_g(A)$ and $E_g(B)$. The semiconductors A and B can be, for example, GaAs and InAs with the corresponding ternary III–V semiconductor $Ga_x In_{1-x}As$.

$$E_g(A_x B_{1-x}) = x \cdot E_g(A) + (1 - x) \cdot E_g(B) - x \cdot (1 - x) \cdot C \qquad (8.2)$$

The coefficient C in Equation (8.2) is the so-called bowing parameter, which is given in the unit of the band gap. The value of C, for example, is equal to 0.477 eV for $Ga_x In_{1-x}As$ (Vurgaftman *et al.*, 2001) (Figure 8.3).

It should be taken into account that Equation (8.2) is valid for the band gaps at the same symmetry points or valleys of the band structure of the given semiconductors (Vurgaftman *et al.*, 2001). Incidentally, the band structure describes the dependence of the energy of electrons on

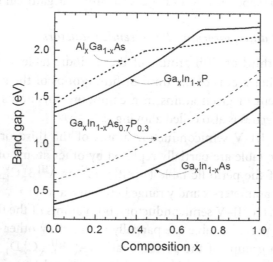

Figure 8.3. Dependence of the band gap on the composition parameter x for the ternary compounds $Ga_x In_{1-x}P$ (thin solid line) and $Ga_x In_{1-x}As$ (thick solid line), for $Al_x Ga_{1-x}As$ (dotted line) and for the quaternary compound $Ga_x In_{1-x}As_{0.7}P_{0.3}$ (thin dashed line).

the wave vector. For interested readers, the band structures of some III–V semiconductors are given in, amongst other sources, Chelikowsky and Cohen (1976).

For GaAs and InAs, both band gaps are related to the so-called Γ-valley. For AlAs and GaP the band gaps are related to the so-called X-valley. The transition energies are 3.0 and 2.9 eV for AlAs and GaP in the Γ-valleys are 1.9 and 1.37 eV for GaAs and InAs in the X-valleys, respectively. The bowing parameters are different for the transition energies in the different valleys and so the transition energies are equal for the Γ- and X-valleys of a certain composition (also called intercrossing point). The intercrossing points are reached for $Al_xGa_{1-x}As$ and $Ga_xIn_{1-x}P$ at compositions of 0.45 and 0.64, respectively. The bowing parameter of $Al_xGa_{1-x}As$ is equal to 0.055 for x larger than 0.45 (X-valleys) and equal to $-0.127 + 1.31x$ for x less than 0.45 (Γ-valleys) (Vurgaftman *et al.*, 2001). At the intercrossing point, the band gap of $Al_{0.45}Ga_{0.55}As$ is about 1.95 eV (Figure 8.3). For $Ga_xIn_{1-x}P$ the bowing parameters are 0.65 for x less than 0.64 (Γ-valleys) and 0.20 for x larger than 0.64 (X-valleys) (Vurgaftman *et al.*, 2001). Figure 8.3 shows the dependence of the band gap on the composition for $Al_xGa_{1-x}As$ and $Ga_xIn_{1-x}P$.

In quaternary compounds, the band gap depends on two composition parameters x and y. The band gap of $Ga_xIn_{1-x}As_yP_{1-y}$ can be calculated, for example, by using the following equation (Nahory *et al.*, 1978):

$$E_g(x, y) = (1.35 + 0.668 \cdot x - 1.17 \cdot y + 0.758 \cdot x^2 + 0.18 \cdot y^2$$
$$- 0.069 \cdot x \cdot y - 0.322 \cdot x^2 \cdot y + 0.03 \cdot x \cdot y^2) \cdot eV \quad (8.3)$$

The dependence of E_g on the composition parameter x is plotted for $Ga_xIn_{1-x}As_{0.7}P_{0.3}$ in Figure 8.3.

A great advantage of alloying III–V semiconductors is that practically any band gap relevant for photovoltaic solar energy conversion can be obtained. This property is very important for the realization of multi-junction solar cells (see Chapter 6) with the highest solar energy conversion efficiencies.

8.1.3 *Optical absorption of III–V semiconductors*

Figure 8.4 shows the optical absorption spectra of some binary III–V semiconductors (InAs, InP, GaAs, GaP and AlAs), $Al_{0.3}Ga_{0.7}As$ and

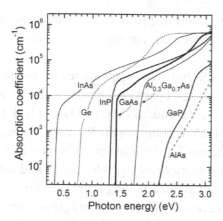

Figure 8.4. Optical absorption spectra of InAs, InP, GaAs, GaP, AlAs, Al$_{0.315}$Ga$_{0.685}$As and Ge. Data taken from Dixon and Ellis (1961), Turner *et al.* (1964), Aspnes and Studna (1983), Aspnes *et al.* (1986), Adachi (1989) and Braunstein *et al.* (1958). The dashed lines mark the absorption lengths of 10 and 1 μm.

germanium. The absorption coefficient increases to values larger than 10^4 cm^{-1} within intervals between E_g and $E_g + 0.1$ eV for InP and GaAs, E_g and $E_g + 0.3$ eV for InAs and Ge and E_g and $E_g + 0.6$ eV for GaP and AlAs. Absorption of photons with energies below E_g can be neglected for epitaxially grown layers of III–V semiconductors due to the extremely low density of defects with electronic states in the band gap.

The optical absorption spectra of ternary and quaternary III–V semiconductors behave very similarly to those of the binary III–V semiconductors. The absorption coefficient of ternary and quaternary III–V semiconductors also increases steeply within an interval between E_g and $E_g + 0.1$ eV when E_g is dominated by the transition in the Γ-valley, for example, for Ga$_x$Al$_{1-x}$As with x < 0.45 (Adachi, 1989; Aspnes *et al.*, 1986). Therefore, absorber layers with large absorption coefficients for photons with energies larger than the band gap can be realized for the whole region of band gaps. This has important implications for multi-junction solar cells (see Chapter 6).

III–V semiconductors are well suited for stacked multi-junction solar cells with integrated series connection due to the variability of band gaps, high absorption coefficients for photons with energies above E_g and excellent transparency for photons with energies below E_g.

The limitation of the quantum efficiency by optical absorption can be obtained for each of the stacked solar cells in a multi-junction solar cell by analyzing the light absorbed in the layers with the larger band gap and transmitted to the following layer with the lower band gap. In the following, the quantum efficiencies of the top, middle and bottom cells (QE_{top}, QE_{middle} and QE_{bottom}) in a triple-junction solar cell are estimated using Equations (2.14) and (1.41). Reflection losses and recombination losses are neglected and $\alpha_{top}(h\nu)$, $\alpha_{middle}(h\nu)$, $\alpha_{bottom}(h\nu)$ and d_{top}, d_{middle} and d_{bottom} correspond to the related absorption coefficients and layer thicknesses, respectively.

$$QE_{top} = 1 - \exp\left(-\alpha_{top} \cdot d_{top}\right) \tag{8.4}$$

$$QE_{middle} = 1 - QE_{top} - \exp\left(-\alpha_{middle} \cdot d_{middle}\right) \tag{8.5}$$

$$QE_{bottom} = 1 - QE_{top} - QE_{middle} - \exp\left(-\alpha_{bottom} \cdot d_{bottom}\right) \tag{8.6}$$

As an example, Figure 8.5 shows approximated optical absorption spectra of $Ga_{0.43}In_{0.57}As$ ($E_g = 0.7\,eV$), $Ga_{0.93}In_{0.07}As$ ($E_g = 1.35\,eV$) and $Ga_{0.5}In_{0.5}P$ ($E_g = 1.86\,eV$), and corresponding spectra of the quantum efficiencies for stacked solar cells in a triple-junction solar cell with different values of layer thicknesses.

The short-circuit current (I_{SC}) density of each cell can be calculated with Equation (1.42). The I_{SC} density of the top cell limits the I_{SC} of the given triple-junction solar cell with integrated series connection. Current matching (see Chapter 6) can be optimized by varying the thicknesses of the absorber layers of the solar cells stacked in a multi-junction solar cell.

8.2 Hetero-Junctions of III–V Semiconductors

8.2.1 *Type I and type II semiconductor hetero-junctions*

Semiconductors with different band gaps are brought into contact with each other in III–V semiconductor solar cells. Contacts between semiconductors with different band gaps are known as semiconductor hetero-junctions.

Barriers for electrons and holes are formed at semiconductor hetero-junctions depending on the energies of the conduction and valence band edges of the two semiconductors at the contact. In the following, the band gaps of the semiconductors with the lower and higher band gaps are denoted

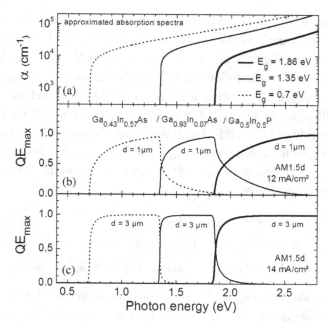

Figure 8.5. Approximated optical absorption spectra of $Ga_{0.43}In_{0.57}As$ (short dashed line), $Ga_{0.93}In_{0.07}As$ (thin solid line) and $Ga_{0.5}In_{0.5}P$ (thick solid line) (a) and spectra of the quantum efficiencies for a stacked triple-junction solar cell with corresponding $Ga_{0.5}In_{0.5}P$ top, $Ga_{0.93}In_{0.07}As$ middle and $Ga_{0.43}In_{0.57}As$ bottom cells (b and c) calculated using Equations (8.4)–(8.6). The layer thicknesses are 1 μm (b) and 3 μm (c). Reflection, internal multiple reflection, and interference and recombination processes are neglected. The short-circuit current densities are limited by the top cell in case of integrated series connection.

by E_{g1} and E_{g2}, respectively. Similarly, the energies of the corresponding conduction and valence band edges are denoted by E_{C1}, E_{C2}, E_{V1} and E_{V2}.

The conduction and valence band offsets (ΔE_C and ΔE_V, respectively) are defined as the differences between E_{C2} and E_{C1} and between E_{V2} and E_{V1}.

$$\Delta E_C \equiv E_{C2} - E_{C1} \tag{8.7'}$$

$$\Delta E_V \equiv E_{V2} - E_{V1} \tag{8.7''}$$

Type I and type II semiconductor hetero-junctions are distinguished with regard to the signs of the valence and conduction band offsets. For type I semiconductor hetero-junctions, the signs of ΔE_C and ΔE_V are different, i.e. positive for ΔE_C and negative for ΔE_V. This means that barriers exist at

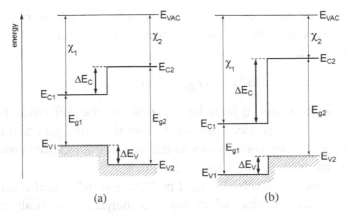

Figure 8.6. Schematic band diagrams of type I (a) and type II (b) semiconductor hetero-junctions. The signs of the conduction and valence band offsets are positive in the case of the type II semiconductor hetero-junction.

the hetero-junction for both free electrons and holes in the semiconductor with the lower band gap (Figure 8.6(a)). Type I semiconductor hetero-junctions are applied for surface passivation of photovoltaic absorbers with wide-gap semiconductors (see Chapter 3).

The signs of the conduction and valence band offsets are identical for type II semiconductor hetero-junctions. A barrier is formed for free electrons whereas the potential energy of free holes in the semiconductor with the lower band gap is reduced at a type II semiconductor hetero-junction in the case of positive signs of ΔE_C and ΔE_V (Figure 8.6(b)). In contrast, a barrier is formed for free holes whereas the potential energy of free electrons in the semiconductor with the lower band gap is reduced at a type II semiconductor hetero-junction in the case of negative signs of ΔE_C and ΔE_V. Therefore type II semiconductor hetero-junctions can serve as charge-selective contacts in solar cells.

In the case that local electric fields are absent at the interface of a semiconductor hetero-junction, i.e. the vacuum level (E_{VAC}) is constant across the interface (Figure 8.6), the values of the conduction band offsets can be obtained from the difference between the electron affinities (see Chapter 5) of both semiconductors (χ_1 and χ_2).

$$\Delta E_C = -(\chi_2 - \chi_1) \tag{8.8}$$

The valence band offset can be calculated from the difference between the band gaps and the conduction band offset at the semiconductor hetero-junction.

$$\Delta E_V = (E_{g2} - E_{g1}) - \Delta E_C \qquad (8.9)$$

The electron affinity can be replaced by the difference between the ionization energy (see Chapter 5) and the band gap. Similarly to Equation (8.8), ΔE_V corresponds to the negative difference between the ionization energies of both semiconductors.

The valence band offset defined by Equation (8.7") is also called the natural valence band offset, which has been analyzed theoretically (see, for example, Wei and Zunger (1998)). The trends of the natural band offsets can be visualized by comparing the natural valence band alignments of the 12 binary III–V semiconductors (Figure 8.7, values taken from Wei and Zunger (1998) and E_V(GaAs) is set to 0 eV). The lowest natural valence band offsets are obtained for the GaSb/InSb ($\Delta E_V = -0.01$ eV) and GaAs/InAs ($\Delta E_V = 0.05$ eV) hetero-junctions. Large natural barriers are formed for free holes at the other III–V semiconductor hetero-junctions. For example, the value of the natural ΔE_V is equal to 0.51 eV for the GaAs/AlAs hetero-junction.

The values of the natural ΔE_C can be calculated from the values of the natural ΔE_V and the differences of the corresponding band gaps (Equation (8.9)). Therefore the nature of a natural binary III–V

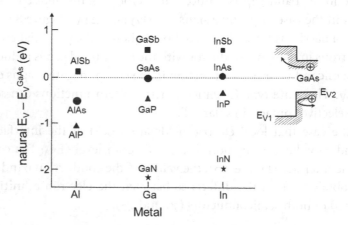

Figure 8.7. Visualization of the natural valence band alignment of binary III–V semiconductors in comparison to GaAs (values taken from Wei and Zunger (1998)).

semiconductor hetero-junction can be found by combining the values of ΔE_V (Figure 8.7) with the values of E_g (Figure 8.2). For example, the natural GaAs/AlAs hetero-junction is a type I hetero-junction, whereas the natural GaAs/InP hetero-junction is a type II hetero-junction (see also task T8.1 at the end of this chapter).

8.2.2 Role of interface dipoles and interface states for band offsets

The density of interface atoms at a hetero-junction is of the order of the reciprocal squared lattice constants of the semiconductors, i.e. of the order of $5 \cdot 10^{14}$ cm^{-2}. Atoms at a hetero-junction can carry a polarization charge in contrast to bulk atoms due to different electronegativity of atoms in both semiconductors. Furthermore, chemical bonds can be distorted and charged defect states can be present at semiconductor hetero-junctions.

The electron density is increased at interface atoms with higher electronegativity, but reduced at interface atoms with lower electronegativity. The polarization charge is of the order of the elementary charge multiplied by the relative change of the electronegativity. Polarization by distorted chemical bonds leads to the formation of interface dipole layers and therefore to the formation of local electric fields. Resulting changes in the band offsets depend on the direction of the interface dipole (Figure 8.8).

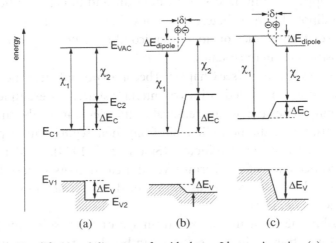

Figure 8.8. Simplified band diagrams of an ideal type I hetero-junction (a) and of type I hetero-junctions with interface dipoles of opposite orientations (b) and (c).

The thickness of interface dipole layers (δ) is of the order of two bond lengths or of the lattice constant, i.e. about 0.5 nm. The polarization charge (ΔQ_{pol}) corresponds only to a certain portion of the charge of an electron in a covalent chemical bond. The drop of the potential energy across a dipole layer (ΔE_{dipole}) can be estimated by using the equation of the parallel plate capacitor (see also task T8.2 at the end of this chapter):

$$\Delta E_{dipole} = q \cdot \frac{\delta}{\varepsilon \cdot \varepsilon_0} \cdot \Delta Q_{pol} \qquad (8.10)$$

Band offsets at intrinsic semiconductor hetero-junctions depend on the orientation of the crystal (Grant *et al.*, 1978). This is not surprising if taking into account, for example, the different densities of atoms of the third and fifth groups of the periodic table at the different surfaces of crystal orientation.

Interface dipoles can be engineered by introducing, for example, additional interfacial doping layers of sub-mono-layer thickness as demonstrated, for example, for a AlGaAs/GaAs hetero-junction (Capasso *et al.*, 1985). The distance between the sheets of donor and acceptor layers at the interface is only several nm for this type of interface dipole engineering.

Electronic defect states can arise at semiconductor hetero-junctions due to imperfections. Interface states are partially charged due to trapping. Charge trapped at interface states causes an additional modification of interface dipole layers at hetero-junctions (Figure 8.9). Depleted barrier layers arise at both sides of doped semiconductor hetero-junctions with large densities of interface states.

Fermi-level pinning (see Chapter 5) becomes possible at semiconductor hetero-junctions for high densities of interface states. For example, Fermi-level pinning arises due to the removal of the elements of the third or of the fifth groups of the periodic table from their regular lattice positions at surfaces of III–V semiconductors (Spicer *et al.*, 1979). Incidentally, the un-pinned behavior of the Fermi-level at a clean GaAs surface has been demonstrated by alternating adsorption of donor or acceptor molecules (Alperovich *et al.*, 1994).

Probably the most important advantage of III–V semiconductors is that the formation of interface states can be almost completely avoided at hetero-junctions formed by lattice-matched epitaxy, for example, at

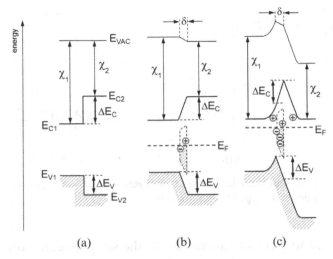

Figure 8.9. Simplified band diagrams of an ideal type I hetero-junction (a) and of type I hetero-junctions with Fermi-level pinning caused by interface states for un-doped and n-type doped type I hetero-junctions ((b) and (c), respectively).

GaAs/AlAs and GaAs/Al$_x$Ga$_{1-x}$As hetero-junctions. The discovery of the GaAs/Al$_x$Ga$_{1-x}$As hetero-junction (Alferov *et al.*, 1969) triggered a burst of development of electronic devices including, for example, semiconductor lasers (Alferov *et al.*, 1970), high-efficiency solar cells (Alferov *et al.*, 1971) and high electron mobility transistors (Mimura *et al.*, 1980).

8.2.3 *Doped type I and type II semiconductor hetero-junctions*

The schematic band diagrams of type I semiconductor hetero-junctions at interfaces of two n-type or two p-typed doped semiconductors are drawn in Figures 8.10(a) and 8.10(b), respectively. Majority charge carriers move from the semiconductor with the higher band gap into the semiconductor with the lower band gap. Therefore, accumulation layers of electrons or holes are formed at the surface of the n-type or p-type doped semiconductor, respectively, with the lower band gap. The surface of the n-type or p-type doped semiconductor with the higher band gap is depleted.

The width of the space charge region of the semiconductor with the higher band gap can be strongly reduced for very high densities of majority charge carriers so that majority charge carriers can tunnel from the semiconductor with the lower band gap into the semiconductor with

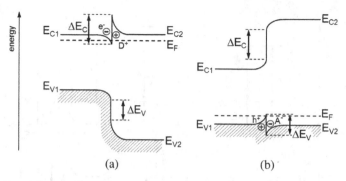

Figure 8.10. Schematic band diagrams of type I hetero-junctions n-type (a) or p-type (b) doped at both sides of the type I hetero-junction.

the higher band gap (see Chapter 5). At the same time, minority charge carriers photo-generated in the semiconductor with the lower band gap can hardly overcome the large barrier at the doped type I semiconductor hetero-junction, which is equal to the sum of the corresponding band offset (ΔE_C for p-type and ΔE_V for n-type doping) and the band bending at the sites of the semiconductors with the higher and lower band gaps.

The highly doped semiconductor with the higher band gap can form an ohmic contact with metals (see Chapter 5) and serve as a barrier layer for minority charge carriers photo-generated in the semiconductor with the lower band gap. Therefore, type I semiconductor hetero-junctions with highly doped semiconductors of a higher band gap than the photovoltaic absorber are nearly ideal contact systems for majority charge carriers if, of course, an extremely low density of interface states is assumed at the type I semiconductor hetero-junction. This behavior is one of the keys for reaching very high solar energy conversion efficiencies with III–V semiconductors.

Ohmic or recombination contacts can be formed at highly doped pn-junctions (see Chapter 5). Solar cells based on III–V semiconductors with different band gaps can be stacked on each other and connected in series by introducing highly pn-doped semiconductor hetero-junctions between the separate solar cells (stacked integrated series connection). Whether the highly pn-doped semiconductor hetero-junction is of type I or of type II is not really important for the formation of recombination contacts (Figure 8.11).

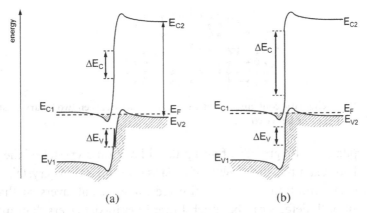

Figure 8.11. Band diagrams of recombination pn-contacts at type I (a) and type II (b) hetero-junctions.

The formation of highly pn-doped semiconductor hetero-junctions for recombination contacts is decisive for the realization of multi-junction concentrator solar cells based on III–V semiconductors with extremely high solar energy conversion efficiency.

8.3 Epitaxial Growth of III–V Semiconductors

8.3.1 *Principle of epitaxy with III–V semiconductors*

Epitaxy is a deposition principle enabling the growth of doped III–V semiconductor layers and layer systems with very low densities of bulk and interface defects. The complex combination of III–V semiconductor absorber layers with charge-selective pn-junctions, with highly doped barrier layers between photovoltaic absorbers and ohmic contacts, and with highly doped recombination contact layer systems is possible only with epitaxy.

Epitaxy means the growth of a crystal layer on a substrate crystal providing orientation and growth direction for the crystalline layer. The normalized difference between the lattice constants of the substrate crystal and of the growing crystal layer is called the lattice mismatch. Figure 8.12 shows a schematic interface between two binary crystals with the same lattice constants, i.e. the lattice mismatch is equal to 0. The crystal lattice of the substrate crystal is continued ideally in the growing crystal layer.

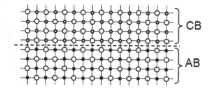

Figure 8.12.   Schematic configuration at an ideal interface between two binary crystals AB and CB with the same lattice constant.

Homoepitaxy is the growth of an epitaxial layer on a crystal of the same material, for example, the growth of a GaAs layer on a GaAs crystal.

Growth conditions can be optimized for epitaxial layers so that the density of bulk defects can be much lower in epitaxial layers than in large crystals of III–V semiconductors. The homoepitaxial growth of a buffering III–V semiconductor layer between a substrate crystal and a layer system of a solar cell is very important for reaching the highest solar energy conversion efficiencies.

The interface between two epitaxial crystal layers is free from interface defect states if both crystal layers have the same lattice constant. Therefore, surface recombination can be neglected at interfaces between epitaxial layers with the same lattice constant or with a very low lattice mismatch. This is the case for interfaces, for example, between epitaxial AlAs and GaAs layers.

Stress arises in growing epitaxial layers if the lattice constants of the substrate and epitaxial crystals are different (Figure 8.13(a)). If the lattice mismatch is not very large, interfacial stress can be compensated by slight distortion of chemical bonds and atoms in the interface region.

Distortion of chemical bonds from their equilibrium causes fluctuations in the periodic lattice potential of the crystals and therefore the formation of exponential tail states at the conduction and valence band edges in the interface region. This type of defect states becomes relevant for surface recombination at concentrated sunlight. Further, distortion of chemical bonds causes electric polarization in the interface region and therefore the formation of an interface dipole layer.

The lattice mismatch between the substrate and epitaxial crystals can be very large and consequently, for compensation of stress, some atoms have to be omitted in the interface region in addition to distortion of chemical bonds (Figure 8.13(b)).The complete removal of interface atoms

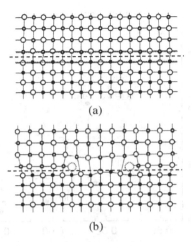

(b)

Figure 8.13. Schematic interface configuration between two binary crystals AB and CB with slightly different lattice constants (a) and with very different lattice constants (b). The narrow dashed lines in (a) characterize the distortion of atoms in the interface region from their lattice positions. The large dashed circles in (b) correspond to missing atoms at the interface of the semiconductor with the higher lattice constant.

from certain lattice positions starts when the lattice mismatch becomes larger than a certain critical lattice mismatch, which also depends on the thickness of the epitaxial layer.

Systems of III–V semiconductor layers with various band gaps are required for high-efficiency solar cells. However, not all III–V semiconductors can be combined with each other for epitaxial growth due to lattice mismatch. The band gaps of several semiconductors — including binary III–V semiconductors and also Si and Ge — are plotted versus their lattice constants in Figure 8.14. The lattice constants of GaAs, GaP, InAs and InP are 0.56532, 0.54505, 0.60583 and 0.58697 nm, respectively, at room temperature (parameters for the calculation of the temperature-dependent lattice constants of III–V semiconductors are given, for example, in Vurgaftman *et al.* (2001)).

The lattice constants of Ge, GaAs and AlAs are very similar and, as such, epitaxial GaAs and AlAs layers can also be grown on Ge crystals. The lattice constants of GaP and AlP are also very close to each other, although their band gaps are rather large for photovoltaic solar energy conversion at maximum efficiency. For most combinations of binary III–V

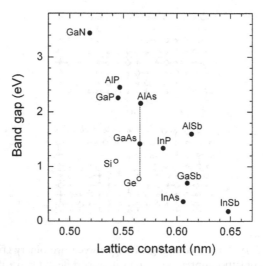

Figure 8.14.   Band gaps of binary III–V semiconductors and Si and Ge plotted against their lattice constants. The vertical line marks epitaxy of GaAs and AlAs on Ge.

semiconductors, the lattice mismatch is too large for epitaxial growth with low density of defects.

Lattice constants can be smoothly changed by alloying for compound crystals. The lattice constant (a_{lc}) of an alloy $A_x B_{1-x}$ is a linear super-position of the composition parameter x and the lattice constants of the compounds $a_{lc}(A)$ and $a_{lc}(B)$ as expressed by the empirical Vegard's law (Vegard, 1921):

$$a_{lc}(A_x B_{1-x}) = x \cdot a_{lc}(A) + (1-x) \cdot a_{lc}(B) \qquad (8.11)$$

For a ternary III–V semiconductor, for example, A and B can correspond to GaAs and InAs. The band gap and the lattice constant of ternary III–V semiconductors can be calculated as a function of the composition parameter by using Equations (8.2) and (8.11). As a result, the band gap can be plotted as a function of the lattice constant for any alloy system of III–V semiconductors. This is shown for the ternary III–V semiconductors $Ga_x In_{1-x}As$, $Ga_x In_{1-x}P$ and $AlAs_x Sb_{1-x}$ and for the quaternary semiconductor $Ga_x In_{1-x}As_{0.7}P_{0.3}$ in Figure 8.15.

Lattice-matched epitaxy is possible along vertical lines of constant lattice constants. GaAs (line (1) in Figure 8.15) and InP (line (2) in Figure 8.15) are important III–V semiconductors for epitaxy. The band

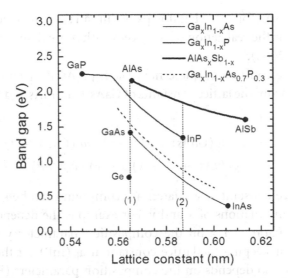

Figure 8.15. Dependence of the band gap on the lattice constant for the ternary III–V semiconductors $Ga_xIn_{1-x}As$, $Ga_xIn_{1-x}P$ and $AlAs_xSb_{1-x}$ (thin, middle and thick solid lines, respectively) and for the quaternary III–V semiconductor $Ga_xIn_{1-x}As_{0.7}P_{0.3}$ (dashed line). The vertical lines mark lattice-matched epitaxy with GaAs (1) and InP (2).

gap between AlAs and GaAs can be smoothly varied by alloying within the $Al_xGa_{1-x}As$ system without changing the lattice constant. The lattice constant of germanium is 0.56575 nm (Dismukes *et al.*, 1964). A perfect lattice matching is obtained for $Ga_{0.99}In_{0.01}As$ and $Ga_{0.51}In_{0.49}P$ epitaxial layers on Ge crystals, which correspond to the combination of band gaps of 1.4 (middle cell), 1.92 (top cell) and 0.7 (bottom cell made from germanium) eV, respectively, for triple-junction solar cells. However, the combination of these band gaps does not correspond to the optimum for reaching maximum energy conversion efficiency (see Chapter 6). A necessary broader variation of the band gaps especially between those of Ge and GaAs is hardly possible at the lattice constant of GaAs.

The ternary III–V semiconductors $Ga_{0.46}In_{0.56}As$ and $AlAs_{0.44}Sb_{0.56}$ have the same lattice constant as InP and band gaps of 0.73 and 1.9 eV, respectively. Therefore a combination of the band gaps of 1.36 (middle cell), 0.73 (bottom cell) and 1.7–1.9 (top cell, variation due to uncertainty in the bowing parameter) eV is possible by lattice-matched epitaxy on InP (line (2) in Figure 8.15).

The high variability of band gaps within a lattice-matched epitaxy is also needed for the realization of solar cells with more than three junctions (Szabo *et al.*, 2008).

The lattice constant of quaternary $Ga_xIn_{1-x}As_yP_{1-y}$ compounds can be calculated from the lattice constants of GaAs, GaP, InAs and InP (Nahory *et al.*, 1978):

$$a_{lc}(x, y) = xy \cdot a_{lc}(GaAs) + x(1 - y) \cdot a_{lc}(GaP) + (1 - x)y$$
$$\cdot a_{lc}(InAs) + (1 - x)(1 - y) \cdot a_{lc}(InP) \qquad (8.12)$$

The lattice constant of a quaternary compound can be kept constant for certain combinations of x and y. For example, the dependence of the composition parameter x on the composition parameter y is shown in Figure 8.16 for keeping the lattice constant at $a_{lc}(InP)$. At the same time the band gap also depends on the composition parameters (Figure 8.16). Therefore, the band gap can be varied smoothly between 0.73 and 1.36 eV along the lattice constant of InP.

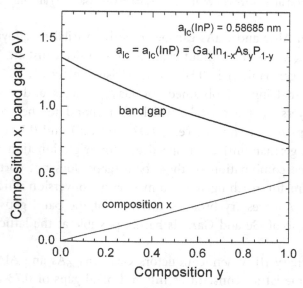

Figure 8.16. Dependence of the metal composition parameter x (thin solid line) and of the band gap (thick solid line) on the composition parameter y for the quaternary III–V semiconductor $Ga_xIn_{1-x}As_yP_{1-y}$ for lattice-matched epitaxy on InP.

The lattice-matched growth of $Ga_xIn_{1-x}As$ with a band gap of about 1.1 eV on InP is impossible but desirable for multi-junction solar cells from a technological point of view. For example, a tandem solar cell with a $Ga_{0.35}In_{0.65}P$ top cell and a $Ga_{0.82}In_{0.18}As$ bottom cell (Dimroth *et al.*, 2000) has a combination of band gaps, i.e. 1.67 and 1.18 eV corresponding to the theoretical optimum (see Chapter 6). However, the lattice constants are quite different for both crystals (0.59853 nm for $Ga_{0.82}In_{0.18}As$ and 0.55972 nm for $Ga_{0.35}In_{0.65}P$). In contrast to lattice-matched epitaxy of III–V semiconductor layers, the combination of lattice-mismatched layers would allow a much higher degree of freedom for optimal choice of band gaps in multi-junction solar cells with extremely high energy conversion efficiency (King *et al.*, 2007).

The lattice mismatch ($\Delta a_{lc}/a_{lc}$) can be calculated by using the following equation:

$$\frac{\Delta a_{lc}}{a_{lc}} = 2 \cdot \left| \frac{a_{lc}(A_xB_{1-x}) - a_{lc}(C_yD_{1-y})}{a_{lc}(A_xB_{1-x}) + a_{lc}(C_yD_{1-y})} \right| \tag{8.13}$$

Additional defects are introduced into epitaxial layers in lattice-mismatched or so-called metamorphic solar cells. It is possible to reduce the density of defects to a certain minimum by introducing, for example, misoriented substrate crystals with composition graded buffers in which stress is reduced (Dimroth *et al.*, 2000; King *et al.*, 2000). Misorientation means that the crystal is cut under a certain angle in relation to the surface at which the epitaxial layer is grown. In this way, additional steps are introduced at the surface of the substrate crystal so that stress or strain can relax in a thin interfacial layer. The relaxation of stress is very important in order to avoid the propagation of extended defects such as threading dislocations into active absorber layers during further crystal growth as well as during operation of the solar cell over its lifecycle (King *et al.*, 2000).

8.3.2 *Molecular beam epitaxy of III–V semiconductors*

Atoms or molecular beams of the third and fifth groups of the periodic table are formed by evaporation through blends from effusion cells into ultra-high vacuum (less than 10^{-8} Pa). The molecular beams are directed onto a substrate crystal kept at a high temperature at which atoms from the molecular beam are incorporated into the lattice of the growing epitaxial layer. A typical growth temperature is, for example, 590°C for epitaxial growth of GaAs (Cho and Arthur, 1975). The evaporation rate is sufficiently low that atoms condensing at the crystal surface can be perfectly incorporated into the growing epitaxial layer before the next atoms arrive. For this reason, the growth rate of molecular beam epitaxy (MBE) is also very low (0.1–1 nm/s).

In ultra-high vacuum, the mean free path of evaporated atoms is much longer than the distance between the effusion cells and the substrate crystal. This means that evaporated atoms do not interact with each other or with rest gas molecules before reaching the substrate surface. In an ultra-high vacuum chamber, the density of rest gas molecules is much lower than the density of evaporated atoms and extremely pure elemental effusion cells can be produced. Therefore, very pure epitaxial layers can be grown by MBE. For example, the background doping of un-doped GaAs layers grown by MBE is less than 10^{14} cm^{-3}, which is two orders of magnitude lower than the maximum density of defects that can be tolerated for the minimum lifetime for Shockley–Read–Hall recombination in GaAs (see Figure 3.7).

Figure 8.17 shows a general set-up for MBE growth of III–V semiconductor layers. Cryogenic traps around the heated sample holder additionally improve the ultra-high vacuum. A mass spectrometer is connected to control the components of the rest gas in the ultra-high vacuum chamber. Effusion cells for evaporation of, for example, Al, Ga, In, P, As, Sb, Be (for p-type doping) and S (for n-type doping) can be closed separately with shutters. The effusion cells, an electron gun and a RHEED (reflection high-energy electron diffraction) detector are all pointed towards the heated sample holder.

For MBE growth, semiconductor crystals are kept in ultra-high vacuum at high temperature. The energy of pure surfaces is minimized by pushing out certain surface atoms from the lattice position of the bulk crystal. In this way, two-dimensional ordered surface structures are formed at very clean semiconductor surfaces in ultra-high vacuum where the geometry of the two-dimensional surface structures depends on the orientation of the crystal and temperature. The formation of two-dimensional ordered

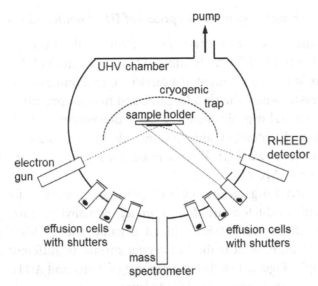

Figure 8.17. Principle set-up for epitaxial layer growth by MBE (UHV — ultra-high vacuum).

surface structures is called surface reconstruction. Surface reconstruction and phase diagrams of surface reconstruction can be investigated by RHEED (see, for example, Däweritz and Hey (1990)).

The epitaxial layer-by-layer growth can be well monitored by analyzing surface reconstruction since the surface becomes rough during the growth of a new mono-layer (reduction of the RHEED signal) and smooth during completion of a growing mono-layer (maximum of the RHEED signal). A very precise control of the growing layer thickness has been reached by connecting computer-controlled shutters of effusion cells with RHEED monitoring of epitaxial layer growth.

Surface chemical bonds and reconstruction influence the optical properties of clean surfaces of III–V semiconductors. The change of optical properties caused by a variation of surface chemical bonds can be monitored very sensitively (see, for example, Samuelson *et al.* (1992)) by measuring the variation of reflected polarized light (the reflection difference — RD) (Aspnes *et al.*, 1988).

Extremely pure epitaxial layers of III–V semiconductors with a defined number of mono-layers and with defined stoichiometry of each mono-layer can be grown by MBE in combination with RHEED and RD.

8.3.3 *Metalorganic vapor phase epitaxy of III–V semiconductors*

Metal organic molecules containing an atom of the third group of the periodic table and hydride molecules containing an atom of the fifth group of the periodic table are diluted in a carrier gas and transported to a heated substrate crystal where the metalorganic and hydride precursor molecules decompose (metal organic chemical vapor deposition — MOCVD). The core atoms of the metal organic and hydride molecules are incorporated into a growing epitaxial III–V semiconductor layer (metal organic vapor phase epitaxy — MOVPE).

Typical metal organic and hydride precursor molecules for MOCVD of III–V semiconductors are, for example, trimethyl gallium (TMG — $Ga(CH_3)_3$) and arsine (AsH_3). Metal organic and hydride precursor molecules decompose near the hot crystal surface by different reactions. As an example, Figure 8.18 shows a scheme of TMG and AsH_3 arriving at the surface of a growing epitaxial GaAs layer.

TMG and AsH_3 start to decompose near the hot surface via reactions such as:

$$AsH_3 \rightarrow\; = AsH + H_2 \tag{8.14}$$

$$Ga(CH_3)_3 + H_2 \rightarrow\; = GaCH_3 + 2\,CH_4 \tag{8.15}$$

Activated radicals such as $=AsH$ and $=GaCH_3$ form bonds with reactive surface sites at the surface of the growing epitaxial layer and

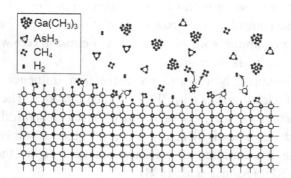

Figure 8.18.   Scheme of trimethyl gallium ($Ga(CH_3)_3$) and arsine (AsH_3) arriving at the surface of a growing GaAs epitaxial layer. Precursor molecules decompose at the hot crystal surface and elements of the third and fifth groups are incorporated into the lattice of the growing epitaxial layer.

continue to decompose. The overall reaction equation of deposition of GaAs by MOCVD is:

$$Ga(CH_3)_3 + AsH_3 \rightarrow GaAs + 3\,CH_4 \tag{8.16}$$

Metal organic precursor molecules such as dimethylzinc (DMZn) are used for doping of III–V semiconductor layers during MOVPE.

A certain amount of carbon, hydrogen and oxygen is incorporated into the growing crystal besides the desired elements of the third and fifth groups of the periodic table. The amount of incorporated carbon, hydrogen and oxygen can be reduced to a minimum by choosing appropriate deposition conditions. The incorporation of carbon can be used for p-type doping of III–V semiconductors. Carbon has a low diffusion coefficient in III–V semiconductors so that highly doped p-type doping is possible with carbon (see, for example, Kuech *et al.* (1988)).

The growth rate of epitaxial III–V semiconductor layers is proportional to the gas flow in the MOCVD reactor. Usual growth rates are of the order of 1–2 nm/s for MOVPE. Gas flows can be well controlled in large-scale production. MOCVD reactors are operated at reduced pressure of the order of 100 mbar.

Hydrides such as AsH_3 and PH_3 are highly toxic gases and metal organics are highly pyrophoric, i.e. they react violently with oxygen and moisture (see, for example, Shenai-Khatkhate *et al.* (2004) and references therein). Therefore, safety measures including gas detection, gas protection and destruction removal of unreacted molecules are very important for the production of III–V semiconductors by MOCVD. Processes of catalytic decomposition, combustion and adsorption are used for the destruction removal of unreacted molecules. This process is called scrubbing.

Figure 8.19 shows a schematic of an MOCVD set-up, including the supply of carrier and hydride gases, the supply of metal organic molecules, the run and venting pipes, the MOCVD reactor with the sample holder, the filters for particles and phosphorus, the processing pump and the scrubber for decomposing unreacted metal organic molecules. The power for substrate heating is coupled into a massive carbon susceptor from a high-frequency power supply (not shown in the figure). Numerous substrates are processed at the same time. Homogeneous deposition is achieved by optimizing the design of the gas flow in the MOCVD reactor and by planetary motion of the substrates during deposition.

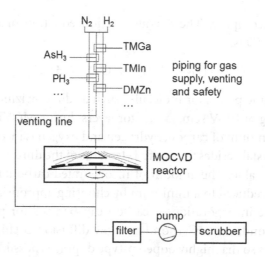

Figure 8.19. Schematic of MOCVD set-up.

8.3.4 *Epitaxial lift-off and exfoliation of III–V semiconductors*

The increase of the diffusion length of photo-generated charge carriers, the reduction of the absorber layer thickness and the realization of a very low surface recombination rate are decisive for reaching very high solar energy conversion efficiencies (see Chapter 4). It is useful to combine GaAs solar cells with back reflectors and/or photonic structures for light trapping in order to reduce the absorber layer thickness (see, for example, Miller *et al.* (2012)). For this purpose, a GaAs solar cell has to be transferred from the substrate crystal to a substrate enabling advanced optical design.

The transfer of GaAs solar cells is possible with the so-called epitaxial lift-off, making use of a sacrificial AlAs layer which is etched away during the lift-off process (Figure 8.20). The etch rate of AlAs in hydrofluoric acid (HF) is several orders of magnitude larger than the etch rate of GaAs in HF (Yablonovich *et al.*, 1987). This property is used to lift-off epitaxial GaAs layers and to recycle the GaAs substrate crystal many times for subsequent epitaxy of GaAs layers (Voncken *et al.*, 2004).

The highest solar energy conversion efficiency of a solar cell with a single band gap has been reached by using epitaxial lift-off of GaAs (Kayes *et al.*, 2011). In the epitaxial lift-off technology of GaAs solar cells, the back contact is deposited onto the complete epitaxial layer system of the GaAs solar cell together with a flexible handle prior to the epitaxial lift-off (Kayes

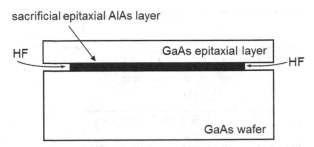

Figure 8.20. Schematic of etching a sacrificial AlAs layer for lifting off a GaAs epitaxial layer.

et al., 2011). Solar cells with relatively large areas of tens of cm^2 can be handled well in this way.

Stacks of many epitaxial sacrificial AlAs layers and epitaxial GaAs solar cells can be grown and etched simultaneously (Yoon *et al.*, 2010). This facilitates ways to assemble flexible lightweight photovoltaic modules with very high solar energy conversion efficiency at moderate costs (Yoon *et al.*, 2010).

Layers of III–V semiconductors can be transferred or exfoliated by combining ion implantation with wafer bonding (see, for example, Singh *et al.* (2006)). First, ions of helium and/or protons are implanted into a wafer of a III–V semiconductor, such as InP (Figure 8.21(a)). A damage layer is formed near the surface during ion implantation.

Second, a handling wafer (crystalline silicon; c-Si) is bonded to the implanted wafer of the III–V semiconductor under pressure (Figure 8.21(b)). An intimate contact is formed between both wafers during wafer bonding. Incidentally, the roughness of wafers should be extremely low for wafer bonding. Voids and blisters are formed during heating and finally the wafer of the III–V semiconductor splits from the handling wafer with the bonded layer of the III–V semiconductor (Figure 8.21(c)). The surface of the bonded layer of the III–V semiconductor has to be polished and etched in the final step (Figure 8.21(d)). Epitaxial layers can be grown on the bonded layer of the III–V semiconductors after removal of a damaged surface layer (see, for example, Zahler *et al.* (2007)). Exfoliation allows epitaxial growth of III–V semiconductors on substrates where lattice mismatch is too large for direct epitaxy.

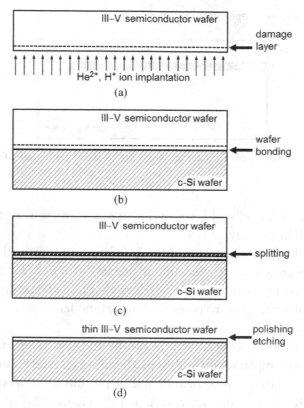

Figure 8.21. Steps of the transfer of a thin III–V semiconductor wafer onto a handling wafer. Crystalline silicon = c-Si.

8.4 Single- and Multi-Junction Solar Cells of III–V Semiconductors

8.4.1 *Design of the pn-junction of solar cells of III–V semiconductors*

III–V semiconductors can be doped n-type and p-type over a wide range. The mobility of free electrons is higher than the mobility of free holes by a factor of 5–10 in III–V semiconductors due to the low effective mass of electrons and the high effective mass of holes. Typical values of electron and hole mobilities are $3000\ \text{cm}^2/\text{Vs}$ and $200\ \text{cm}^2/\text{Vs}$, respectively, for GaAs doped with donor or acceptor densities of about $5 \cdot 10^{17}\ \text{cm}^{-3}$ (Sze 1981). The diffusion coefficients of free electrons and holes can be obtained from the corresponding mobility by using the Einstein equation (Equation (9.13)).

The diffusion length of photo-generated minority charge carriers can be calculated if the diffusion constant and the lifetime are known (Equation (3.5)). The minimum lifetime condition (Equation (3.6)) resulted in a minimum radiative lifetime for GaAs of about 10 ns, which corresponds to a density of majority charge carriers of the order of $5 \cdot 10^{17}$ cm^{-3} (see Chapter 3). The measured diffusion length of electrons in p-type doped GaAs is about 8 and 5 μm for $p_0 = 3 \cdot 10^{17}$ and $8 \cdot 10^{17}$ cm^{-3} (Casey *et al.*, 1973). The measured diffusion length of holes in n-type doped GaAs is about 1.5 μm for $n_0 = 1 \cdot 10^{18}$ cm^{-3} (Casey *et al.*, 1973). The diffusion potential of a pn-junction of a conventional GaAs solar cell ranges between 1.38 and 1.40 v.

The sum of the thicknesses of the n-type and p-type doped regions of a solar cell should be larger by a factor of three than the optical absorption length. More than 95% of incoming photons with energy above the band gap of GaAs is absorbed within a GaAs layer with a thickness of 3 μm (see Chapter 2). Therefore, a GaAs solar cell with high solar energy conversion efficiency usually contains an n-type doped layer with a thickness of about 0.5 μm and a p-type doped layer with a thickness of about 2.5 μm.

The absorption lengths and the diffusion coefficients of electrons and holes are of the same order in III–V semiconductors. Therefore doping and extension of n-type and p-type doped regions are rather similar for solar cells based on different III–V semiconductors.

8.4.2 *Architecture and band diagram of single-junction solar cells*

Figure 8.22 shows an example for the layer structure of a GaAs solar cell with very high solar energy conversion efficiency grown on a highly n-type doped wafer of GaAs(100).

First, a so-called buffer layer with a thickness of the order of 1 μm was grown on top of the GaAs(100) wafer in order to reduce the density of extended defects. The buffer layer is highly n-type doped as well. Second, a highly n-type doped $Al_xGa_{1-x}As$ layer was grown on top of the defect-free buffer layer to form the ohmic contact with the n-type doped GaAs layer and with a barrier for holes photo-generated in the n-type doped GaAs layer. Third, the n-type and p-type doped absorber layers were grown. Fourth, a p-type doped $Al_xGa_{1-x}As$ passivation layer was grown on the p-type doped GaAs layer. Fifth, the p-type doped $Al_xGa_{1-x}As$ buffer was partially opened by a mesa etching and the p-type doped GaAs layer was contacted

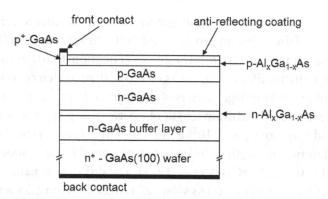

Figure 8.22. Schematic cross section of a GaAs solar cell grown on a highly n-type doped GaAs wafer.

with a highly doped GaAs layer. The $Al_xGa_{1-x}As$ buffer layer was coated with a silicon nitride antireflection coating. Finally, metal contacts were deposited and ohmic contacts were formed by annealing steps between 100 and 300°C.

The idealized band diagram of the solar cell shown in Figure 8.22 is depicted in Figure 8.23 for the cross section across the antireflection coating and the following passivation, absorber and barrier layers and the highly n-type doped GaAs wafer. The band gap of the antireflection coating is much larger than the band gap of the absorber layer. Numerous defects at the interface between the antireflection coating and the passivation layer are not important for the performance of the solar cell since the $p-Al_xGa_{1-x}As$ passivation layer keeps photo-generated electrons and holes away from defects at the interface between the antireflection coating and the passivation layer (see Chapter 3).

Electrons can be transferred from the n-type doped region of the GaAs absorber through the highly n-type doped $Al_xGa_{1-x}As$ barrier layer to the ohmic back contact whereas holes are reflected at the interface between the n-type doped region of the GaAs absorber and the barrier layer. Holes can be transferred from the p-type doped region of the GaAs absorber through the highly p-type doped GaAs layer to the ohmic front contact.

The combination of absorber layers with pn-homo-junctions as charge-selective contacts and hetero-junctions for formation of barrier and passivation layers is applied in all kinds of solar cells based on III–V semiconductors.

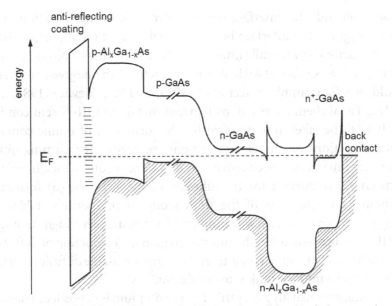

Figure 8.23. Idealized band diagram of a GaAs solar cell in the dark across the anti-reflection coating, the passivation, absorber and barrier layers and the highly n-type doped GaAs wafer with the ohmic back contact.

Solar cells based on GaAs can also be prepared on p-type doped GaAs wafers. In this case, the thickness of the p-type doped GaAs absorber layer is increased to about 3 μm concomitant with a reduction of p_0 to $5 \cdot 10^{16}$ cm^{-3} whereas the thickness of the n-type doped GaAs layer is reduced to about 0.1 μm concomitant with an increase of n_0 to about $1–2 \cdot 10^{18}$ cm^{-3} (see, for example, Lu *et al.* (2011)).

The principle architecture of solar cells based on III–V semiconductors does not change if complete stacks of epitaxial layers for a solar cell are transferred onto a substrate serving as back contact and if the crystal of the III–V semiconductor is removed from the stack of epitaxial layers by etching of a sacrificial AlAs layer (Kayes *et al.*, 2011). In the case of epitaxial growth on exfoliated layers bonded to a silicon wafer, the back contact has to be realized at the front side of the solar cell (Zahler *et al.*, 2007).

8.4.3 Ohmic metal contacts with III–V semiconductors

The formation of ohmic metal contacts with low contact resistance demands a highly doped semiconductor and a low barrier height at the metal–semiconductor interface (see Chapter 5). A strong reduction of the barrier height at the interface between metal and differently doped III–V semiconductors is practically impossible due to Fermi-level pinning during covering a clean surface of a III–V semiconductor with one given metal such as gold (see, for example, Spicer *et al.* (1979) and Walukiewicz (1988)).

Interfacial chemical reactions between metals and III–V semiconductors have to be taken into account for the formation of ohmic contacts. Various functions of interfacial reactions are needed for forming ohmic contacts at interfaces between metals and III–V semiconductors. Interfacial reactions are important for (i) adhesion of metal layers, (ii) formation of vacancies of elements of the third group of the periodic table, (iii) occupying these vacancies with dopants for creating very high doping of the III–V semiconductor in the interface region, (iv) reduction of defects by recrystallization of contact regions and (v) creating stable diffusion barriers of metal atoms into the bulk semiconductor.

In a first step, usually a very thin layer of titanium is deposited. Titanium getters oxygen very efficiently in a high vacuum chamber. Titanium also forms an adhesion layer for following metal layer systems on top of III–V semiconductors.

The formation of ohmic contacts with III–V semiconductors requires the deposition of several metals causing different interfacial reactions. Here the formation of ohmic contacts at n-type doped III–V semiconductors is considered. Layer systems of, for example, Au/Ge and Ni/Ge/Au (Kim and Holloway, 1997) or Pd/Ge/Ti/Pt/Au (Zide *et al.*, 2006) are deposited and the ohmic contacts are formed during annealing steps. Arsenic (As) is partially replaced in the presence of nickel (Ni) or palladium (Pd) due to the formation of stable intermetallic compounds such as NiAs at the interface between the metal and GaAs (see also Figure 8.24).

Ohmic metal contacts can be formed on p-type doped GaAs, for example, with an Au/BeAu/Cr contact system (Algora *et al.*, 2001). Beryllium atoms are acceptors in GaAs crystals. Beryllium atoms are incorporated into the surface region of the GaAs layer during the formation of the ohmic contact. The ohmic contact is formed between the metal layer and a highly p-type doped GaAs(Be) layer at the contact.

The Fermi-energy at metal–semiconductor interfaces is often pinned near the middle of the band gap of the semiconductor. Therefore the barrier

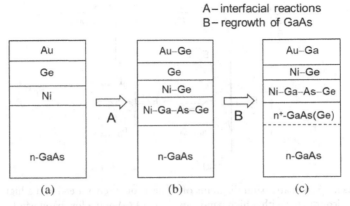

Figure 8.24. Example sequence for the formation of an ohmic metal contact on n-type doped GaAs consisting of the as-deposited Ni/Ge/Au layer system (a), a reacting system leading to the formation of an Au–Ge eutectic, a Ni–Ge alloy and Ni–Ga–As–Ge phases (b) and of the final contact system containing a highly n-type doped GaAs layer regrown by solid-state epitaxy, a layer of Ni–Ga–As–Ge phases, a Ni–Ge layer and an Au–Ga layer (c). After Kim and Holloway (1997).

heights at contacts between metals and III–V semiconductors usually increase with the increasing band gap of the III–V semiconductor. Very thin layers of highly doped III–V semiconductors with low band gaps are grown on III–V semiconductors with high band gaps for forming ohmic metal contacts with very low contact resistance (see, for example, Kim and Holloway (1997)). For contacting GaAs, a highly doped germanium layer can also be grown to form an ohmic contact (Devlin *et al.*, 1980). The contact resistance has been reduced to 10^{-7} Ωcm^2 by implementing a highly doped germanium (Stall *et al.*, 1981), InAs (Kumar *et al.*, 1989) or graded InGaAs (Mehdi *et al.*, 1989) contact layer between the metal contact and the GaAs absorber. This is shown schematically in Figure 8.25 for a p-type doped semiconductor.

The interface at ohmic metal contacts with III–V semiconductors is rough due to interfacial chemical reactions and several phase transitions and due to inter-diffusion of atoms between layers.

The thickness of the complete ohmic contact layer system is of the order of 100–200 nm. After the formation of the ohmic contact, electroplating is applied for increasing the thickness of the contact pad so that external metal wires can be bonded to the ohmic metal contact of the solar cell.

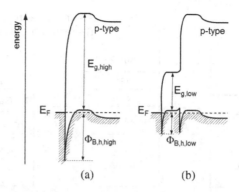

Figure 8.25.   Schematic band diagrams of contacts between a metal and a highly p-type doped semiconductor with a high band gap ($E_{g,high}$) (a) and a low band gap ($E_{g,low}$) (b) for contacting a moderately p-type doped semiconductor with a high band gap. The barrier height for holes has been reduced from $\Phi_{B,h,high}$ to $\Phi_{B,h,low}$ in addition to a reduction of the width of the space charge region. Fermi-level pinning around the middle of the band gap at the metal–semiconductor contact and a type I hetero-junction between the semiconductors with high and low band gaps are assumed.

The contact resistance of ohmic metal contacts formed on GaAs with Au/Ge can be as low as 10^{-5} Ωcm^2 (Kim and Holloway, 1997). This value has to be related to the tolerable series resistance of solar cells based on III–V semiconductors and operated under concentrated sunlight by taking into account a reduced area of the front contact for minimized shading. A contact resistance of about $3 \cdot 10^{-6}$ Ωcm^2 would be required for a GaAs solar cell with open-circuit voltage (V_{OC}) = 1.107 V and I_{SC} = 29.6 mA/cm^2 at AM1.5 (Kayes *et al.*, 2011) with an area of the front contact reduced by 100 times and operated at a maximum concentration factor of 1000. This makes the contact resistance very important for solar cells based on III–V semiconductors operated under highly concentrated sunlight.

8.4.4 *Tandem solar cells with III–V semiconductors*

In comparison to single-junction solar cells, the solar energy conversion efficiency can be increased with tandem solar cells due to the reduction of thermalization losses while the theoretical limit of the efficiency is about 44% for a combination of band gaps of 1.1 and 1.7 eV (see Chapter 6).

GaAs and $Al_xGa_{1-x}As$ have nearly identical lattice constants (Figure 8.13) and, as such, lattice-matched epitaxial layer growth can be used for fabrication of tandem solar cells. The band gap of GaAs is 1.42 eV and the corresponding optimum band gap of the top cell is 1.91 eV with regard to Figure 6.12. A band gap of 1.91 eV is obtained for $Al_{0.36}Ga_{0.64}As$ with regard to Figure 8.3. The maximum solar energy conversion efficiency of a tandem solar cell with band gaps of 1.91 and 1.42 eV for the top and bottom cells, respectively, is about 40% (see Figure 6.15).

Figure 8.26 shows a schematic cross section of a $GaAs/Al_{0.36}Ga_{0.64}As$ tandem solar cell grown by MOVPE on an n-type doped GaAs wafer with a layer system used by Takahashi *et al.* (2005). For reaching a very high efficiency of 28.85% under illumination at air mass (AM)1.5, the authors (i) increased the lifetime of holes in n-type doped $Al_{0.36}Ga_{0.64}As$ layers from about 1.7 ns to 4.1 ns by exchanging Si for Se donors and (ii) reduced the thermal diffusion of acceptors into the highly n-type doped region of the tunneling junction by exchanging Zn for carbon

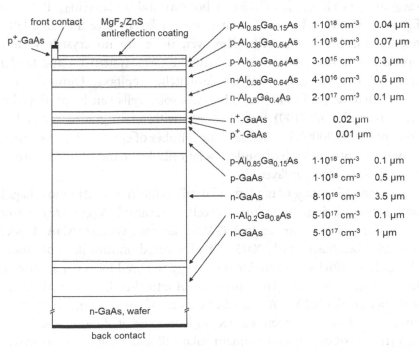

Figure 8.26. Schematic cross section of a $GaAs/Al_{0.36}Ga_{0.64}As$ tandem solar cell grown on an n-type doped GaAs wafer. After Takahashi *et al.* (2005).

acceptors, which have a low diffusion coefficient at increased temperatures (Takahashi *et al.*, 2005). This example demonstrates the importance of a dedicated increase of the lifetime of minority charge carriers in the absorber and the reduction of the tunneling resistance of the recombination contact.

A GaAs/$Al_{0.36}Ga_{0.64}As$ tandem solar cell consists of at least 12 epitaxial layers including the n-type and p-type doped absorber layers, barrier layers, passivation layers and the highly doped layers of the tunneling contact (see also Bedair *et al.* (1979)). Layers with thicknesses between 10 nm and 3.5 μm and a doping range between $3 \cdot 10^{15}$ cm^{-3} and $5 \cdot 10^{19}$ cm^{-3} have been grown. In addition, highly doped contact layers were formed for ohmic contacts. Finally, a MgF_2/ZnS double antireflection coating and ohmic contact layer systems were deposited.

A combination of band gaps close to the maximum of solar energy conversion efficiency can be obtained for tandem solar cells with a $Ga_{0.83}In_{0.17}As$ ($E_{g,bottom} = 1.18$ eV) bottom and a $Ga_{0.35}In_{0.65}P$ top cell ($E_{g,top} = 1.67$ eV). The III–V semiconductor layers have the same lattice constant for both solar cells. However, there are no crystals with an appropriate lattice constant on which the layer system for the tandem solar cell can be grown by lattice-matched epitaxy. Dimroth *et al.* showed that very efficient $Ga_{0.83}In_{0.17}As$ solar cells can be produced on a misoriented GaAs(100) wafer covered with an In-graded buffer layer (Dimroth *et al.*, 2000). Incidentally, the number of epitaxial layers increases for metamorphic growth of tandem solar cells due to the additional growth of the In-graded buffer layer.

A strong advantage of tandem solar cells is the increased photovoltage in comparison to a single-junction solar cell. For example, V_{OC} increased from 1.047 V for a GaAs solar cell to 2.42 V for an $Al_{0.36}Ga_{0.64}As$/GaAs tandem solar cell (Takahashi *et al.*, 2005). The increased photovoltage of tandem solar cells is sufficient to provide the energy required for water splitting by electrolysis, at least 1.23 eV minimum (see, for further discussion, the review by Walter *et al.* (2010)). A record efficiency of solar energy to hydrogen conversion of 18% has been reached with an integrated system including a $Ga_{0.83}In_{0.17}As$/$Ga_{0.35}In_{0.65}P$ tandem solar cell and a cell for electrolysis of water with a proton exchange membrane (Dimroth, 2006).

8.4.5 *Triple-junction concentrator solar cells*

Very high solar energy conversion efficiencies, of more than 40% (King *et al.*, 2007), were reached with triple-junction concentrator solar cells. Conventional triple-junction solar cells consist of a germanium bottom cell, a $Ga_xIn_{1-x}As$ middle cell and a $Ga_xIn_{1-x}P$ top cell (see Figures 8.27(a) and 8.27(b)).The band gap of germanium is about 0.7 eV.

The optimum band gaps of the middle ($E_{g,middle}$) and top ($E_{g,top}$) cells arc 1.15 and 1.73 eV, respectively, for a triple-junction solar cell with a band gap of the bottom cell equal to 0.7 eV (see Figure 6.12). As already

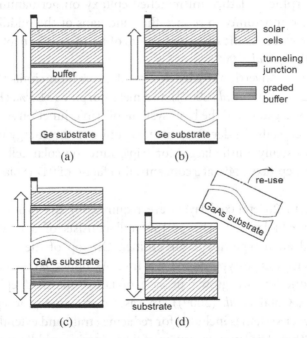

Figure 8.27. Simplified architectures of triple-junction solar cells with very high solar energy conversion efficiency grown on Ge substrates by lattice-matched (a) and metamorphic (b) epitaxy, grown on a GaAs substrate by lattice-matched epitaxy of the middle and top cell at one side and metamorphic epitaxy of the bottom cell at the other side of the GaAs substrate (c) and grown on a GaAs substrate by lattice-matched epitaxy of the top and middle cell and metamorphic epitaxy of the bottom cell followed by transfer on a substrate and removal of the GaAs substrate (d). The middle and top cells are made from $Ga_xIn_{1-x}As$ and $Ga_xIn_{1-x}P$, respectively. The bottom cells are made from Ge ((a) and (b)) and from $Ga_xIn_{1-x}As$ ((c) and (d)). The arrows denote the growth direction.

mentioned, the values of $E_{g,middle}$ and $E_{g,top}$ are 1.4 eV ($Ga_{0.99}In_{0.01}As$) and 1.92 eV ($Ga_{0.51}In_{0.49}P$), respectively, for ideal lattice matching. These values are almost close to the optimum values. Incidentally, the short-circuit current density of the top cell of the lattice-matched triple-junction solar cell limits I_{SC} so that the maximum solar energy conversion efficiency is reduced in comparison to the maximum (see also task T8.6).

A certain lattice mismatch can be tolerated for lattice-matched epitaxy, and so values of $E_{g,middle}$ and $E_{g,top}$ of about 1.38 and 1.86 eV, respectively, have been used (King et al., 2007) (see also task T8.7). By this measure, losses due to limitation of I_{SC} by the top cell were reduced.

Metamorphic or lattice-mismatched epitaxy on germanium crystals provides the opportunity to reduce the band gaps of the middle and top cells to come closer to the optimum values of band gaps in triple-junction solar cells (King et al., 2000).

The energy conversion efficiencies at AM1.5 are quite similar for lattice-matched (32%, (King et al., 2007)) and metamorphic (31.3%, (King et al., 2007)) epitaxial growth of the layer systems of triple-junction solar cells on germanium crystals. Under concentrated sunlight, the energy conversion efficiency is usually a little larger for triple-junction solar cells grown by metamorphic epitaxy (44% at a concentration factor of 947 (Solar Junction, 2012)).

More than 20 epitaxial layers are required for the production of a triple-junction solar cell. The bottom cell consists of a p-type doped Ge base and an n-type doped Ge emitter. On top of the Ge emitter a nucleation layer, an n-type doped $Ga_xIn_{1-x}As$ buffer layer and the first tunneling junction are grown in case of lattice-matched epitaxy (see, for example, Cotal et al. (2009)). In case of metamorphic epitaxy, an additional layer system is included for reducing strain and extended defects (Figure 8.27(b)). The layer structures of the following middle and top solar cells are similar for lattice-matched and metamorphic epitaxy and include layers for absorption, barrier layers, the second tunneling junction and passivation layers (see, for example, Cotal et al. (2009)).

A bottom cell with a band gap of 1.0 eV has an optimum combination with band gaps of a top cell and middle cell of 1.9 ($Ga_{0.5}In_{0.5}P$) and 1.4 eV ($Ga_{0.99}In_{0.01}As$), respectively (Figure 6.12). The maximum efficiency in the limit of radiative recombination is about 50% for a triple-junction solar cell with $E_{b,bottom} = 1.0$, $E_{g,middle} = 1.4$ and $E_{g,top} = 1.9\,\mathrm{eV}$ (Figure 6.15). For this combination of band gaps, the top cell and the middle cell can be grown on a GaAs crystal or on a germanium crystal by lattice-matched epitaxy. A band gap of $1.0\,\mathrm{eV}$ can be realized with a layer of $Ga_{0.68}In_{0.32}As$ which has a high lattice mismatch with GaAs of about 2.3%.

A layer of $Ga_{0.68}In_{0.32}As$ can be grown on a GaAs substrate or a $Ga_{0.99}In_{0.01}As$ layer by metamorphic epitaxy with a system of graded buffer layers for reducing strain and for reducing the density of extended defects (see, for example, Cotal *et al.* (2009)). The transparency of a thin GaAs substrate is very high for photons with energy below the band gap of GaAs and so the GaAs substrate can be used for lattice-matched epitaxial growth of the $Ga_{0.99}In_{0.01}As$ middle and of the $Ga_{0.5}In_{0.5}P$ top cell at one side and of the $Ga_{0.68}In_{0.32}As$ bottom cell at the other side of the GaAs substrate (Figure 8.27(c)).

It is possible to start the epitaxial growth with the $Ga_{0.5}In_{0.5}P$ top cell and the $Ga_{0.99}In_{0.01}As$ middle cell by lattice-matched epitaxy on a GaAs substrate and to finish the layer growth by metamorphic epitaxy of the $Ga_{0.68}In_{0.32}As$ bottom cell (inverted growth). In this case, the solar cell has to be transferred to a mechanically stable substrate and the GaAs substrate has to be removed, for example, by introducing a sacrificial AlAs layer at the GaAs substrate for epitaxial lift-off so that the GaAs substrate can be re-used for epitaxial growth (Figure 8.27(d)).

Values of V_{OC} larger than $3\,\mathrm{V}$ and a solar energy conversion efficiency as high as 37.7% have been be reached for a triple-junction solar cell with $E_{b,bottom} = 1.0$, $E_{g,middle} = 1.4$ and $E_{g,top} = 1.9\,\mathrm{eV}$ at AM1.5 (Green *et al.*, 2013).

8.4.6 *Quadruple-junction solar cells*

The further increase of the number of solar cells with different band gaps allows a further increase of the solar energy conversion efficiency of multi-junction solar cells. With regards to Figures 6.15 and 6.12, the maximum efficiency in the limit of radiative recombination is about 54% for a quadruple-junction solar cell with band gaps of about 0.7, 1.0, 1.4 and 1.9 eV, respectively, illuminated at AM1.5.

Taking into account Figure 8.15, the layer structure of a quadruple-junction solar cell can probably be grown by lattice-matched epitaxy on an

InP substrate. For example, a $Ga_{0.2}In_{0.8}As_{0.43}P_{0.57}$ solar cell with a band gap of about 1.03 eV followed by a $Ga_{0.43}In_{0.57}As$ bottom cell with a band gap of 0.73 eV can be grown on one side of an InP substrate (Szabo *et al.*, 2008), where an InP solar cell and an $AlAs_xSb_{1-x}$ top cell with a band gap of 1.86 eV can be realized on the other side of the same InP substrate (Figure 8.28(a)).

A GaAs solar cell with a band gap of 1.42 eV and a top $Ga_{0.5}In_{0.5}P$ solar cell with a band gap of 1.9 eV can be realized by lattice-matched epitaxial growth on a GaAs substrate. The solar cells grown on the InP substrate and the solar cells grown on the GaAs substrate can be connected with each other by wafer bonding (Figure 8.28(b)). However, extremely smooth and clean surfaces and a relatively high pressure are required for wafer bonding.

To date, the absolutely highest solar energy conversion efficiency has been achieved with a quadruple-junction solar cell (44.7% under concentrated sunlight; see Dimroth *et al.* (2014)).

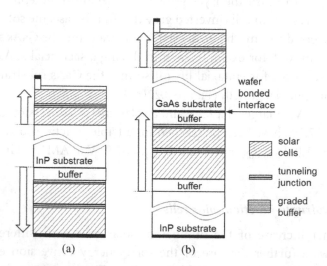

Figure 8.28. Simplified architectures of quadruple-junction solar cells grown by lattice-matched epitaxy consisting of an InP solar cell and an $Al_xAs_{1-x}Sb$ top cell on one side, and of a $Ga_xIn_{1-x}As_yP_{1-y}$ solar cell and an $In_xGa_{1-x}As$ bottom cell on the other side of the InP substrate (a) and consisting of an GaAs solar cell and a $Ga_xIn_{1-x}P$ top cell on a GaAs substrate, and of a $Ga_xIn_{1-x}As_yP_{1-y}$ solar cell and an $In_xGa_{1-x}As$ bottom cell on an InP substrate (b). The InP and GaAs substrates are connected by wafer bonding. The arrows denote the growth direction.

8.5 Summary

Binary, ternary and quaternary III–V semiconductors are practically ideal photovoltaic absorbers. Their band gaps can be varied over the whole range of the sun spectrum, and their diffusion lengths of minority charge carriers can be much longer than the optical absorption length. Additionally, III–V semiconductors can be doped over a wide range so that pn-junctions for practically ideal charge-selective contacts with very low densities of defects at the contact can be realized and ohmic contacts can be formed with metal contacts. Hetero-junctions that are free from interface defects can be formed between layers of different III–V semiconductors by epitaxial growth due to engineering of lattice constants of ternary and quaternary III–V semiconductors. This allows for the implementation of (i) passivation layers without introducing remarkable surface recombination at hetero-junctions and (ii) barrier layers for minimizing the density of minority charge carriers at ohmic contacts. Therefore, single-junction solar cells based on III–V semiconductors include n-type and p-type doped absorber layers of the same band gap, barrier and passivation layers made from III–V semiconductors with a higher band gap than the absorber layer and metal layer systems for forming ohmic contacts to the two external metal leads. The absolute highest solar energy conversion efficiency of a single-junction solar cell illuminated at AM1.5 was reached for a GaAs absorber with $Al_xGa_{1-x}As$ barrier and passivation layers. A solar energy conversion efficiency larger than 28% has been reached with a GaAs solar cell illuminated at AM1.5 (Green *et al.*, 2013).

Layers of III–V semiconductors are transparent for photon energies below the band gap of the given III–V semiconductor. This means that solar cells based on III–V semiconductors with different band gaps can be stacked on top of each other without additional optical losses. Band gaps of absorber layers can be tuned for reaching the optimum combination for the maximum efficiency of multi-junction solar cells. At the same time, lattice constants of III–V semiconductors can be engineered in such a way that sophisticated layer systems including different absorber layers, barrier layers, passivation layers and contact layers can be grown with a minimum of defects in the different absorbers and at the numerous interfaces. The

Table 8.1. Short-circuit current density (I_{SC}), open-circuit voltage (V_{OC}), fill factor (FF) and solar energy conversion efficiency (η) for different single-junction and multi-junction solar cells based on III–V semiconductors for given concentration factors of sunlight (X). References are given.

	X (suns)	I_{SC} (mA/cm^2)	V_{OC} (V)	FF (%)	η (%)	Reference
GaAs/wafer	1	29.1	1.04	86	26	Lu *et al.* (2011)
GaAs transferred	1	29.68	1.122	86.5	28.8	Green *et al.* (2013)
Ga$_{0.44}$In$_{0.56}$P (1.875 eV)	1	13.5	1.37	88	16.4	Lu *et al.* (2011)
GaAs/AlGaAs	1	13.44	2.42	88.7	28.85	Takahashi *et al.* (2005)
GaInP/GaAs/Ge	1	14.7	2.69	86	34.1	Bett *et al.* (2009)
GaInP/GaAs/GaInAs	1	14.57	3.014	86	37.7	Green *et al.* (2013)
GaInP/GaInAs/Ge	240	3830	2.911	87.5	40.7	King *et al.* (2007)
GaInP/GaAs/GaInAs	302	—	—	—	44.4	Sharp (2013)
GaInP/GaInAs/Ge	454	7475	2.867	87.2	41.1	Bett *et al.* (2009)
GaInP/GaAs/ GaInNAs	947	—	—	—	44	Solar Junction (2012)
GaAs/wafer	2509	67340	1.184	75	23.1	Algora *et al.* (2001)
quadruple	297	192.1	4.165	86.5	44.7	Dimroth (2014)

properties of the tuning of band gaps, of ideal transparency for photon energies below the band gap and of the realization of tunneling contacts between highly doped layers of III–V semiconductors are additional properties, which are decisive for the realization of very efficient stacked multi-junction solar cells. A solar energy conversion efficiency larger than 37% has been reached with an Ga$_x$In$_{1-x}$P/GaAs/Ga$_x$In$_{1-x}$As triple-junction solar cell illuminated at AM1.5 (Green *et al.*, 2013). Table 8.1 compares basic characteristics of different single-junction and multi-junction solar cells.

The solar energy conversion efficiency of solar cells can be increased under illumination with highly concentrated sunlight. However, illumination of solar cells with highly concentrated sunlight demands very low series resistances for minimizing resistive losses. The resistance of layers of III–V semiconductors is very low and can be neglected under illumination with highly concentrated sunlight. A very low resistance of the tunneling contact between metals and III–V semiconductors can be realized with appropriate systems of metal layers taking into account interfacial reactions and epitaxial regrowth after contact formation. Tunneling resistances between highly n-type and p-type doped layers of III–V semiconductors usually limit the series resistance of multi-junction solar cells under illumination with highly concentrated sunlight. Inter-diffusion of dopants at highly doped pn-hetero-junctions leads to a partial compensation of doping and therefore to an increase of the width of the corresponding space charge regions. As a consequence, the tunneling resistance increases. Inter-diffusion of dopants at highly doped pn-hetero-junctions has to be minimized by choosing dopants with low diffusion coefficients and by reducing the thermal impact during the subsequent epitaxy and other processing steps.

A solar energy conversion efficiency as high as 44.4% has been reached with an $Ga_xIn_{1-x}P/GaAs/Ga_xIn_{1-x}As$ triple-junction solar cell illuminated with highly concentrated sunlight with a concentration factor of 302 (Sharp, 2013). To date, the absolutely highest solar energy conversion efficiency has been achieved with a quadruple-junction solar cell with an efficiency of 44.7% at a concentration factor of 297 (Fraunhofer, 2013).

8.6 Tasks

T8.1: Type I and type II semiconductor hetero-junction

Establish whether the natural GaP/InP, InN/AlAs and GaAs/InAs interfaces form type I or type II semiconductor hetero-junctions.

T8.2: Influence of an interface dipole on the band offset

Estimate the change of the potential energy at a semiconductor hetero-junction with a polarization charge of 10% of the elementary charge per unit cell of the lattice. Discuss the consequences for the natural valence and conduction band offsets at the GaAs/AlAs and GaAs/InAs interfaces.

T8.3: Tuning of band gaps in lattice-matched epitaxy

Calculate the minimum and maximum band gaps that can be realized with lattice-matched epitaxy of $Ga_xIn_{1-x}As_yP_{1-y}$ at a lattice constant of 0.580 nm.

T8.4: Absorber resistance of solar cells based on III–V semiconductors

Estimate the resistance of a GaAs absorber in a GaAs solar cell and discuss the role of the resistance of the absorber in solar cells based on III–V semiconductors operated under highly concentrated sunlight.

T8.5: Epitaxial layers in tandem solar cells with III–V semiconductors

Explain the function of each epitaxial layer in a GaAs/$Al_{0.36}Ga_{0.64}As$ tandem solar cell as shown in Figure 8.24.

T8.6: Maximum efficiency of a lattice-matched triple-junction solar cell

Ascertain the maximum efficiency within the limit of radiative recombination of a lattice-matched triple-junction solar cell with $E_{g,bottom} = 0.7$ eV, $E_{g,middle} = 1.38$ eV and $E_{g,top} = 1.86$ eV illuminated at AM1.5. Use the nomograms given in Chapter 6. Compare the result with the maximum efficiency of a triple-junction solar cell with optimum band gaps.

T8.7: Lattice mismatch for lattice-matched epitaxy

Ascertain the lattice mismatch for lattice-matched epitaxy realized for a triple-junction solar cell on a Ge substrate with an $Ga_xIn_{1-x}As$ middle cell and an $Ga_xIn_{1-x}P$ top cell with band gaps of 1.38 and 1.86 eV, respectively.

T8.8: Lattice mismatch for lattice-mismatched epitaxy

Ascertain the lattice mismatch for the growth of a triple-junction solar cell with $Ga_xIn_{1-x}As$ middle ($E_g = 1.28$ eV) and $Ga_xIn_{1-x}P$ top ($E_g = 1.80$ eV) cells on c-Ge ($a_{lc} = 0.56575$ nm) substrate. Compare with the lattice mismatch for an inverted metamorphic triple-junction solar cell including a $Ga_xIn_{1-x}As$ bottom cell with $E_g = 1.0$ eV grown on GaAs.

T8.9: Type II hetero-junctions for charge separation

Is it useful to apply type II hetero-junctions instead of pn-homo-junctions as charge-selective contacts in multi-junction solar cells?

T8.10: *Finger grid of concentrator solar cells*

Estimate the distance between grid fingers at the front contact of a triple-junction solar cell operated at concentrated sunlight with a concentration factor of 2000.

9

Thin-Film Solar Cells

Thin-film solar cells and photovoltaic (PV) thin-film modules are based on inorganic PV absorber layers, which have high absorption coefficients and can be deposited at a high quality on foreign substrates at reduced temperatures. A thin-film absorber layer has a high quality if the lifetime of photo-generated charge carriers is longer than the transport time to the charge-selective contact. The concept of thin-film photovoltaics and two examples of typical thin-film deposition techniques are given at the beginning of this chapter. Transparent conducting oxides (TCO) are key for thin-film photovoltaics. Resistive and optical losses caused by TCO layers in thin-film solar cells are illuminated. The main part of this chapter is devoted to principle properties of absorber layers and charge-selective contacts for the different absorber classes of thin-film solar cells. The three classes of thin-film solar cells are distinguished with respect to the formation of the absorber layer by (i) direct deposition (amorphous and micro-crystalline hydrogenated silicon: a-Si:H, μc-Si:H), (ii) phase transformation by selenization or sulfurization of metal precursors (chalcopyrites: $CuIn_{1-x}Ga_x(Se_{1-y}S_y)_2$, kesterites: $Cu_2ZnSn(Se_{1-x}S_x)_4$) and (iii) congruent sublimation and post-treatment (cadmium telluride: CdTe). Chalcopyrite and CdTe thin-film solar cells have the potential for very high energy conversion efficiencies. Amorphous silicon and kesterite absorbers are produced from earth abundant elements.

9.1 Concept of Thin-Film Photovoltaics

9.1.1 *About the architecture of thin-film solar cells and modules*

Thin-film absorbers have a thickness of the order of 0.5–5 μm. Therefore, the absorption length of thin-film absorbers should be of the order of

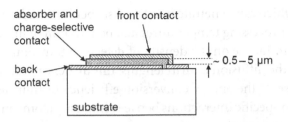

Figure 9.1. Principle cross section of a thin-film solar cell.

0.1–2 μm. Thin-film absorber layers are deposited directly onto a foreign substrate, such as glass sheets, metal foils or polyimide foils. Properties of foreign substrates can be adjusted according to application demands including costs, weight and flexibility of solar cells and PV modules.

Figure 9.1 shows a principle cross section of a thin-film solar cell. The back-contact layer is deposited onto the foreign substrate and the thin-film absorber and the charge-selective contact layers are deposited onto the back-contact layer. Charge-selective contact layers at thin-film absorbers have a thickness of only several tens of nm. Therefore, the front-contact layer should completely cover the absorber and charge-selective contact layers. Depending on whether the thin-film solar cell is illuminated through the front contact (strate configuration) or through the back contact (super-strate configuration), the front- or back-contact layers, respectively, have to be transparent in the absorption range of the absorber. Transparent contact layers are realized with transparent conducting oxides (TCOs).

Thin-film absorbers can be based on hydrogenated amorphous and micro-crystalline silicon (a-Si:H, μc-Si:H), chalcopyrites ($CuIn_{1-x}Ga_x$ $(Se_{1-y}S_y)_2$), kesterites ($Cu_2ZnSn(Se_{1-x}S_x)_4$) or cadmium telluride (CdTe). These absorbers have high absorption coefficients and can be deposited with sufficient diffusion lengths or drift lengths on appropriate substrates coated with back-contact layers. Charge-selective contact layers are called buffer layers in thin-film solar cells with chalcopyrite, kesterite or cadmium telluride absorbers.

In thin-film technology, the maximum processing temperature is often limited by the foreign substrate, for example, by the softening temperature of glass. In this way, a foreign substrate can limit the formation temperature of absorber layers, contacts and contact layers. Furthermore, the foreign substrate and the back-contact layer are sources for uncontrolled impurities, which can penetrate into the absorber layer and contact regions. Both limited processing temperatures and penetration of impurities have a tremendous influence on the density of defects in thin-film absorbers and therefore on the diffusion or drift lengths. Impurities can cause a reduction or an increase of the energy conversion efficiency of thin-film solar cells depending on specific interactions between impurity atoms and host atoms in absorber layers and/or contact regions.

Thin-film deposition technologies are optimized for a homogeneous covering of substrates with thin-metal, absorber and TCO layers on areas as large as several m^2. This enabled a lateral integrated series connection of thin-film solar cells in a monolithic process of a PV thin-film module. For this purpose, the large area of the deposited back contact is separated into numerous back-contact stripes determining the areas of the separate solar cells. The separation of the back-contact stripes is called the P1 cut (Figure 9.2(a)). The P1 cut is realized with a focused laser beam (laser scribing). The absorber layer and charge-selective contact layers are deposited in the subsequent step. The absorber and charge-selective contact layers are separated with a narrow overlap between two neighboring solar cells during the P2 cut (Figure 9.2(b)). After the P2 cut, the front-contact layer is deposited. The front-contact layer is separated into front-contact stripes during the P3 cut in such a way that the front contact of each solar cell is connected with the back contact of the following solar cell (Figure 9.2(c)). Depending on the absorber layer, the P2 and P3 cuts can be performed by laser or mechanical scribing. Incidentally, the area of the P1, P2 and P3 cuts and the area between the P1 and P3 cuts are lost for photocurrent generation. Conventional P1, P2 and P3 cuts have a width of the order of 100 μm.

The values of the photovoltage of thin-film solar cells are added together for integrated series connection in a PV thin-film module (Figure 9.2(d)). The solar cell with the lowest photocurrent limits the photocurrent of the whole PV thin-film module. Therefore, the current-matching condition demands identical photocurrents for all solar cells in a PV thin-film module in the ideal case. As a consequence, all solar cells in a PV thin-film module must be very homogeneous.

The current-matching condition requires thin-film deposition techniques capable of producing very homogeneous absorber and TCO layers with respect to photo-generation, optical losses and resistive losses. The homogeneity of optical losses, for example, is limited by the thickness of

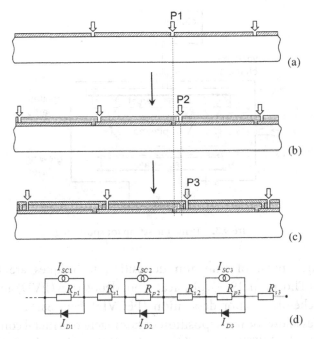

Figure 9.2. Separation of back contacts (a), absorber layers (b) and front contacts (c) for lateral integrated series connection of thin-film solar cells. The arrows mark the corresponding P1 (a), P2 (b) and P3 (c) cuts, and the dashed lines mark the alignment between them. The equivalent circuit of thin-film solar cells connected in series is given in (d).

the TCO layer, which should be controlled with an accuracy of only a few nanometers. Resistive losses are mainly caused by the resistance of the TCO layers and local shunts.

Thin-film technologies are scalable so that the area of the thin-film PV modules can easily be varied from several cm^2 and less (low-power mobile applications) to several m^2 (high-power energy production). Furthermore, PV thin-film modules can be combined with additional functionalities given by the choice of the substrate. This broadens degrees of freedom for applications of thin-film solar cells, for example, in consumer goods and architecture.

9.1.2 *Sputtering and plasma-enhanced chemical vapor deposition*

The concept of thin-film photovoltaics is closely related to the development of technological tools for thin-film deposition. In thin-film

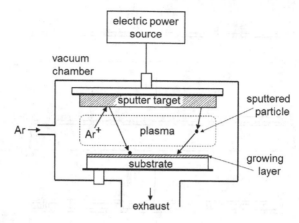

Figure 9.3.　Principle set-up for sputtering.

photovoltaics, most of thin-film deposition techniques are related to sputtering (Thornton, 1974), physical vapor deposition (PVD) and plasma-enhanced chemical vapor deposition (PECVD) (van Sark, 2002). These technologies allow for the deposition of homogeneous metal contact layers (PVD, sputtering), TCO layers (sputtering (Minami *et al.*, 1985)) and/or absorber layers (sputtering, PECVD) on large areas, such as PECVD of a-Si:H (for example, Otero *et al.* (2012)).

　　Figure 9.3 shows a principle set-up for sputtering based on a vacuum chamber, an electric power supply, a sputter target and a substrate. Energetic ions are used for the transfer of a material from a sputter target onto a substrate at well-controlled low substrate temperature of the order of 100–400°C. Argon atoms are ionized for sputtering. Trajectories of ions and electrons can be enhanced in a magnetron so that ionization of argon is very efficient. Argon ions are accelerated in a high electrical field towards the sputter target where the argon ions kick out or sputter target atoms. Sputtered target atoms have a high kinetic energy and move towards the substrate where they form a growing layer. Sputtered target atoms passing the plasma can be ionized and undergo numerous chemical reactions before reaching the substrate.

　　Constant or variable electric fields are applied in DC (direct current) or RF (radio frequency, 13.56 MHz) sputtering from conducting or insulating sputter targets, respectively. Sputter targets used for the production of thin-film solar cells consist of (i) metals such as molybdenum, copper or indium,

(ii) metal oxides such as ZnO or (iii) sulfides such as ZnS. The rates of sputtering are specific for each element in mixed or multi-component sputter targets. Oxygen as a reactive species is added to the argon gas in reactive ion sputtering for the deposition of TCO layers such as ZnO:Al. Reactive selenium or sulfur species can also be added to the argon gas for depositing sulfide or selenide layers. For example, $CuInSe_2$ has been sputtered by Thornton *et al.* in an Ar/H_2Se working gas (Thornton *et al.*, 1988).

The stoichiometry of sputter targets can be varied by the alloying of metals or by combining different sputter targets. It should be pointed out that deposition by sputtering is far from chemical equilibrium. Therefore, practically any stoichiometry can be adjusted in a layer deposited by sputtering. This is practically impossible by other deposition techniques working close to chemical equilibrium.

Layers deposited by sputtering are multi-crystalline. The deposition rate depends mainly on the argon pressure and on the power of the electric power source. Deposition rates are of the order of hundreds of nm per minute for sputtering.

A PVD process is performed in a high vacuum chamber with a base pressure below 10^{-4} mbar or less. Elements such as metals, sulfur or selenium are evaporated from heated evaporation sources. The stoichiometry and deposition sequences are controlled by opening or closing shutters covering the evaporation sources. Deposition rates are of the order of nanometers per second for PVD processes. Sulfide- and/or selenide-based chalcopyrite or kesterite absorber layers can be deposited by PVD, such as $CuInS_2$ by a co-evaporation process (Scheer *et al.*, 1993). Metal layers deposited by PVD or sputtering can be transformed by post-treatment steps into chalcopyrite absorber layers, such as by the sulfurization of Cu/In layers (Klenk *et al.*, 2005).

PECVD is used for the deposition of a-Si:H and μc-Si:H thin-film absorbers. A PECVD set-up contains a gas supply system besides the electric power source and the high vacuum chamber with the substrate (Figure 9.4.). The gas supply system is connected with several gas sources such as argon as a carrier gas and silane as the main precursor gas. The gas flow is controlled with a flowmeter. A precise temperature control is important for the quality of growing a-Si:H or μc-Si:H layers.

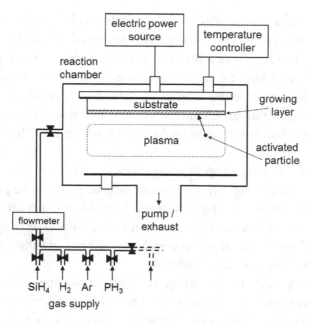

Figure 9.4. Schematic of a PECVD set-up.

Precursor molecules are activated in the plasma by electron impact excitation, dissociation and ionization and undergo different chemical reactions in the plasma and at the surface of a substrate (Perrin, 1995). The chemical surface reactions are related to bonding of $-SiH_3$ at reactive surface sites and to the transformation of $\equiv Si\text{-}SiH_3$ surface species to $=Si=SiH_2$, $-Si\equiv SiH$ and $\equiv Si\text{-}Si\equiv$ bond configurations under the release of hydrogen.

Substitutional doping of absorber layers based on a-Si:H can be well controlled in PECVD processes by adding gases such as phosphine or diborane for n-type or p-type doping, respectively (Spear and Le Comber, 1975).

Deposition rates can be controlled in PECVD processes by modifying gas flow rates and substrate temperatures, by changing the geometry of PECVD reactors, by diluting precursor gases with argon and hydrogen and by the power of the RF power source. Layers of a-Si:H, layers of mixed phases of a-Si:H and very small silicon crystallites (μc-Si:H) and layers of silicon micro-crystals can be formed by PECVD depending mainly on the silane dilution in hydrogen (Vetterl *et al.*, 2000). The deposition rate of a-Si:H absorber layers by PECVD is of the order of 1–10 nm/min, which is very slow in comparison to sputtering.

9.2 Transparent Conducting Oxides

9.2.1 *Doping and electron mobility in transparent conducting oxides*

Optical and electrical properties of TCOs are very important for the energy conversion efficiency of thin-film solar cells. Some metal oxides have large forbidden band gaps so that most of sunlight can be transmitted. For example, the band gaps of ZnO, SnO_2 and In_2O_3 are 3.2 (Thomas, 1960), 3.6 (Summitt *et al.*, 1964) and 2.98 (Haines and Bube, 1978) eV, respectively. Un-doped transparent metal oxides are insulators. The electric conductivity of transparent metal oxides can be increased by many orders of magnitude by intrinsic and extrinsic doping. Intrinsic doping is caused by deviations in the stoichiometry. An excess of oxygen causes p-type doping and a deficiency of oxygen leads to n-type doping (see, for example, Kim *et al.* (1986)). Defined intrinsic doping demands a well-controlled partial pressure of oxygen over several orders of magnitude during processing at high temperature. Therefore, extrinsic doping is much more suitable for practical reasons.

Knowledge about the density and mobility of free electrons (n_0, μ_n) in TCOs is important for minimizing resistive losses in thin-film solar cells. Free carrier absorption also causes optical losses in TCO layers depending on the doping.

Metal atoms are oxidized and oxygen atoms are reduced (oxidation state O^{2-}) in metal oxides. Host metal atoms can be replaced by metal atoms in a higher oxidation state. For example, zinc is in the oxidation state Zn^{2+} in ZnO and zinc atoms can be replaced by aluminum atoms, which are in the oxidation state Al^{3+} (Minami *et al.*, 1985). The additional positive charge at the metal side is compensated by a free electron in the conduction band of ZnO. Aluminum-doped ZnO:Al is the most important TCO for thin-film photovoltaics.

In SnO_2, oxygen atoms can be replaced by fluorine atoms, which are in the oxidation state F^- (Bhardwaj *et al.*, 1981). The missing negative charge is compensated by a free electron in the conduction band of SnO_2:F (also called FTO — fluorine-doped tin oxide). FTO is a standard glass in architecture for reflecting infrared light but transmitting visible light. It can be also applied in solar cells. Incidentally, FTO can be produced, for example, by spray pyrolysis on glass sheets (see, for example, Zhang *et al.* (2011)).

In In_2O_3:Sn (also called ITO — indium tin oxide) a part of indium atoms, which are in the oxidation state In^{3+} are replaced by tin atoms, which are in the oxidation state Sn^{4+} (Haines and Bube, 1978). The additional positive charge at the metal side is compensated by a free electron in the conduction band of ITO, which is the standard TCO for thin-film displays.

The density of free electrons can be as high as 10^{21} cm^{-3} in TCO layers. The electron mobility can be as high as 130 cm^2 /(Vs) in moderately doped ZnO crystals; see, for example, Wagner and Helbig (1974). The electron mobility decreases with increasing density of free electrons in highly doped TCOs or semiconductors due to scattering at ionized impurities. For example, the electron mobility is up to 22 cm^2 /(Vs) in highly doped sputtered ZnO:Al layers (Minami *et al.*, 1985).

Sputtered TCO layers are polycrystalline. Therefore, the electron mobility is limited not only by transport processes within one grain (intra-grain transport) but also by charge transfer across grain boundaries (inter-grain electron transfer), as depicted in Figure 9.5.

A part of the free electrons from the bulk of a TCO crystallite can be trapped at interface states in grain boundaries. The negative charge of electrons trapped at grain boundaries leads to an increase of the potential energy of free electrons at the conduction band edge and therefore to the formation of a barrier between two neighboring grains. The barrier

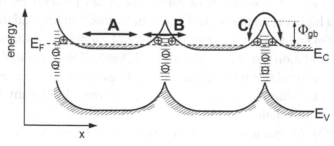

Figure 9.5. Band diagram across charged grain boundaries in a TCO layer. A, B and C denote intra-grain transport, inter-grain electron transfer by tunneling and inter-grain electron transfer by thermionic emission, respectively.

height depends on the density of interface states at grain boundaries (of the order of 10^{12}–10^{13} cm^{-3}). The negative charge of electrons trapped at grain boundaries is compensated by positive charge in the space charge region.

Electrons transferred from one to the other grain have to overcome the barrier at the grain boundary by thermionic emission or by tunneling (see Chapter 5). Therefore, the electron mobility increases with increasing grain size in TCO layers. Tunneling becomes important for a density of free electrons larger than 10^{19} cm^{-3}. The electron mobility usually increases with increasing tunneling rate for densities of free electrons up to about 10^{20} cm^{-2} and decreases for higher densities due to the increase of scattering at ionized impurities in the bulk of a grain. Typical values of the electron mobility in a TCO layer are of the order of 10–40 cm^2/(Vs) for n_0 between 10^{21} and 10^{20} cm^{-3} (Minami *et al.*, 1985).

9.2.2 *Resistance of transparent conducting oxide layers*

The specific resistance of an electron conductor is given by:

$$\rho = \frac{1}{q \cdot n_0 \cdot \mu_n} \tag{9.1}$$

The electron mobility in conventional TCO layers is about 5–40 cm^2/(Vs) for n_0 between 10^{21} and 10^{20} cm^{-3}. Therefore, the specific resistance of a sputtered polycrystalline TCO layer is of the order of 10^{-3} Ωcm, about three orders of magnitude larger than the specific resistance of copper.

The resistance of the TCO layer gives the major contribution to the series resistance of thin-film solar cells. The height (H, of the order of 0.5–1 μm) and the width (Figure 9.6) of the TCO layer (L) have to be adjusted in accordance with the tolerable series resistance of a thin-film solar cell (see Chapter 1). Typically, thin-film solar cells in thin-film PV modules have the shape of a long stripe (length B of the order of 1 m) with a width of about 0.5–2 cm. The resistance of a TCO stripe can be calculated by the following equation:

$$R_{TCO} = \frac{1}{q \cdot n_0 \cdot \mu_n} \cdot \frac{L}{B \cdot H} \tag{9.2}$$

The short-circuit current of a thin-film solar cell is equal to the short-circuit current density (I_{SC}) multiplied by the area of the solar cell which

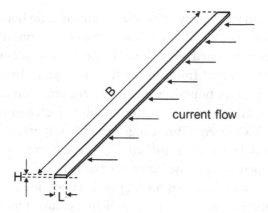

Figure 9.6. TCO stripe of a thin-film solar cell.

is practically equal to the area of the TCO stripe. The ratio between the squared width and the thickness of the TCO stripe can be estimated by the following expression if taking into account Equation (9.2) and the condition for the tolerable series resistance (see Chapter 1):

$$R_{TCO,tol} \leq \frac{1}{100} \cdot \frac{V_{OC}}{I_{SC} \cdot B \cdot L} \tag{9.3}$$

$$\frac{L^2}{H} \leq \frac{q \cdot n_0 \cdot \mu_n}{100} \cdot \frac{V_{OC}}{I_{SC}} \tag{9.3'}$$

Expression (9.3′) gives an idea of the minimization of resistive losses in thin-film modules caused by the series resistance of the TCO layer. If assuming values of 10^{21} cm^{-3}, 10 cm^2/(Vs), 0.6 V and 0.03 A/cm^2 for n_0, μ_n, V_{OC} (open-circuit voltage) and I_{SC}, respectively, the ratio between the squared width and the thickness of the TCO stripe should be less than about 320 cm. For a thickness of the TCO layer of 1 μm, this would mean that the width of the TCO stripe should be equal to or less than only 2 mm. This value has to be compared with the losses in the area due to the P1–3 cuts taking about 0.4 mm away from the width of a TCO stripe, i.e. losses due to the reduced active area would be 20 times larger than the losses due to the series resistance. The compromise is that higher series resistances are tolerated in thin-film PV modules. Therefore, the width of TCO stripes is usually between 0.3 and 2 cm in thin-film PV modules (for example, L is about 0.3–0.6, 1.2 and 2.0 cm for $Cu(In,Ga)Se_2$, CdTe and a-Si:H pin–pin thin-film solar cells, respectively). The resistance of the TCO layer is one of the reasons why fill factors are lower for thin-film solar cells compared to solar cells based on crystalline silicon (c-Si; see Chapter 7) or gallium arsenide (see Chapter 8). Furthermore, the resistance of a TCO layer generally avoids the efficient application of concentrated sunlight in thin-film photovoltaics.

The thickness of TCO layers in thin-film solar cells is usually fixed to minimize optical losses. The sheet resistance ($R_\square$) of a TCO layer is defined as the specific resistance divided by the thickness of the TCO layer:

$$R_\square = \frac{\rho}{H} \tag{9.4}$$

TCO layers with sheet resistances between 2 and 8 $\Omega\square$ are used for thin-film solar cells. As an example, a TCO layer with $R_\square$ of about 2 $\Omega\square$ would be required for a high-efficiency (single) a-Si:H thin-film solar cell with I_{SC} and V_{OC} of about 0.8 V and 16 mA/cm^2, respectively, and with L of about 0.5 cm. However, a value 2 $\Omega\square$ for a sheet resistance is a harsh condition, which demands a thickness of the TCO layer above 1 μm. Incidentally, specially designed finger grids of Ni:Al are deposited onto the TCO layer for reducing the series resistance in small area record thin-film solar cells.

The resistance of ZnO:Al layers increases under wet atmosphere due to the formation of defects compensating the charge of mobile electrons. The degradation of TCO layers of thin-film PV modules is minimized by careful sealing. Accelerated degradation tests (so-called damp heat test, relative humidity of 85%, temperature of 85°C and time of 1000 h) have been developed for testing thin-film PV modules.

9.2.3 *Optical losses in transparent conducting oxides*

In a thin-film solar cell, the sunlight has to penetrate the TCO layer before it reaches the PV absorber. The optical transmission of TCO layers has to be as large as possible in the range of the part of the sun spectrum which is absorbed by the thin-film absorber. Therefore, optical losses caused by reflection, free carrier absorption and fundamental absorption in TCO layers have to be minimized for the absorption range of the thin-film absorber. The minimization of optical losses in thin-film solar cells demands simulations with sophisticated optical models including the spectra of refractive indices and absorption coefficients of all materials involved, as well as light scattering in the near and far fields. Here the principal role of reflectivity, free charge carriers and absorption of photons with energies above the band gap of a TCO is demonstrated.

The reflection coefficient of a thin-film solar cell depends on the differences between the refraction indices of the ambience (n_{air}), of the TCO layer (n_{TCO}) and of the PV absorber (n_S). A system with reflection between the ambience and the TCO layer and between the TCO layer and the substrate will be considered for simplicity (see also Figure 2.6). The squared refractive index of a transparent material depends on the reciprocal squared wavelength (λ) and can be well approximated by the Sellmeier equation:

$$n^2_{TCO} = A + \frac{B \cdot \lambda^2}{\lambda^2 - C^2} + \frac{D \cdot \lambda^2}{\lambda^2 - E^2} \tag{9.5}$$

For ZnO thin films, the values of the coefficients A, B, C, D and E are 2.0065, $1.5748 \cdot 10^6$, $1 \cdot 10^8$, 1.5868 and 2.606, respectively (for wavelength given in Å) (Sun *et al.*, 1999). As an example, Figure 9.7 demonstrates the spectra for the refractive index of ZnO and glass. The refractive indices of ZnO and glass are about 1.9 or 1.5 at longer wavelengths and range between 2.3 and 1.9 and between 1.7 and 1.5 at shorter wavelengths, respectively.

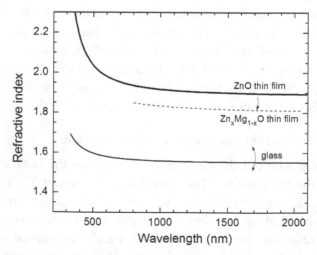

Figure 9.7. Spectra of the refractive index for ZnO thin films and glass. The trend for the reduction of the refractive index by alloying ZnO with MgO and the variability of the refractive index of glass are also shown.

The consideration of reflection leads to the following equation for the transmission coefficient (T_{otr}) under normal incidence of light.

$$T_{otr} = \frac{1 + r_{air,TCO}^2 \cdot r_{TCO,S}^2 - r_{air,TCO}^2 - r_{TCO,S}^2}{1 + r_{air,TCO}^2 \cdot r_{TCO,S}^2 + r_{air,TCO} \cdot r_{TCO,S} \cdot \cos(\theta)} \tag{9.6}$$

The reflectivity at the interfaces between air and the TCO layer ($r_{air,TCO}^2$) and between the TCO layer and the substrate ($r_{TCO,S}^2$) are given by the following equations:

$$r_{air,TCO}^2 = \left(\frac{n_{TCO} - n_{air}}{n_{TCO} + n_{air}} \right)^2 \tag{9.7'}$$

$$r_{TCO,S}^2 = \left(\frac{n_S - n_{TCO}}{n_S + n_{TCO}} \right)^2 \tag{9.7''}$$

The phase shift (θ) due to the optical path depends on the reciprocal wavelength of incident light, on the thickness of the TCO layer and on the refractive index of the TCO layer:

$$\theta = \frac{4 \cdot \pi}{\lambda} \cdot n_{TCO} \cdot H \tag{9.8}$$

The product of the doubled thickness and of the refractive index of TCO layers is larger than the wavelength of incident sunlight. This leads to the appearance of characteristic interference patterns in the transmission spectrum of a transparent TCO layer, as is shown in Figure 9.8. The transmission of the TCO layer in the given configuration is reduced over the whole spectral range. The reflection losses in the TCO layer are usually between 10 and 25%.

In order to minimize reflection losses in thin-film solar cells, the differences between the refractive index of the TCO and air (Equation (9.7')) and/or between the refractive index of the substrate and the TCO (Equation (9.7'')) should be reduced. This is possible by (i) adding an additional antireflection coating on top of the TCO layer (see Chapter 2) such as MgF_2, (ii) reducing the thickness of the TCO layer (Equation (9.8)) and (iii) reducing the refractive index of the TCO layer. The reduction of the thickness of the TCO layer leads to an increase of its sheet resistance. The resulting increase of the series resistance can be compensated by increasing the specific conductivity of the TCO or by reducing the width of TCO

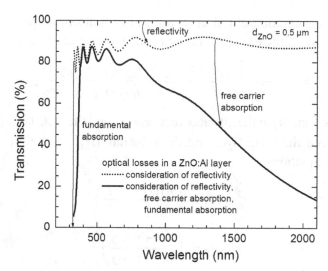

Figure 9.8. Transmission spectra of a ZnO layer (thickness 0.5 μm) deposited on glass under consideration of reflectivity only (dotted line) and of reflectivity, fundamental absorption and free carrier absorption (solid line).

stripes. The increase of the specific conductivity of a given TCO is rather limited. A reduction of the refractive index of a TCO is possible in a certain range by alloying, for example, ZnO with MgO, the product of which has a lower refractive index than ZnO (Teng *et al.*, 2000). However, the alloying of different metal oxides often leads to an increase of the specific resistance of the TCO.

Free electrons can be excited to higher energetic states in the conduction band with photons of smaller energies than the band gap of the TCO (free carrier absorption, Figure 9.9).

For free carrier absorption, the absorption coefficient is proportional to the density of free charge carriers (n_0) and to the wavelength to the power of 2 (Drude model) or 3 (Equation (9.9), highly doped ZnO (Weiher, 1966)):

$$\alpha(\lambda, n_0) = \alpha_0(n_0) \cdot \left(\frac{\lambda}{\lambda_0}\right)^3 \tag{9.9}$$

The absorption coefficient of ZnO amounts to $\alpha_0 = 200 \, \text{cm}^{-1}$ at a wavelength of $\lambda_0 = 1 \, \mu\text{m}$ for $n_0 = 5 \cdot 10^{19} \, \text{cm}^{-3}$ (Weiher, 1966).

Figure 9.8 shows an example of the transmission of a TCO layer. Losses caused by free carrier absorption increase with increasing wavelength.

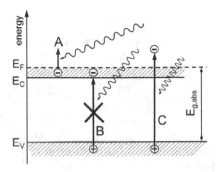

Figure 9.9. Photo-excitation of free charge carriers (A) and fundamental absorption (C) in a TCO. Transitions from occupied to occupied states are forbidden (B). The absorption gap ($E_{g,abs}$) is larger than the band gap of a degenerated TCO.

Therefore, the influence of free carrier absorption of TCO layers is less important for thin-film solar cells based on PV absorbers with larger band gaps such as a-Si:H (see also task T9.2 at the end of this chapter).

The density of free electrons in TCOs is larger than the effective density of states at the conduction band edge. Therefore, the Fermi-energy is shifted into the conduction band (Figure 9.9). The electronic states between the conduction band edge and the Fermi-level are occupied. Light is absorbed by excitation of electrons from occupied into unoccupied states. Consequently, the absorption gap of TCO layers is increased by the difference between the conduction band edge and the Fermi-energy. This effect is known as the Burstein shift (Burstein, 1953). For example, fundamental absorption of ZnO:Al (doped at $3 \cdot 10^{20}$ cm^{-3}) sets in at photon energies of about 3.7 eV (see also task T9.3). Therefore, fundamental absorption in the TCO layer does not play an important role for optical losses in thin-film solar cells.

9.3 Amorphous and Micro-Crystalline Silicon Solar Cells

9.3.1 *Disorder and mobility of charge carriers in amorphous silicon*

Amorphous silicon is a disordered semiconductor, i.e. a unit cell and translational symmetry are missing. The amorphous structure is caused by large variations in bond angles and bond lengths (see Figure 9.10). The probability of the formation of unsaturated or dangling bonds (dbs) is

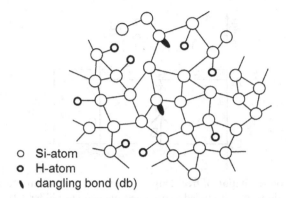

○ Si-atom
◑ H-atom
❱ dangling bond (db)

Figure 9.10. Schematic of disordered bond configuration and components in a-Si:H.

very high in amorphous silicon. Dangling bonds have electronic states in the middle of the band gap, i.e. the Shockley–Read–Hall recombination rate (see Chapter 3) is at a maximum (lifetime killers). The density of dbs in amorphous silicon is reduced by hydrogen passivation. Silicon thin-film solar cells are based on hydrogenated amorphous silicon (a-Si:H and μc-Si:H).

Each atom in a-Si:H has a specific number of bonds to its nearest neighbors (coordination number). In a relaxed amorphous structure, the preferred coordination number corresponds to the valence of the atoms, i.e. the coordination numbers are 4, 1 and 3 for Si, H and P, respectively. Incorporation of atoms with different coordination into amorphous structures is possible. Coordination defects are the elementary defects in a-Si:H. All bond configurations are equivalent in a-Si:H. The variability of bond configurations causes a natural variability of the properties of a-Si:H, depending on the content of hydrogen and on preparation parameters. For example, the band gap of a-Si:H increases with the increasing content of hydrogen.

The disorder of bond angles and bond lengths causes a random distribution of atom potentials and interatomic distances (Figure 9.11) (Mott, 1985). The disorder potential (V_0) describes the variation of atom potentials. The average interatomic distance is described by a_0. The uncertainty of localization (Δx) is of the order of the interatomic distance in a-Si:H. Therefore, with respect to the Heisenberg uncertainty principle, the uncertainty of the momentum (Δk) in a-Si:H is of the same order as

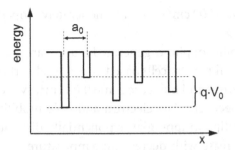

Figure 9.11. Schematic of the random distribution of atom potentials and interatomic distances in a-Si:H.

the momentum (k).

$$\Delta k = \frac{\hbar}{\Delta x} \approx \frac{\hbar}{a_0} = k \qquad (9.10)$$

As a consequence, the momentum is not conserved in a-Si:H and localization in space is most important. Optical transitions are possible between occupied and unoccupied electronic states, which overlap in space. Furthermore, disorder has strong influence on the transport of free charge carriers (Anderson, 1958).

Due to disorder, the motion of free charge carriers in a-Si:H is changed by scattering at each atom passed by a free charge carrier. The mobility of free charge carriers is given by the ratio between the time interval between two scattering events of a moving particle (δt) and the mass of the moving particle ($m^*_{e(h)}$) (Grimsehl, 1990).

$$\mu = q \cdot \frac{\delta t}{m^*_{e(h)}} \qquad (9.11)$$

The mobility of free charge carriers in a-Si:H can be estimated by taking into account a scattering length of a_0, which is about 0.025 nm and the maximum velocity, which is the thermal velocity v_{th} of 10^7 cm/s.

$$\mu \approx \frac{q \cdot a_0}{m^*_{e(h)} \cdot v_{th}} \qquad (9.12)$$

The field-effect mobility of free electrons, for example, is about 2 cm^2/(Vs) in un-doped a-Si:H (Han *et al.*, 2009). This agrees very well with the value estimated from Equation (9.12). For comparison, the electron mobility is about $1000\,cm^2/(Vs)$ in moderately doped silicon crystals (Jacoboni *et al.*, 1977).

The drift mobility of photo-generated charge carriers in electric fields is important for thin-film solar cells based on a-Si:H absorber layers. The drift mobilities of electrons and holes are 2 and $0.02\,cm^2/(Vs)\,cm^2/(Vs)$ at room temperature, respectively. The electron and hole mobilities in a-Si:H are controlled by multiple trapping at exponentially distributed defect states and therefore decrease with decreasing temperature (Tiedje *et al.*, 1981). Localization of charge carriers at electronic states causes time-dependent drift mobilities of photo-generated charge carriers (see, for example, Nebel *et al.* (1992)). Electronic transport with time-dependent mobility is called dispersive transport. The electron drift mobility in a-Si:H depends on the electric field and increases with increasing temperature up to values of $2-5\,cm^2\,/(Vs)$ (Marshall *et al.*, 1986).

There are numerous electronic states in the band gap of a-Si:H. Defects are part of the chemical equilibrium between different bond configurations in a-Si:H and cannot be avoided in general. This is a principal difference from ideal crystalline semiconductors. Extended and localized states are distinguished in a-Si:H (Figure 9.12). Delocalized states overlap in space so that charge carriers are mobile. Disorder and stress in bonds cause the so-called exponential tail states at the valence and conduction band edges. The bonding and anti-bonding states of dbs are in the region of the middle

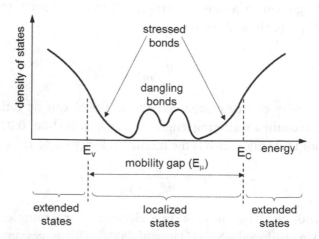

Figure 9.12. Schematic of the distribution of the density of states in a-Si:H.

of the band gap of a-Si:H (Cody *et al.*, 1981). The high density of electronic states in the band gap of a-Si:H is responsible for low mobilities and very short lifetimes of photo-generated charge carriers in a-Si:H and μc-Si:H.

The so-called mobility gap of a-Si:H is defined as the difference between energies of holes in the valence and of electrons in the conduction bands at which the holes and electrons are still mobile. Electrons and holes trapped at localized states are not mobile (Figure 9.11). The mobility gap of a-Si:H is not well-defined as the band gap of crystalline semiconductors. The mobility gap of a-Si:H decreases with increasing density of defects and with increasing temperature.

9.3.2 *Optical absorption and optical band gap in amorphous silicon*

Electrons can be excited from occupied into unoccupied electronic states. Therefore, the distribution of the density of states is reflected in the absorption spectrum. Figure 9.13 shows the schematic absorption spectrum of a-Si:H. It can be separated into three distinct sections. At lower photon energies, optical absorption is caused by dbs (Cody *et al.*, 1981; Stutzmann *et al.*, 1987). At high photon energies, the absorption coefficient is proportional to the square root of the photon energy. This part of the absorption spectrum defines the optical band gap (E_{og}) of a-Si:H (Brodsky *et al.*, 1970). For absorption coefficients larger than 10^4 cm^{-1}, the absorption spectrum has been fitted by the following equation (Cody

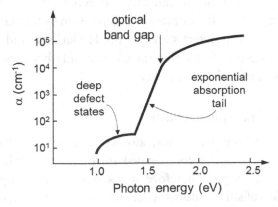

Figure 9.13. Schematic absorption spectrum of a-Si:H.

et al., 1981):

$$\sqrt{\alpha \cdot h\nu} = C_{a-Si:H} \cdot (\hbar\omega - E_{og}) \tag{9.13}$$

The parameter $C_{a-Si:H}$ is independent of temperature and amounts to 6.9 $(eV\mu m)^{-0.5}$ (Cody *et al.*, 1981).

The value of E_{og} increases with an increasing amount of hydrogen in a-Si:H, for example, from about 1.5 eV (2% of hydrogen) to about 1.8 eV (10% of hydrogen) (van Sark, 2002). The optical band gap of a-Si:H, which is used in thin-film solar cells, is about 1.7 eV (7% of hydrogen). The band gap of a-Si:H is controlled by the amount of disorder (Cody *et al.*, 1981). The mobility gap is lower than the optical band gap of a-Si:H. The absorption length is several hundred nm in a-Si:H absorber layers (Cody *et al.*, 1981).

At photon energies in the region of the exponential absorption tail, the absorption coefficient can be described by the following expression:

$$\alpha = \alpha_0 \cdot \exp\left(\frac{h\nu - E_1}{E_t}\right) \tag{9.14}$$

The parameters α_0 and E_1 are equal to about $1.3 \cdot 10^6$ cm^{-1} and 2.17 eV, respectively (Cody *et al.*, 1981). The characteristic energy describing the exponential increase of the absorption coefficient with increasing photon energy (tail energy, E_t) depends on temperature and structural disorder. The value of E_t is of the order of 50–100 meV (Cody *et al.*, 1981). The value of E_{og} increases with decreasing E_t (Cody *et al.*, 1981).

The absorption coefficient can vary between $3 \cdot 10^{-1}$ and 10^2 cm^{-1} in the spectral range below the influence of absorption by tail states depending on the density of deep defect states in a-Si:H (Jackson and Amer, 1982). The density of deep defect states ranges in a-Si:H between $3 \cdot 10^{15}$ and 10^{18} cm^{-3} (Jackson and Amer, 1982).

9.3.3 *Doping of amorphous silicon*

In relaxed amorphous structures, atoms are usually in their preferred coordination, i.e. fourfold coordinated silicon or threefold coordinated phosphorus, for example. Therefore, it was expected that doping of a-Si:H by a substitution of silicon atoms would not work. However, in 1975, Spear and Le Comber found a correlation between the increase of the conductivity

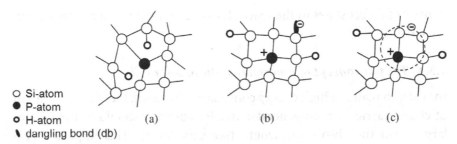

O Si-atom
● P-atom
o H-atom   (a)   (b)   (c)
❧ dangling bond (db)

Figure 9.14. Some bond configurations of phosphorus in a-Si:H in inactive configuration (a), defect-compensated donor configuration (b) and active configuration (c).

of a-Si:H in correlation and the content of phosphine or diborane added to the silane gas during deposition of a-Si:H layers (Spear and Le Comber, 1975).

Atoms can also appear in other than preferred coordinations due to local chemical reactions. Figure 9.14 shows examples for a phosphorus atom in inactive (threefold coordination) and active (fourfold coordination) configurations. It has to be considered that an electron leaving a phosphorus atom in fourfold coordination can be trapped at defects. The densities of phosphorus atoms in threefold and fourfold coordination are balanced by chemical reactions involving silicon and phosphorus atoms in different configurations, hydrogen atoms and dbs.

Figure 9.14 suggests that only a part of phosphorus or boron atoms is in active configuration in a-Si:H. The density of phosphorus or boron atoms in active configuration increases with increasing amount of phosphine or diborane in the silane gas. However, the relative amount of phosphorus or boron atoms in active configuration decreases with increasing amount of phosphine or diborane in the silane gas. The density of dbs increases with increasing density of phosphorus or boron atoms in a-Si:H. Therefore, the lowest density of dbs is achieved in un-doped a-Si:H (Stutzmann *et al.*, 1987).

The so-called doping efficiency (Stutzmann *et al.*, 1987) of a-Si:H is about 10^{-2}–10^{-3} at high concentration of phosphorus or boron, i.e. only one from 100–1000 phosphorus or boron atoms is in an active configuration. This means that densities of free charge carriers up to about 10^{18} cm^{-3} can be reached in a-Si:H. These high densities of activated dopants are sufficient to form ohmic contacts with a-Si:H due to the high density of defect states in the space charge region (trap-assisted tunneling, see Chapter 5).

9.3.4 *The pin concept of amorphous silicon solar cells*

In order to realize a high energy conversion efficiency, the diffusion length of charge carriers photo-generated in a PV absorber should be three times larger than the absorption length (see Chapter 3). This implies that a diffusion length of more than 1 μm is required for a-Si:H absorbers.

The diffusion length of photo-generated charge carriers can be calculated if the diffusion coefficient and the lifetime of photo-generated free charge carriers are known (Equation (3.5)). The diffusion coefficient and the mobility are connected by the Einstein equation:

$$D = \frac{k_B \cdot T}{q} \cdot \mu \qquad (9.15)$$

The longest lifetime of photo-generated charge carriers in a-Si:H can be estimated by taking into account the minimum possible density of dbs of about 10^{15}–10^{16} cm^{-3} (un-doped a-Si:H) and by using Equation (3.21). The resulting lifetime of photo-generated free charge carriers in un-doped a-Si:H is about 1–10 ns.

The diffusion length of photo-generated charge carriers in un-doped a-Si:H can be calculated by using the following equation:

$$L = \sqrt{\frac{k_B \cdot T}{q} \cdot \mu \cdot \tau} \qquad (9.16)$$

The diffusion length of photo-generated charge carriers is of the order of 100 nm in un-doped a-Si:H and decreases even further for doped a-Si:H. Therefore, the diffusion length of photo-generated charge carriers in a-Si:H is not long enough to realize a-Si:H thin-film solar cells with high energy conversion efficiency.

Photo-generated charge carriers can reach charge-selective contacts by drift in electric fields. The drift length (L_{drift}) of free charge carriers photo-generated in a-Si:H can be calculated by considering a certain electric field and by assuming that free charge carriers can drift until they recombine. The value of L_{drift} can be estimated from the product of the lifetime and the average drift velocity of photo-generated free charge carriers. The average

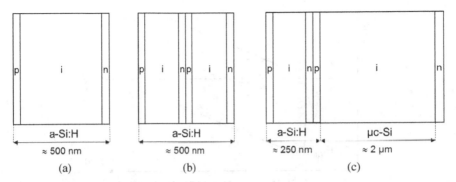

Figure 9.15. Layer structures of silicon thin-film absorbers with pin a-Si:H (a), pin–pin tandem a-Si:H (b) and tandem pin–pin a-Si:H/μc-Si:H (c) configuration, where "i" stands for un-doped or intrinsic.

drift velocity is equal to the product of the drift mobility and the average electric field (ε).

$$L_{drift} = \mu \cdot \tau \cdot \varepsilon \tag{9.17}$$

A homogeneous built-in electric field can be realized in un-doped a-Si:H layers if sandwiching the un-doped a-Si:H layer between highly p-type and n-type doped a-Si:H layers (Figure 9.15(a)). This is the so-called pin configuration of a a-Si:H thin-film solar cell.

The electric field in the un-doped a-Si:H absorber can be estimated if taking into account a built-in potential between the p-type and n-type doped regions of the order of 1 V and a thickness of the a-Si:H absorber layer of about 1 μm. Then the value of L_{drift} is of the order of 1 μm for an un-doped a-Si:H absorber layer, i.e. a-Si:H absorbers are suitable for thin-film solar cells in the pin configuration. The pin concept has been proposed for thin-film solar cells based on a-Si:H absorber layers by Carlson and Wronski (1976).

Figure 9.16 shows the schematic band diagram of an a-Si:H solar cell in pin configuration in the dark (a) and the distribution of dbs across the pin structure (b). The density of dbs (N_{db}) is of the order of 10^{17} cm^{-3} in the very thin n-type and p-type doped layers, where N_{db} is of the order of 10^{15} cm^{-3} in the un-doped or intrinsic layer (Stutzmann *et al.*, 1987). The built-in potential in the intrinsic region corresponds to the difference of the conduction band edges and the Fermi-level in the p-type and n-type

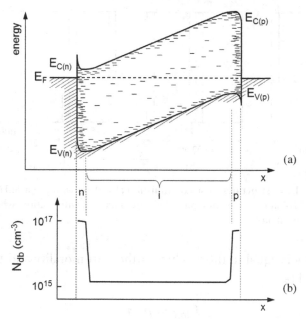

Figure 9.16. Schematic band diagram of an a-Si:H thin-film solar cell in pin configuration (a) and density of dbs (b).

doped regions. The potential energy of electrons or holes in the conduction or valence bands, respectively, decreases across the un-doped region nearly linearly towards the n-type or p-type doped layers, respectively. Electrons or holes photo-generated in the un-doped layer are accelerated towards the very thin n-type or p-type doped layers, respectively, where they become majority charge carriers and are collected at the ohmic contacts.

The maximum built-in potential in a-Si:H solar cells can be estimated if considering an effective density of states of the order of 10^{20}–10^{21} cm^{-3} at the valence and conduction band edges and a saturation limit of phosphorus and boron of the same order. Then, considering a doping efficiency of 0.1–1%, the differences between the conduction or valence band edges and the Fermi-level in the n-type or p-type doped regions, respectively, correspond to about 0.12–0.18 eV with respect to Equation (2.37). Therefore, the built-in potential is about 1.3–1.4 V in an a-Si:H solar cell with a pin structure. However, the upper limit of V_{OC} has to be significantly lower than the maximum built-in potential in a-Si:H solar cells since a certain electric field has to remain in order to realize carrier collection by drift (see also task T9.4). As a consequence, it is not useful to operate solar cells based on a-Si:H absorber layers under concentrated sunlight, and it is practically impossible to reach very high energy conversion efficiencies with a-Si:H solar cells.

The p-type doped contact layer often consists of amorphous silicon carbide (a-SiC:H), which has a larger band gap and enables additional blocking of electrons at the p-type doped contact (Konagai, 2011).

The maximum thickness of the absorber layer in a-Si:H solar cells is limited by the minimum drift length in relation to the absorption length. This makes the estimation of the minimum diode saturation or thermal generation current density of a-Si:H solar cells complicated since a-Si:H solar cells cannot be treated as ideal absorbers. Furthermore, the mobility gap of a-Si:H is lower than the optical band gap and photo-generation takes place due to absorption of blackbody radiation by defect states. In comparison, with an ideal solar cell, these two factors lead to a strong increase in the thermal generation current density, thereby limiting V_{OC}. Therefore, the solar energy conversion efficiency of an ideal solar cell with a band gap of 1.7 eV in the Shockley–Queisser limit (28%) is significantly higher than the highest energy conversion efficiency that can be reached with a-Si:H solar cells in reality (see also Figure 6.6).

9.3.5 Staebler–Wronski effect in amorphous silicon solar cells

The solar energy conversion efficiency of a-Si:H solar cells decreases during the first two to three days of operation of the a-Si:H solar cell and then stabilizes after this degradation. The energy conversion efficiency can be recovered by heating the a-Si:H solar cell to 160–170°C. The effect of reversible degradation of a-Si:H solar cells under illumination is called the Staebler–Wronski effect and was described by Staebler and Wronski in terms of reversible conductivity changes (Staebler and Wronski, 1977).

The Staebler–Wronski effect is caused by light-induced metastable defects (Stutzmann *et al.*, 1985). The energy conversion efficiency of a-Si:H solar cells is reduced from about 10% to about 7% due to the Staebler–Wronski effect. Degradation by the Staebler–Wronski effect cannot be avoided in a-Si:H solar cells. The role of the Staebler–Wronski effect for degradation of a-Si:H solar cells can be reduced by increasing the built-in electric field in the un-doped a-Si:H absorber. However, the a-Si:H

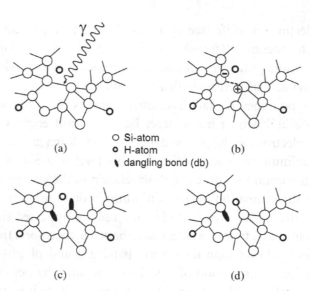

O Si-atom
o H-atom
\ dangling bond (db)

(a)   (b)   (c)   (d)

Figure 9.17.   Possible mechanism of photo-induced generation of a db in a-Si:H consisting of the steps of photon absorption (a), weakening of a Si–Si bond due to localization of a photo-generated electron-hole pair (b), breaking of a weakened Si–Si bond (c) and stabilization of one of the dbs (d).

absorber thickness has to be reduced for this purpose, which in turn leads to a decrease of absorption in the a-Si:H layer. The reduced absorption in a thinner a-Si:H layer can be compensated by stacking two a-Si:H solar cells in series (pin–pin structure of an a-Si:H solar cell, Figure 9.15(b)). It should be mentioned that the thicknesses of the stacked a-Si:H solar cells with integrated series connection have to be adjusted in accordance with the current-matching condition.

The origin of the Staebler–Wronski effect is still under discussion and there exist different models for explaining the effect. Figure 9.17 gives an idea of possible mechanisms of photo-induced generation of dbs in a-Si:H (close to the bond breaking model of Stutzmann (Stutzmann *et al.*, 1985)).

Hydrogen plays a decisive role in the formation and annealing of dbs in a-Si:H. It is assumed that a photon is absorbed at a Si–Si bond near a hydrogen atom. The Si–Si bond can be weakened by localization of the electron-hole pair at neighboring Si atoms and the weakened Si–Si bond can break by forming two dbs. Both dbs will recombine within a very short time. However, one of the dbs can be stabilized by a hydrogen atom saturating the other db leading to an increase of N_{db} in a-Si:H.

The Staebler–Wronski effect can be avoided in μc-Si:H consisting of nanoparticles of c-Si with diameters of tens of nm surrounded by a-Si:H (Meier *et al.*, 1994). The a-Si:H provides hydrogen for passivation of surface states at the c-Si nanoparticles. The potential energy of photo-generated electrons and holes is lower in the c-Si nanoparticles so that photo-generated electrons and holes are delocalized within a very short period of time after photo-generation. Recombination of delocalized electrons and holes does not cause bond breaking.

9.3.6 *a-Si:H/μc-Si:H tandem solar cells*

The absorption coefficient of μc-Si:H is higher than the absorption coefficient in c-Si even though the band gaps are the same (Vetterl *et al.*, 2000). This revealed the opportunity to apply a PV absorber with the band gap of c-Si in thin-film solar cells. Solar cells based on μc-Si:H absorbers also have a pin architecture like a-Si:H but do not show degradation due to the Staebler–Wronski effect (Meier *et al.*, 1994). Values of V_{OC} of the order of 0.526 V can be reached in solar cells based on μc-Si:H absorbers (Konagai, 2011).

A tandem solar cell with the combination of band gaps of 1.7 eV (top cell) and 1.1 eV (bottom cell) can increase the solar energy conversion efficiency of silicon thin-film solar cells. In a-Si:H/μc-Si:H tandem solar cells the pin–pin concept is applied (Figure 9.15(c)). It is important to note that the values of I_{SC} reduced from about 17.5 or 25.3 mA/cm^2 for solar cells based on a-Si:H or μc-Si:H absorbers to 14.4 mA/cm^2 for a solar cell based on a-Si:H/μc-Si:H (Konagai, 2011), which is even larger than half of the I_{SC} of the solar cell based on a μc-Si:H absorber, i.e. optical losses can be further reduced in a-Si:H/μc-Si:H multi-layer structures.

Figure 9.18 shows a schematic cross section of a high-efficiency a-Si:H/μc-Si:H tandem solar cell. Light trapping is optimized by introducing a silver back reflector, a textured TCO layer on top of the silver back reflector and a TCO interlayer. The TCO interlayer has been introduced to increase the reflection of photon with energy above the band gap of a-Si:H, i.e. for optimizing the current-matching condition at a thickness of the un-doped a-Si:H layer of only 0.3 μm. Tandem solar cells based on a-Si:H/μc-Si:H are also called micromorph (micro-crystalline-amorphous)

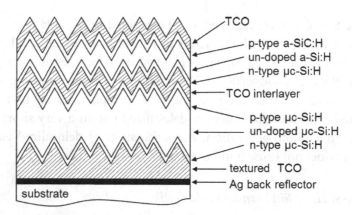

Figure 9.18. Schematic cross section of a high-efficiency a-Si:H/μc-Si:H tandem solar cell.

solar cells. Solar energy conversion efficiencies of the order of 14% have been reached with the a-Si:H/μc-Si:H concept (Konagai, 2011).

9.4 Chalcopyrite and Kesterite Solar Cells

9.4.1 *The chalcopyrite and kesterite absorber families*

Chalcopyrite is a copper-based mineral with the chemical formula $CuFeS_2$. Chalcopyrite consists of Cu^I (copper can have the valence of Cu^I and Cu^{II}), Fe^{III} (Fe can have the valence of Fe^{II} and Fe^{III}) and S having a valence of 2. The structure of chalcopyrite is shown in Figure 9.19(a) (see, for example, Jaffe and Zunger (1984)). All crystals with the structure of the mineral chalcopyrite are members of the chalcopyrite family keeping in mind that Fe atoms can be replaced by other atoms with a valence of 3 (In, Ga, Al) and that S atoms can be replaced by Se atoms with a valence of 2. Therefore, members of the chalcopyrite family have the chemical formula $A^I B^{III} X_2^{VI}$ where A^I corresponds to Cu^I, B^{III} corresponds to In or Ga and X^{VI} is related to S or Se atoms.

Atoms are bonded by a polar covalent bonding in chalcopyrites. Each atom is coordinated in a chalcopyrite with four other atoms. The metal atoms have four sulfur or selenium atoms as neighbor atoms, where sulfur or selenium atoms have two metal atoms with the valence of 1 and 2 metal atoms with the valence of 3 as neighbor atoms. The electronegativities of sulfur and selenium are 2.58 and 2.55, respectively, after Pauling (see, for

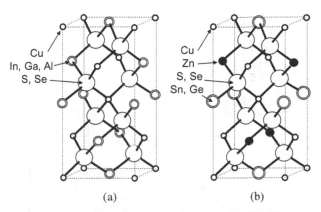

Figure 9.19. Chalcopyrite ($A^IB^{III}X_2^{VI}$) and kesterite ($A_2^IC^{II}D^{IV}X_4^{VI}$) structures (a) and (b), respectively, and some elements with the valence of A^I (Cu), B^{III} (Fe^{III}, In, Ga, Al), C^{II} (Zn), D^{IV} (Sn, Ge) and X^{VI} (S, Se).

example, Allred (1961)). The electronegativity of the metal atoms (1.9, 1.78, 1.81 for Cu, In and Ga, respectively) is less than the electronegativity of sulfur or selenium atoms. Therefore, the center of positive charge is partially shifted to the metal atoms and the center of negative charge is partially shifted to sulfur or selenium atoms in a chalcopyrite.

The structure of kesterite is shown in Figure 9.19(b) (see, for example, Persson (2010)). It can be derived from the structure of chalcopyrite if replacing half of the atoms with valence 3, for example, In atoms by atoms with a valence of 2 such as zinc atoms, and the other half by atoms with a valence of 4 such as Sn^{IV} or Ge atoms. The chemical formula of kesterites is $A_2^IC^{II}D^{IV}X_4^{VI}$ where A^I corresponds to Cu^I, C^{II} corresponds to Zn, D^{IV} means Sn^{IV} or Ge and X^{VI} is related to S and/or Se atoms, i.e. $Cu_2ZnSn(Se_{1-x}S_x)_4$.

9.4.2 *Band gaps and absorption in chalcopyrites and kesterites*

Chalcopyrites and kesterites are semiconductors. The values of the forbidden band gaps are 1.04, 1.53 and 1.68 eV for $CuInSe_2$ (Shay *et al.*, 1973), $CuInS_2$ (Tell *et al.*, 1971) and $CuGaSe_2$ (Shay *et al.*, 1972), respectively. The band gaps of $Cu_2ZnSnSe_4$ and Cu_2ZnSnS_4 are about 1.0 and 1.5 eV, respectively (Persson, 2010). These band gaps are in the range of PV absorbers for very high solar energy conversion efficiency (see Chapter 6). Incidentally, the stoichiometry between Cu and In or Ga can vary within

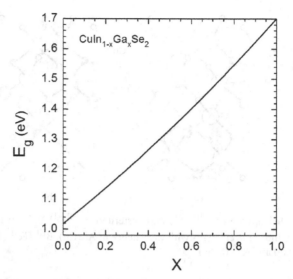

Figure 9.20. Dependence of the band gap of $CuIn_{1-x}Ga_xSe_2$ on the stoichiometry following from Equation (9.14).

the chalcopyrite phase so that the band gap of a certain chalcopyrite can also vary depending on the Cu:In or Cu:Ga ratio.

The band gap of chalcopyrite thin films can be smoothly tuned by changing the stoichiometry, for example, in $CuIn_{1-x}Ga_xSe_2$ (Dimmler *et al.*, 1987) (Figure 9.20):

$$E_g[eV] = 1.018 + 0.575 \cdot x + 0.108 \cdot x^2 \tag{9.18}$$

It has to be mentioned that the parameters in Equation (9.18) can vary depending on deposition, probably due to little deviations of the content of copper (Paulson *et al.*, 2003). The tuning of the band gap of chalcopyrites is important for the optimization of the energy conversion efficiency of thin-film solar cells based on chalcopyrite absorbers. In a similar way, band gaps can be varied in other quaternary ($CuIn(Se_{1-x}S_x)_2$) or penternary chalcopyrite ($CuIn_{1-x}Ga_x(Se_{1-y}S_y)_2$) or kesterite ($Cu_2ZnSn(Se_{1-x}S_x)_4$) compounds.

Chalcopyrite and kesterite PV absorbers have large absorption coefficients where the absorption coefficient increases strongly within a quite narrow range of photon energies above the band gap (Paulson *et al.*, 2003). Figure 9.21 shows the absorption spectrum of $CuInSe_2$ as a function of the

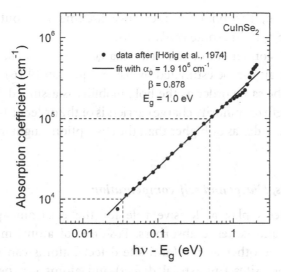

Figure 9.21. Absorption spectrum of CuInSe$_2$ as a function of the photon energy reduced by the band gap. Data points taken after (Hörig *et al.*, 1974), the line is a fit with an empirical power law (power coefficient $\beta = 0.878$, pre-factor $\alpha_0 = 1.9 \cdot 10^5$ cm^{-1}).

photon energy reduced by the band gap in a double logarithmic scale. The data points were taken from Hörig *et al.* (1974). The absorption coefficient follows an empirical power law for photon energies up to 1 eV above the band gap. A large absorption coefficient of 10^4 cm^{-1} and a very large absorption coefficient of 10^5 cm^{-1} are reached at photon energies of about 35 meV and 0.45 eV above the band gap, respectively, i.e. at 1.035 and 1.45 eV. CuInSe$_2$ is the strongest absorber used in thin-film photovoltaics.

9.4.3 *Lifetime and mobility for the example of* CuIn$_{1-x}$Ga$_x$Se$_2$

Epitaxial layers of CuIn$_{1-x}$Ga$_x$Se$_2$ are usually intrinsically p-type doped with densities of free holes (p$_0$) between $4 \cdot 10^{16}$ and $2 \cdot 10^{19}$ cm^{-3} where two different acceptors and compensating donors exist (Schroeder *et al.*, 1998). The mobility of free holes is between 167 and 311 cm^2/(Vs) in epitaxial layers of CuIn$_{1-x}$Ga$_x$Se$_2$ independent of x and p$_0$ (Schroeder *et al.*, 1998). The mobility of electrons is in the range of 20–1100 cm^2/(Vs) (Schumann *et al.*, 1980). The high densities of acceptors and donors in chalcopyrite epitaxial layers point to the importance of defects.

Wagner *et al.* measured response times of p-type doped CuInSe$_2$/CdS detectors between 5 and 150 ns (Wagner *et al.*, 1974). For means of comparison, the photoluminescence lifetime is about 5–50 ns for $CuIn_{1-x}Ga_xSe_2$ (Ohnesorge *et al.*, 1998).

The diffusion length of minority charge carriers in chalcopyrite and kesterite crystals can be estimated by using Equation (9.16) if an electron mobility of the same order as the hole mobility is assumed. The resulting diffusion length of minority charge carriers is of the order of 1–5 μm, which is of the same order as or higher than the absorption length in chalcopyrite absorbers.

9.4.4 *Defects, phases and self-compensation*

Intrinsic defects play a decisive role for the electronic properties of chalcopyrite and kesterite absorbers. Positions of atoms in a lattice can be occupied by other atoms (anti-site defect), atoms can occupy space between lattice sites (interstitial defect) and atoms can be missed at a lattice site (vacancy defect) as shown schematically in Figure 9.22 for a two-dimensional lattice with two different atoms.

The number of elementary defect configurations in a lattice increases with the number of different atoms. In addition, defects can form complexes, thus further increasing the number of possible defect configurations. A high probability of defect formation makes the precise control of stoichiometry and phases in chalcopyrite and kesterite thin-film absorbers complicated.

There are two prominent defects in chalcopyrites (Figure 9.23). The anti-site defect pair $(Cu_{In}^{2-} + In_{Cu}^{2+})$ consists of a Cu atom on a lattice

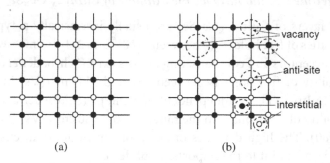

(a)   (b)

Figure 9.22. Two-dimensional defect-free lattice with two different atoms (a) and two-dimensional lattice with vacancy, anti-site and interstitial defects (b).

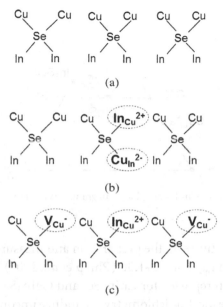

Figure 9.23. Undisturbed configurations of Se atoms in CuInSe$_2$ (a) and configurations with anti-site defect pair (b) and neutral Cu vacancy complex (c) defects in CuInS$_2$.

site of In (Cu$_{In}^{2-}$) and of an In atom on a lattice site of Cu (In$_{Cu}^{2+}$). The formation energy of the anti-site pair defect in chalcopyrite is relatively low and consequently heavy randomization sets in at temperatures much lower than the melting temperature (Zhang *et al.*, 1998). The neutral Cu vacancy complex (2 V$_{Cu}^{-}$ + In$_{Cu}^{2+}$) consists of two Cu vacancies (V$_{Cu}^{-}$) and one In atom on a lattice site of Cu (In$_{Cu}^{2+}$) compensating the negative charge of the Cu vacancies (Zhang *et al.*, 1998).

The presence of a neutral Cu vacancy complex changes the local stoichiometry of a chalcopyrite. The replacement of Cu atoms by In atoms can be expressed by the following equation:

$$m \cdot CuInSe_2 + n \cdot In \rightarrow Cu_{m-3n}In_{m+n}Se_{2m} + 3n \cdot Cu \qquad (9.19)$$

The values of the numbers m and n have to be equal to or larger than 3 or 1, respectively. The compounds Cu$_{m-3n}$In$_{m+n}$Se$_{2m}$ can form phases (also known as ordered vacancy compounds (OVCs)) in the phase diagram of Cu, In and Se (Figure 9.24), for example, CuIn$_5$Se$_8$, CuIn$_3$Se$_5$. Ordered vacancy compounds are also semiconductors. The band gaps of

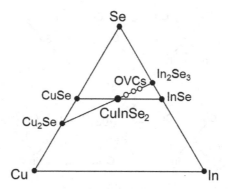

Figure 9.24. Simplified Cu–In–Se phase diagram. OVC stands for ordered vacancy compound.

OVCs depend sensitively on the preparation and measurement conditions. For example, band gaps of 1.3–1.36 (Philip *et al.*, 2005) and 1.21 (Bodnar, 2008) eV have been reported for $CuIn_5Se_8$ and $CuIn_3Se_5$.

Changes in the local stoichiometry can induce uncontrolled local phase transitions. The control of phases during the formation of chalcopyrite and kesterite thin-film absorbers is a key technological issue. The formation and transformation of various phases and the distribution of elements have been analyzed in detail by Mainz and Klenk (2011). Chalcopyrites are usually grown in a Cu-rich regime (Klenk *et al.*, 2005) so that copper sulfide can form on top of a chalcopyrite layer. The highly conductive copper sulfide layer has to be removed in KCN (potassium cyanide) solution (see, for example, Scheer *et al.* (1993)). It has been shown by neutron scattering that the lowest densities of defects are reached when there is a slight excess of copper in the chalcopyrite layer (Stephan *et al.*, 2011).

The density of each type of defect is determined by a complex of chemical reactions. The equilibrium of chemical reactions involving certain defects and phase transformation can be strongly influenced by additives. For example, layers of $CuIn_{1-x}Ga_xSe_2$ absorbers are usually produced by co-evaporation onto substrates of soda lime glass or in the presence of additionally deposited sodium increasing the density of acceptors and passivating grain boundaries (see, for example, Ård *et al.* (2000)). The performance of thin-film solar cells based on $CuIn_{1-x}Ga_xSe_2$ improved strongly when sodium was present during the formation of the absorber. The highest absorber energy conversion efficiencies of a thin-film solar cell ($\eta = 20.3\%$) were achieved on soda lime glass with a $CuIn_{1-x}Ga_xSe_2$ absorber (Jackson *et al.*, 2011). Furthermore, the treatment of the $CuIn_{1-x}Ga_xSe_2$ absorber with potassium from a very thin deposited KF surface layer allowed for the increase of the solar energy conversion efficiency to 20.4% on flexible substrates (Chirilă *et al.*, 2013).

Thin-film solar cells based on a $CuIn_{1-x}Ga_xSe_2$ absorber are stable over a long period of time despite the large variety of possible reactions of defect formation. Chalcopyrite and kesterite thin films are excellent PV absorbers in which even the presence of grain boundaries has no influence on the solar energy conversion efficiency (Abou-Ras *et al.*, 2008) and diffusion lengths of minority charge carriers above 1 μm are easily reached (Scheer *et al.*, 1995). Therefore, defect reactions and/or local phase transitions can stabilize chalcopyrite and kesterite absorbers and avoid the formation of recombination active defects. However, a detailed analysis of defects and phase diagrams is rather complex for $CuIn_{1-x}Ga_xSe_2$ absorbers (Guillemoles, 2000).

Doping by variation of the stoichiometry in $CuInSe_2$ has been demonstrated by Noufi *et al.* where the Cu to In and the Se to metal ratios are important (Noufi *et al.*, 1984). Well-controlled doping of chalcopyrite and kesterite thin films is practically impossible. Chalcopyrite and kesterite absorber layers are usually p-type doped in thin-film solar cells.

Different defect configurations have been analyzed for $CuInSe_2$. It has been shown by the group of Zunger *et al.* that the Cu vacancy is a shallow acceptor with very low formation energy, which can explain intrinsic p-type doping of $CuInSe_2$ thin films (Zhang *et al.*, 1998).

$$V_{Cu} \leftrightarrow V_{Cu}^- + h^+ \qquad (9.20)$$

Low formation energies of defects cause self-compensation and prevent controlled doping. For example, the formation energy of the Cu vacancy becomes even negative for a Fermi-energy close to the conduction band edge of $CuInSe_2$ so that compensating Cu vacancies are formed together with incorporation of donor atoms (Zhang *et al.*, 1998). Self-compensation avoids the formation of pn-homo-junctions (see Chapter 4) as well as of ohmic tunneling contacts (see Chapter 5) in thin-film solar cells based on chalcopyrite or kesterite absorbers. On the other hand, self-compensation leads to an exceptional radiation hardness of thin-film solar cells based on $CuIn_{1-x}Ga_xSe_2$ (Jasenek and Rau, 2001).

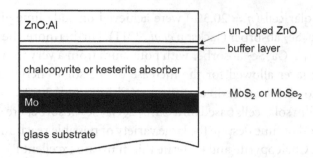

Figure 9.25. Schematic cross section of a thin-film solar cell based on a chalcopyrite or kesterite absorber.

9.4.5 *Contacts in chalcopyrite and kesterite solar cells*

Charge-selective contacts are realized with hetero-junctions (see Chapter 4) in thin-film solar cells based on chalcopyrite and kesterite absorbers. Figure 9.25 shows a schematic cross section including all principal layers in related solar cells.

Ohmic back contacts of chalcopyrite or kesterite absorbers are prepared on sputtered molybdenum layers (thickness of the order of 0.5–1 μm) where the ohmic contact with the absorber layer is formed with an interfacial MoS_2 or $MoSe_2$ layer (see, for example, Scheer *et al.* (1993)). Incidentally, MoS_2 or $MoSe_2$ are layered materials well known for use as lubricants. From this point of view, MoS_2 or $MoSe_2$ interface layers avoid stress caused by different thermal expansion of the substrate and the absorber layer.

Ohmic front contacts are formed between an un-doped ZnO layer and a highly doped ZnO:Al layer (thickness of the order of 0.5–1 μm). Reflection losses in the TCO layer can be reduced by depositing an additional MgF_2 antireflection coating layer on top of the ZnO:Al layer.

Charge-selective contacts are realized by a hetero-junction between the chalcopyrite or kesterite absorber and a so-called buffer layer, which is a semiconductor with a wide band gap. The buffer layer enables transfer of photo-generated electrons from the conduction band of the p-type doped absorber into the conduction band of the wide-gap semiconductor whereas the transfer of photo-generated holes from the absorber into the buffer layer is blocked. Band gaps of materials that are used as buffer layers are 2.4 eV (CdS (Shin *et al.*, 1991)), 2.1 eV (In_2S_3 (Allsop *et al.*, 2006)) and between 2.6 eV ($ZnO_{0.5}S_{0.5}$) and 3.4 eV ($ZnO_{1-x}S_x$ (Meyer *et al.*, 2004)). The feasibility of $CuInS_2/CdS$ charge-selective contacts has been demonstrated by Wagner *et al.* in their pioneering works (Wagner *et al.*, 1974; Shay *et al.*, 1975).

Contacts with CdS are usually prepared in a chemical bath containing cadmium acetate or $CdSO_4$, ammonia and thioruea (see, for example, Britt and Ferekides (1993)). The range of the charge-selective contact layer is extended by a thin un-doped ZnO layer leading to a smaller spread of device parameters (Scheer *et al.*, 2011). The highest solar energy conversion efficiency reached with a thin-film solar cell ($\eta = 20.3\%$) has been obtained for the $CuIn_{1-x}Ga_xSe_2/CdS$ system in which the Ga content is increased at the charge-selective front contact and at the ohmic back contacts but reduced near the front contact (Jackson *et al.*, 2011). Photo-generation in the buffer layer does not contribute to the short-circuit current in solar cells based on chalcopyrite or kesterite thin-film absorbers due to a short value of L_{diff}. Therefore, optical absorption in the buffer layer has to be minimized by reducing the thickness of the buffer layer to 40–60 nm.

The advantages of using In_2S_3 for charge-selective contacts are that In_2S_3 does not contain the poison Cd and that In_2S_3 can be deposited by a spray ILGAR® (ion layer gas reaction) process (Fischer *et al.*, 2011). The spray ILGAR® process consists of cycles of spraying a chemisorbed layer of precursor metal ions, a purge, sulfurization step and a purge. Each cycle gives a thickness control of about 1 nm and additives such as chlorine or oxygen can be incorporated via the precursor salt or water molecules during the purge step. High energy conversion efficiencies comparable to those with a CdS buffer layer can be reached for solar cells based on $CuIn_{1-x}Ga_xSe_2/In_2S_3$ (Fischer *et al.*, 2011).

$ZnO_{1-x}S_x$ can be used to align both the valence (oxygen-rich) and conduction (sulfur-rich) band offsets at hetero-junctions with chalcopyrite or kesterite absorber layers due to the strong bowing (Persson *at al.*, 2006). Figure 9.26 shows the dependencies of the calculated energies of the conduction (E_V) and valence (E_C) band edges and of the calculated (Persson *at al.*, 2006) and measured (Meyer *et al.*, 2004) band gaps on x.

The values of E_V and E_C can be varied over 0.9 and 1.1 eV for x < 0.5 and x > 0.5, respectively (Persson *at al.*, 2006). The calculated and measured band gaps of $ZnO_{1-x}S_x$ are in good agreement. The controlled band alignment with $ZnO_{1-x}S_x$ buffer layers makes in-line processing of highly

Materials Concepts for Solar Cells

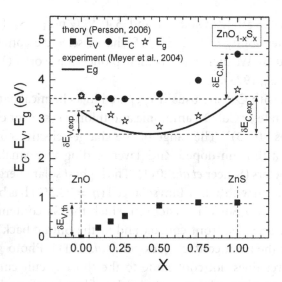

Figure 9.26. Dependence of the energies of the conduction (filled circles) and valence (filled squares) band edges and of the band gap (open stars and solid line) in $ZnO_{1-x}S_x$ on the stoichiometry. The symbols and the solid line correspond to theoretical analysis (Persson, 2006) and to experimental results (Meyer *et al.*, 2004), respectively. The arrows mark the theoretical and experimental variability of the energies of the conduction and valence band edges.

efficient chalcopyrite solar cells possible by applying sputtering (Grimm *et al.*, 2012). Atomic layer deposition (ALD) by pulsing gaseous precursors allows as well for a controlled variation of the stoichiometry in $ZnO_{1-x}S_x$ layers. Solar energy conversion efficiencies above those of reference samples with a CdS buffer layer have been demonstrated for $ZnO_{1-x}S_x$ buffer layers deposited by ALD (Platzer-Björkman *et al.*, 2006).

An energy conversion efficiency of 18% has been reached for thin-film solar cells with a $CuIn_{1-x}Ga_xSe_2$ absorber and a ZnS buffer layer prepared by chemical bath deposition (Nakada and Mizutani, 2002).

Hetero-junctions between chalcopyrite or kesterite absorbers and buffer layers cannot be considered as metallurgically sharp contacts. The contact is extended to an intermixed region with a thickness of the order of 2 nm as demonstrated directly for a $CuInSe_2$/CdS interface (Cojacaru-Mirédin *et al.*, 2011). Metallurgical intermixing at interfaces makes a theoretical description of charge-selective contacts with chalcopyrites or kesterites rather complicated. Nevertheless, useful information can be obtained from models with sharp hetero-junctions (Klenk, 2001).

The value of V_{OC} increases with increasing band gap as expected, but only within a range of band gaps between 1 and about 1.2 eV for a $CuIn_{1-x}Ga_xSe_2$ absorber with a CdS charge-selective contact. For larger band gaps, the value of V_{OC} increases much less than expected with increasing E_g and tends to saturate for larger band gaps (Herberholz *et al.*, 1997). For example, V_{OC} is up to 0.74 V for a $CuIn_{1-x}Ga_xSe_2$ absorber with a band gap of 1.2 eV (Jackson *et al.*, 2011) and only up to 0.88 V for a $CuInS_2$ absorber having a band gap of 1.5 eV (Merdes *et al.*, 2013). This phenomenon is also known as the V_{OC} problem of thin-film solar cells.

Figure 9.27 depicts the simplified band diagrams of chalcopyrite solar cells based on $CuInSe_2$/CdS (a) and $CuInS_2$/CdS (b) charge-selective contacts. It is commonly agreed that a type I hetero-junction is formed at the $CuInSe_2$/CdSe contact (Herberholz *et al.*, 1997; Klenk, 2001). Electrons can be partially transferred from the ZnO:Al surface layer into the un-doped ZnO interfacial and CdS buffer layer. Electrons crossing the CdS/CuInSe$_2$ interface accumulate in the interface region due to the positive conduction band offset of the order of +0.31 eV (theoretical analysis (Wei and Zunger, 1993)). Therefore, free electrons form an inversion layer at the $CuInSe_2$ side. As a consequence, the density of holes is strongly reduced at the

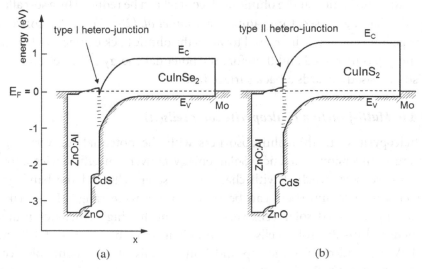

Figure 9.27. Simplified band diagrams of thin-film solar cells based on $CuInSe_2$ (a) and $CuInS_2$ (b) with metallurgically sharp junctions and interface states.

charge-selective contact, and a solar cell based on $CuInSe_2/CdS$ is limited by bulk recombination in the $CuInSe_2$ absorber (Turcu *et al.*, 2002).

The conduction band offset at the $CdS/CuIn_{1-x}Ga_xSe_2$ interface reduces with an increasing amount of Ga. A type II hetero-junction (see Chapter 8) is formed at the $CdS/CuInS_2$ contact (Herberholz *et al.*, 1997). Surface recombination dominates in this case, and the limit of V_{OC} reduces to the difference between the E_C reduced by the conduction band offset at the $CdS/CuInS_2$ interface and E_V. The activation energy of the recombination process (E_a) depends on the type of the hetero-junction. It has been found by the group of Rau that E_a follows E_g for solar cells based on copper-poor $CuIn_{1-x}Ga_x(Se_{1-y}S_y)_2/CdS$ hetero-junctions, but also that E_a is independent of E_g for copper-rich $CuIn_{1-x}Ga_x(Se_{1-y}S_y)_2/CdS$ hetero-junctions (Turcu *et al.*, 2002).

The ideality factor of thin-film solar cells based on chalcopyrite and kesterite absorbers depends on light intensity and temperature (see, for example, Turcu *et al.* (2002)). The solar energy conversion efficiency of chalcopyrite and kesterite solar cells can increase under illumination due to light soaking (Haug *et al.*, 2003). This shows that the charge-selective interface is not optimized in such cases, and that saturation of defect states under illumination can become important.

Recombination at the ohmic back contact can be reduced by a so-called graded band gap (see, for example, Jackson *et al.* (2011)). In this concept, the gallium content is increased towards the ohmic back contact so that the band gap is increased and therefore the potential energy of free electrons is also increased towards a back surface field.

9.4.6 Multi-junction chalcopyrite solar cells(?!)

Chalcopyrites are thin-film absorbers with the potential for very high energy conversion efficiency. Solar energy conversion efficiencies above 20% have been reached with chalcopyrite solar cells and the band gap of chalcopyrite absorbers can be tuned over a wide range. This makes chalcopyrite-based solar cells very interesting for high-efficiency multi-junction thin-film solar cells. The combination of band gaps of 1.7 and 1.1 eV are optimal for the top and bottom cells of a tandem solar cell (see Chapter 6). These values can be realized with $CuIn_{0.85}Ga_{0.15}Se_2$ ($E_g = 1.1$ eV) and $CuGaSe_2$ ($E_g = 1.68$ eV).

Besides the tunability of the band gap, three other conditions have to be fulfilled in order to attain highly efficient multi-junction thin-film solar cells based on chalcopyrites.

First, the V_{OC} of solar cells with the different band gaps should follow the value of the band gap. This may be problematic with respect to the V_{OC} problem of thin-film solar cells.

Second, the top cell of a tandem solar cell has to be transparent for photons with photon energies below the band gap of the top cell since these photons are required for photo-generation in the bottom cell. The relatively high absorption coefficient of chalcopyrite absorber layers for photon energies below the band gap as well as the increasing influence of free carrier absorption in the TCO layer with decreasing photon energy are problematic. The thickness of the chalcopyrite absorber and the density of defects and/or inclusions of phases other than the desired chalcopyrite phase have to be reduced in order to increase the transmittance for photons with energies below the band gap of the top cell. It has been demonstrated that the thickness of chalcopyrite absorbers transferred onto a gold reflector can be reduced to 0.3 μm without significant losses in I_{SC} (Naghavi *et al.*, 2012). At the same time, however, V_{OC} decreased with decreasing thickness of the absorber layer (Naghavi *et al.*, 2012) due to the increasing role of recombination at the ohmic back contact (see Chapter 5).

Third, a transparent recombination contact for integrated series connection is needed. For this purpose, the molybdenum layer has to be avoided. It seems possible to transfer chalcopyrite solar cells with ohmic molybdenum selenide or sulfide back contacts onto a highly doped and very thin TCO layer.

The realization of highly efficient tandem or multi-junction chalcopyrite thin-film solar cells is challenging. It is important to mention that the energy conversion efficiency of chalcopyrite solar cells can be further increased in microcells with micro-concentrators (Paire *et al.*, 2012).

9.5 Cadmium Telluride Solar Cells

9.5.1 *Properties of CdTe*

CdTe belongs to the family of $A^{II}B^{VI}$ (A: Zn, Cd, Hg; B: S, Se, Te) semiconductors. The cubic zincblende structure of CdTe crystals is shown

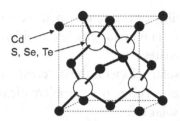

Figure 9.28. Structure of cubic zincblende.

in Figure 9.28. The band gap of CdTe ranges between 1.5 eV (Adachi *et al.*, 1993) and 1.43 eV (Yamada, 1960), which is in the optimum for PV solar energy conversion (see Chapter 6). The absorption coefficient of CdTe (Adachi *et al.*, 1993) is high enough so that a thickness of a CdTe layer of several μm is sufficient for light absorption.

The electron or hole mobility is up to 1100 cm^2/(Vs) in n-type doped (Triboulet *et al.*, 1974) or 80 cm^2/(Vs) in p-type doped (Yamada, 1960) CdTe crystals, respectively. Controlled n-type and p-type doping as well as high doping is difficult or practically impossible for CdTe thin films due to enhanced compensation and segregation effects at grain boundaries (Birkmire and Eser, 1997). CdTe thin films are intrinsically p-type doped. Therefore, pn-homo-junctions as charge-selective contacts and ohmic contacts cannot be formed by controlled doping of CdTe thin films.

No indication has been found for recombination of minority charge carriers at grain boundaries in thin-film solar cells based on CdTe (Scheer, 1999).

9.5.2 *Contacts in CdTe solar cells*

CdTe thin-film solar cells are produced in super-strate configuration (Figure 9.29). The TCO layer (SnO$_2$:F or Cd$_2$SnO$_4$) is deposited on a glass substrate for high-efficiency CdTe solar cells. The mobility of Cd$_2$SnO$_4$ is relatively large and so a thickness of the TCO layer of 0.2–0.3 μm is sufficient (Wu, 2004). A layer of un-doped ZnSnO$_x$ is deposited on the TCO layer. The CdS buffer layer is about 70-nm thick. The CdTe absorber layer is several-μm thick.

A hetero-junction with a highly doped p-type semiconductor has to be prepared on top of the p-type doped CdTe thin-film solar cell in order to form an ohmic contact. Deposition of Cu, Hg, Pb or Au followed

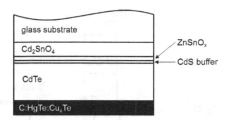

Figure 9.29. Schematic cross section of a thin-film solar cell based on CdTe absorber.

by heat treatment at 150°C leads to the formation of a highly p-type doped surface region of CdTe thin films and therefore to ohmic contacts (Birkmire and Eser, 1997). Ohmic contacts of CdTe thin films are also realized by depositing ZnTe–Cu, HgTe or PbTe thin films (Birkmire and Eser, 1997). Ohmic contacts with CdTe thin films can also be formed with alloys containing tellurium and copper such as HgTe:Cu_xTe (Wu, 2004).

9.5.3 *Technological aspects of CdTe solar cells*

CdTe sublimates or evaporates in correct stoichiometry as Cd and Te_2 (congruent evaporation) and forms only one phase (de Nobel, 1959).

$$2CdTe(s) \leftrightarrow 2Cd(g) + Te_2(g) \tag{9.21}$$

CdTe is the only semiconductor that can be directly transferred by a very simple sublimation process from a CdTe source onto a substrate within a short time of only 3–5 min. The short time of the deposition of a PV CdTe absorber is a decisive key for the low production costs of CdTe thin-film solar cells. The distance between the source and the substrate is only 2 mm in the CSS (close space sublimation (Britt and Ferekides, 1993; Birkmire and Eser, 1997)) process so that losses of CdTe are very low during the deposition process. Furthermore, the CSS process of CdTe deposition is performed in Ar or O_2/He atmosphere at a reduced pressure of 1000 Pa without need for high vacuum. The deposition temperature of between 550 and 700°C is not critical and standard glass sheets can be used as substrates (Wu, 2004).

As-deposited CdTe thin films have to be post-treated with $CdCl_2$ in order to reduce recombination losses and to increase the lifetime of photo-generated charge carriers. The $CdCl_2$ is deposited, for example, by spraying a $CdCl_2$/methanol solution onto the CdTe thin film (McCandless

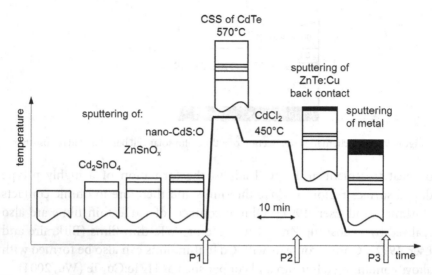

Figure 9.30. Basic technological steps of a CdTe thin-film solar module in time and temperature scales. The arrows mark the P1–3 cuts after deposition of the TCO, after post-treatment with CdCl₂ and after final metallization, respectively. For more details see Wu (2004).

and Birkmire, 1991). Annealing of CdTe thin films at about 400°C for 10 min in the presence of $CdCl_2$ leads to an increase of grains, to a passivation of grain boundaries and to alloying at the CdS/CdTe interface. The post-treatment of CdTe thin films with $CdCl_2$ is key for high solar energy conversion efficiencies (Birkmire and Eser, 1997).

Figure 9.30 gives an overview of the basic technology of thin-film solar cells and modules based on CdTe absorber. The amazing points are the simplicity and the short time period of the complete technological process. The production of CdTe solar cells is by far the simplest and fastest technology in photovoltaics in general.

Layers are deposited by sputtering, CSS and spraying, all of which do not demand high energies. Therefore, the application of CdTe thin-film PV modules results in the shortest energy payback time (0.7 years under conditions in Catania, Spain) and the lowest carbon footprint (15 g CO_2-eq/kWh under conditions in Catania) for PV systems (Wild-Scholten *et al.*, 2010).

Despite the high energy conversion efficiency and despite the fast and relatively simple technology, CdTe thin-film solar cells are not suitable for a global energy supply due to the limitation of Te resources (George, 2008) (see also task T9.8). Furthermore, broken or degraded CdTe PV modules have to be completely recycled in order to recover Te and to avoid emission of the toxic Cd into the environment.

9.6 Summary

Thin-film solar cells consist of a thin-film absorber as well as of charge-selective and ohmic contact layers, which can be deposited on large foreign substrates. Table 9.1 gives an overview of the main materials and deposition methods in thin-film photovoltaics. There are three classes of materials which have been successfully applied for absorbers in thin-film solar cells: (i) a-Si:H (hydrogenated amorphous silicon) and μc-Si:H (hydrogenated micro-crystalline silicon), (ii) $CuIn_{1-x}Ga_x(Se_{1-y}S_y)_2$ (chalcopyrites) and $Cu_2ZnSn(Se_{1-x}S_x)_4$ (kesterites) and (iii) CdTe. Each material of these classes can be produced only with its very specific deposition method being able to minimize the density of defects and to maximize the drift

Table 9.1. Overview of the main materials and deposition methods in thin-film photovoltaics (PECVD — plasma-enhanced chemical vapor deposition, CSS — close space sublimation, CBD — chemical bath deposition, ILGAR® — ion layer gas reaction, ALD — atomic layer deposition).

Function	Material	(Scalable) deposition method
absorber layer	a-Si:H, μc-Si:H (un-doped)	PECVD
	$CuIn_{1-x}Ga_x(Se_{1-y}S_y)_2$	co-evaporation and selenization/
	$CuIn_{1-x}Ga_x(Se_{1-y}S_y)_2$	sulfurization of precursors
	CdTe	CSS, $CdCl_2$ post-treatment
ohmic contact	ZnO, ZnO:Al	sputtering
layer	Mo	sputtering
	(other) metals	sputtering, evaporation
	SnO_2:F	sputtering, sol-gel
	ZnTe:Cu	sputtering
charge-selective	a-Si:H, a-SiC:H, μc-Si:H	PECVD
contact layer	CdS	CBD
	In_2S_3	ILGAR®, CBD
	$ZnO_{1-x}S_x$	sputtering, CBD, ALD

or diffusion length of photo-generated charge carriers, i.e. (i) PECVD for a-Si:H and μc-Si:H, (ii) co-evaporation and selenization/sulfurization of metallic precursors with sophisticated annealing regimes for chalcopyrites and kesterites and (iii) CSS with $CdCl_2$ post-treatment for CdTe.

Ohmic contact layers (TCOs or metal layers) are usually sputtered. Evaporation for metals or sol-gel deposition of metal oxides are alternative deposition methods for some cases. Optical losses are caused in TCO layers by reflection, fundamental absorption and free carrier absorption. The resistance of TCO layers is a major reason for resistive losses in thin-film solar cells.

Charge-selective p-type and n-type a-Si:H and μc-Si:H contact layers are prepared by PECVD. A relatively large variety exists for the deposition of CdS (CBD), In_2S_3 (ILGAR®, CBD) and $ZnO_{1-x}S_x$ (sputtering, CBD, ALD) charge-selective contacts.

Table 9.2 summarizes the main materials and principles with respect to the classes of thin-film absorbers: (i) silicon thin-film, (ii) chalcopyrite and kesterite thin-film and (iii) CdTe thin-film solar cells.

Each class of thin-film absorbers has its specific measures for maximizing the drift or diffusion length of photo-generated charge carriers. The implementation of a large amount of hydrogen of about 7% is decisive for the passivation of silicon dbs in a-Si:H and μc-Si:H. Substitional doping is possible for a-Si:H and μc-Si:H where only a part of the dopants is activated and the density of dbs increases with increasing density of dopants. The realization of pin structures enables a drift length of photo-generated charge carriers larger than the absorption length in a-Si:H and μc-Si:H absorbers.

The interplay between enhancement of crystallization (for example, by $CdCl_2$ post-treatment of CdTe or by introduction of Na or K into $CuInSe_2$ absorbers) and segregation of impurity atoms or excess host atoms at inactive grain boundaries is very important for reaching long diffusion lengths in chalcopyrite, kesterite and CdTe absorber layers. The formation of secondary phases in chalcopyrite and kesterite absorber layers has to be avoided by applying well-controlled annealing regimes.

Ohmic front and back contacts with ZnO:Al are not critical for silicon thin-film solar cells. Ohmic contacts with ZnO:Al or the more stable SnO_2:F are not critical for CdS, In_2S_3 or $ZnO_{1-x}S_x$ buffer layers. Ohmic contacts for majority charge carriers in chalcopyrite, kesterite or CdTe absorber layers

Table 9.2. Overview of materials for absorber, charge-selective and ohmic contacts as well as of the passivation principle of defects and of the type of charge-selective contact for silicon thin-film, chalcopyrite and kesterite thin-film and CdTe thin-film solar cells.

Function	Silicon thin-film	Chalcopyrite and kesterite thin-film	CdTe thin-film
absorber	a-Si:H μc-Si:H	$CuIn_{1-x}Ga_x(Se_{1-y}S_y)_2$ $Cu_2ZnSn(Se_{1-x}S_x)_4$	CdTe
passivation	H-passivation of dbs	segregation of impurities at electrically inactive grain boundaries	
charge-selective contact	pin by substitutional doping	electron extracting hetero-junction	
		CdS In_2S_3:Cl, :O $ZnO_{1-x}S_x$	CdS
ohmic front contact	ZnO:Al	ZnO/ZnO:Al	SnO_2:F Cd_2SnO_4
ohmic back contact	ZnO:Al	$Mo/Mo(Se_{1-x}S_x)_2$	C:Cu_xTe ZnTe-Cu HgTe PbTe

are more critical since effective surface passivation strategies are missing. Therefore, the density of photo-generated minority charge carriers has to be reduced at the ohmic contacts by increasing the thickness of the chalcopyrite, kesterite or CdTe absorber layers to values larger than the absorption and diffusion lengths.

Segregation avoids the well-controlled doping of chalcopyrite, kesterite and CdTe thin-film absorbers, which are usually p-type. Electron extracting hetero-junctions are formed with CdS for chalcopyrite, kesterite and CdTe absorber layers. Charge-selective contacts were successfully tested with In_2S_3 and $ZnO_{1-x}S_x$ buffer layers deposited on chalcopyrite absorber layers.

Table 9.3 summarizes values of E_g, I_{SC}, V_{OC}, fill factor (FF) and η of record thin-film solar cells for the most representative thin-film solar cells. A very high energy conversion efficiency (20.3%) has been reached with a chalcopyrite $CuIn_{1-x}Ga_xSe_2$ absorber layer across which the Ga content

Table 9.3. Values of the band gap of thin-film absorbers (E_g), short-circuit current density (I_{SC}), open-circuit voltage (V_{OC}), fill factor (FF) and solar energy conversion efficiency (η) for different thin-film world record solar cells representing different technologies. References are given.

	E_g (eV)	I_{SC} (mA/cm^2)	V_{OC} (V)	FF (%)	η (%)	Reference
a-Si:H	1.7	16.75	0.886	67.8	10.1	Benagli *et al.* (2009)
a-Si:H/μc-Si	1.7/1.1	14.4	1.41	72.8	14.7	Konagai (2011)
CuIn$_{1-x}$Ga$_x$Se$_2$/CdS	1.17–1.24	35.4	0.740	77.5	20.3	Jackson *et al.* (2011)[a]
CuIn$_{1-x}$Ga$_x$Se$_2$/CdS	1.17–1.24	35.1	0.736	78.9	20.4	Chirilă *et al.* (2013)[b]
CuIn$_{1-x}$Ga$_x$Se$_2$/ In$_2$S$_3$	1.2	35.5	0.630	72.0	16.1	Sáez-Araoz *et al.* (2012)
CuIn$_{1-x}$Ga$_x$Se$_2$/ ZnS$_{1-x}$O$_x$	1.1	38.4	0.654	78.3	18.3	Klenk *et al.* (2014)
CuInS$_2$/CdS	1.5	21.2	0.835	73	12.9	Merdes *et al.* (2013)
CuGaSe$_2$/CdS	1.67	16.4	0.795	69.2	9.0	Contreras *et al.* (2012)
Cu$_2$ZnSn(Se$_{1-x}$S$_x$)$_4$ /CdS	1.15	30.8	0.517	63.7	10.1	Bag *et al.* (2012)
CdTe/CdS	1.4	28.59	0.857	80.0	19.6	Gloeckler *et al.* (2013)

[a]on soda lime glass, [b]KF treated on flexible substrate.

was varied in such a way that the potential energy of free electrons and holes increased towards the ohmic back contact and towards the charge-selective hetero-junction with CdS, respectively (Jackson *et al.*, 2011). It has been demonstrated that sprayed In$_2$S$_3$ (Sáez-Araoz *et al.*, 2012) and sputtered ZnS$_{1-x}$O$_x$ (Klenk *et al.*, 2014) buffer layers also have a potential for high energy conversion efficiency for thin-film solar cells based on chalcopyrites. The great technological advantage of the use of ZnS$_{1-x}$O$_x$ buffer layers is that in-line processing without wet chemical treatments becomes possible. A high energy conversion efficiency of 18.3% was reached for thin-film solar cells based on CdTe/CdS (Gloeckler *et al.*, 2013).

For wide-gap thin-film solar cells based on a-Si:H, CuInS$_2$ or CuGaSe$_2$, the energy conversion efficiencies are reduced due to insufficient values of V_{OC}, i.e. the diode saturation current densities are quite high. Here, an improved understanding of chemical processes at charge-selective and ohmic contacts and their relation to defect formation is required.

The reserves of indium and tellurium are rather limited in relation to global application of thin-film solar cells based on $CuIn_{1-x}Ga_xSe_2$ or CdTe absorbers (Andersson, 2000). Kesterite-based solar cells contain only earth abundant elements and kesterite PV modules can be produced with similar technologies as chalcopyrite solar cells. It seems that kesterite absorber layers have the potential for high solar energy conversion efficiency. Different technological approaches are under development to ascertain materials limitations in kesterite-based solar cells (Delbos, 2012).

Thin-film solar cells have the highest potential for sustainable energy production due to short energy payback times (0.7 years for CdTe, 1.1 years for Si thin-film and 1.2 years for $CuIn_{1-x}Ga_xSe_2$ for PV systems installed in the south of Italy; for comparison, 1.5 years for c-Si PV systems (Wild-Scholten *et al.*, 2010)).

The transformation of printed metal (Ungelenk *et al.*, 2010) and metal sulfide nanoparticles (Guo *et al.*, 2010) into chalcopyrite or kesterite thin-film absorbers, respectively, as well as the sintering of printed CdTe (Gur *et al.*, 2005) or $CuInSe_2$ (Guo *et al.*, 2008) nanoparticle layers can further reduce the energy needed for the production of thin-film PV modules.

Thin-film PV modules based on a-Si:H have a high shunt resistance so that the energy conversion efficiency is also high at reduced light intensity. Thin-film PV modules based on a-Si:H also have the lowest decrease of the energy conversion efficiency with increasing temperature so that they provide the highest amount of energy produced per kW_p installed in tropical regions, for example.

9.7 Tasks

T9.1: Bulk and tunneling resistances at grain boundaries in a TCO layer

Compare the bulk and tunneling resistances at grain boundaries in highly doped TCO layers ($n_0 = 5 \cdot 10^{20} \, cm^{-3}$, $\mu_n = 20 \, cm^2/Vs$) with an average grain size of $d_g = 200 \, nm$ and a surface band bending at grain boundaries of (a) $\varphi_0 = 1.2 \, V$ (Fermi-level pinning due to high density of interface defects) and (b) $\varphi_0 = 0.3 \, V$ (reduced density of interface defects). The effective electron mass (m_e^*) is 0.26 times the free electron mass (m_e).

T9.2: TCO contact stripe of thin-film solar cells

Find the optimal width of a TCO ($R_\square = 5\,\Omega\square$) contact stripe for a high efficiency pin–pin a-Si:H solar cell.

T9.3: Optical losses in TCO layers

Calculate I_{SC} as a function of the wavelength corresponding to the band gap of PV absorbers on the basis of the AM1.5d spectrum. The absorber layer has a surface ZnO:Al layer ($n_0 = 10^{21}\,cm^{-3}$) with a thickness of 0.5 or 1 μm. The refractive index of the absorber layer is approximately 3. Compare the maximum I_{SC} with I_{SC} obtained for both surface layers in the cases of a-Si:H and CuInSe$_2$ absorber layers by taking into account losses due to reflection and due to free carrier absorption.

T9.4: Burstein shift in ZnO:Al

Calculate the increase of the optical band gap in highly doped ZnO:Al for a density of free electrons of 10^{20} and $10^{21}\,cm^{-3}$ by taking into account the energy dependence of the density of states (see Chapter 2) and an effective electron mass of $0.265 \cdot m_e$ (Weiher, 1966).

T9.5: Open-circuit voltage in an a-Si:H solar cell

Estimate the V_{OC} in an a-Si:H solar cell by taking into account the densities of majority charge carriers in the p- and n-type doped regions and the density of photo-generated charge carriers in the un-doped a-Si:H absorber layer. Discuss the importance of light trapping for thin-film solar cells with un-doped a-Si:H absorber layers.

T9.6: Chalcopyrite tandem solar cell

Estimate the maximum possible energy conversion efficiency of a chalcopyrite tandem solar cell by taking into account data given in Table 9.3 and compare the value with the world record chalcopyrite solar cell. Discuss the importance of the V_{OC} problem of thin-film solar cells.

T9.7: Advantages of CdTe thin-film technology

Explain why the thin-film technology of CdTe solar cells is so stable and robust with respect to temperature regimes and purity standards of materials and ambient environment during production.

T9.8: Strategic potential of CdTe, chalcopyrite and kesterite solar cells

Estimate (a) the mass of Te, In and Sn in solar cells based on CdTe, $CuIn_{0.75}Ga_{0.25}Se_2$ and $Cu_2ZnSn(Se_{1-x}S_x)_4$ absorbers needed for 1 W_p of installed power and (b) the mass of Te, In and Sn that would be required if all electricity needs of mankind (2 kW per capita) would be satisfied with only one of these technologies. Compare with the annual production of Te (80 t in 2012; George (2013)), In (640 t in 2011; Tolcin (2012)) and Sn (240000 t in 2012; Carlin (2013)).

10

Nano-Composite Solar Cells

Materials with different properties are combined in nano-composites to achieve functionalities needed in solar cells but which cannot be realized with only one of these materials. Nano-composite absorbers offered the opportunity for application of highly absorptive materials with very short diffusion or drift lengths of photo-generated free charge carriers in solar cells. The concept of nano-composite materials is explained and an example of folded absorbers is shown. Next, the quantum confinement effect, the concept of quantum dot absorbers and the concept of surface conditioning are introduced. The principles of organic photovoltaic absorbers and the realization of local charge separation with donor–acceptor hetero-junctions in organic solar cells are illuminated, the exceptional role of C_{60} molecules as acceptors is highlighted, and examples for conjugated polymers and oligomers are given. The principles of dye-sensitized solar cells and their realization with sintered networks of TiO_2 nanoparticles sensitized with dye molecules and contacted with redox couples of an electrolyte are given in detail. Passivation concepts and the inherent stability of dye-sensitized solar cells are mentioned. The concept of the solid-state methyl-ammonium-lead-iodide-sensitized solar cell is given. The basic idea of nano-photonic concepts for solar cells is briefly explained and examples are given. The potential of nano-composite solar cells for sustainable solar energy conversion is discussed.

10.1 The Concept of Nano-Composite Materials for Photovoltaics

10.1.1 *New quality of properties in nano-composite materials*

A nano-composite is a material in which two or more materials interpenetrate each other on a scale of 1–500 nm without changing the phase of

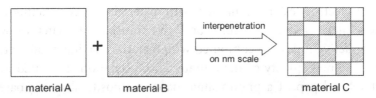

material A   material B   material C

Figure 10.1. Schematic of a nano-composite formed by interpenetration of two materials A and B with different properties on a scale of about 1–500 nm resulting in qualitatively new properties of material C.

each individual material (Figure 10.1) such that the individual properties of each component are conserved. The combination of very different properties of individual materials results in qualitatively new properties of nano-composite materials. For example, materials with very different mechanical and thermal properties can interpenetrate each other and form a new material, such as aerogels, which can be used for ultra-light and heat-insulating structures. Actually, micro-crystalline hydrogenated silicon (μc-Si:H; see Chapter 9) is a thin-film absorber consisting of silicon nano-crystals embedded in amorphous hydrogenated silicon (a-Si:H) where silicon nano-crystals provide delocalization of photo-generated charge carriers and a-Si:H provides a high absorption coefficient and hydrogen for passivation of dangling bonds.

Each component contributes with a specific functionality to the new quality of a nano-composite. Functionalities important for solar cells are, amongst others, light absorption, photo-generation, charge separation, charge transport, surface passivation and series connection of nanoparticles. A new quality can be, for example, an extremely low thermal conductivity (in the case of aerogels) or a stable silicon thin-film absorber with a band gap of 1.1 eV based on silicon (in the case of μc-Si:H). A qualitatively new property of a nano-composite cannot be achieved with the homogeneous individual materials.

In nano-composites for photovoltaics, (i) materials transparent to most of the sun spectrum such as surfactant molecules are combined with materials absorbing light very well such as semiconductor quantum dots (QDs), (ii) materials accepting electrons very well such as C60 molecules are combined with materials donating electrons such as phthalocyanine molecules, (iii) materials with high conductivity of charge carriers such as

TiO$_2$ or electrolytes are combined with materials not conducting charge carriers such as dye molecules and (iv) transparent materials with a high refractive index are combined with materials having a low refractive index. The new quality of these material combinations is related to (i) a tunable band gap of a photovoltaic nano-composite absorber based on semiconductor QDs, (ii) charge separation at donor–acceptor bulk hetero-junctions in organic solar cells, (iii) application of dye molecules for light absorption in dye-sensitized solar cells and (iv) possible incorporation of photonic structures into nano-composite solar cells.

A certain individual material can have several functionalities in a nano-composite solar cell. For example, the structure of a nano-composite is often defined by the morphology of a nano-structured skeleton of a transparent metal oxide such as nanoporous TiO$_2$ or ZnO nanorods, which also form charge-selective contacts and connect individual solar cells with a locally very thin absorber layer in parallel with each other. Furthermore, the morphology and the refractive index of nano-structured transparent metal oxides can contribute to light scattering and light trapping.

The optimization of a certain functionality of a given material in a nano-composite absorber is, in general, a complex task since the variation of any local parameter in a nano-composite absorber can decisively influence other functionalities. The challenge of photovoltaic nano-composite absorbers is to combine materials in such a way that light absorption, local charge separation and transport of photo-generated charge carriers are optimized at a minimum of optical, recombination and resistive losses limiting the short-circuit current density, the diode saturation current density and the fill factor of a solar cell.

10.1.2 *Photovoltaic absorbers with very short diffusion or drift lengths*

Several concepts of nano-composite absorbers aim for the incorporation of absorber materials with very short or even extremely short diffusion or drift lengths of photo-generated charge carriers into solar cells. Related absorber materials are, for example, dye molecules that do not form conducting or semiconducting layers. A nano-composite absorber can be treated in general as a photovoltaic absorber with an effective absorption coefficient and an effective diffusion or drift length of photo-generated charge carriers.

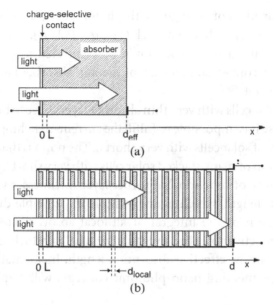

Figure 10.2. Principle of solar cells with an absorber material having a diffusion or drift length (L) much shorter than the absorption length for a compact absorber layer (a) and for locally thin absorber layers with local charge-selective contacts (b). The individual solar cells in the stack of many solar cells are connected in parallel.

The thickness of a photovoltaic absorber should be three times larger than the absorption length in order to absorb more than 95% of incoming light (see Chapter 2). Figure 10.2(a) shows the schematic cross section of a solar cell with a photovoltaic absorber material having a diffusion or drift length of photo-generated charge carriers (L) much shorter than the absorption length. Such an absorber is called an absorber with very short diffusion or drift length. It is obvious that efficient solar cells cannot be prepared from homogeneous absorbers with very short diffusion or drift lengths since the vast majority of photo-generated charge carriers recombine before reaching the charge-selective contact.

Absorber materials with very short diffusion or drift lengths can only be applied in efficient solar cells if the local thickness of the absorber (d_{local}) is shorter than L. In this case, most of the photo-generated charge carriers can reach the charge-selective contact. However, only a small fraction of photons in the sun spectrum is absorbed in an optically thin absorber layer. Therefore, a stack of many optically thin absorber layers is needed for complete absorption of sunlight with photon energies above the optical band gap of the absorber material (Figure 10.2(b)). Each of the thin absorber layers in such a stack should be connected individually with a charge-selective contact so that each of the thin absorber layers is part of a complete solar cell.

Stacked solar cells with very thin absorber layers have to be connected in parallel because it is impossible to fulfill the current-matching condition for a large number of solar cells with very short L. The point is that an integrated series connection of many stacked solar cells with very short L would require a large variability of individual absorber layer thicknesses with respect to the absorption length, which is not possible in principle due to the very short L. The sum or the integral of all local absorber layer thicknesses results in the effective absorber layer thickness (d_{eff}), which should be three times larger than the effective absorption length. Incidentally, there is the hope that implementing nano-photonic concepts will help to drastically reduce d_{eff}.

The values of d_{local} and the geometry of individual solar cells with very thin absorber layers can vary depending on the kind of materials used and technologies applied. A solar cell consisting of many individual solar cells with locally very thin absorbers and which are connected in parallel can be treated as one solar cell with one effective nano-composite absorber.

An absorber with a very short L can be folded by using, for example, the surface of ZnO nanorods where the ZnO nanorods additionally serve as the electron conductor. A related system has been demonstrated for a solar cell based on ZnO nanorods/In_2S_3:Cu/CuSCN (see Dittrich *et al.* (2011) and references therein) where the In_2S_3:Cu absorber has a band gap of about 1.5 eV and CuSCN serves as the hole conductor.

An absorber layer folded between the electron and hole conductors is illustrated in Figure 10.3. The value of d_{eff} increases with increasing surface area of a supporting matrix. For example, the short-circuit current density (I_{SC}) increased with increasing length of ZnO nanorods in the ZnO nanorod/In_2S_3:Cu/CuSCN system (Kieven *et al.*, 2008). However, the increase of the contact area causes an increase of surface recombination and therefore of the diode saturation current density (I_0). The increase of I_0 leads to a decrease of the open-circuit voltage (V_{OC}) with increasing d_{eff}. Furthermore, the decrease of d_{local} to about 10 nm also led to an increase

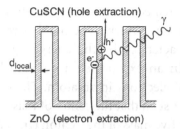

Figure 10.3. Schematic nano-composite absorber based on ZnO nanorod/In$_2$S$_3$:Cu/CuSCN.

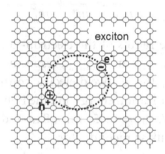

Figure 10.4. Schematic of an exciton.

of I_{SC} in the ZnO nanorod/In$_2$S$_3$:Cu/CuSCN system due to increasing drift of photo-generated charge carriers in the increased electrical field between the electron and hole conductors. However, V_{OC} and FF decreased with decreasing d_{local} due to an increase of local shunts (Belaidi *et al.*, 2008). This example shows that very high solar energy conversion efficiencies cannot be reached with related systems without a drastic reduction of surface recombination and/or without a drastic improvement in light trapping.

10.2 Quantum Dot-Based Nano-Composite Solar Cells

10.2.1 *Quantum confinement in semiconductor nano-crystals*

Electrons and holes influence each other due to Coulomb attractive force. The attraction between a free electron and a free hole leads to the formation of a bond state. The bond state between a free electron and a free hole is called exciton (Figure 10.4). Excitons (see, for example, Scholes and

Rumbles (2006)) are treated as electrically neutral quasi-particles, which can diffuse in photovoltaic absorbers before they recombine.

Excitons are characterized by the energy and by the radius of the electron-hole orbit. In an exciton, electrons and holes move on orbits similarly to orbitals of electrons in an atom. The masses of electrons and holes are rather similar such that the center of mass is between the electron and the hole. The effective mass of the exciton (μ_{eh}) is given by the effective masses of the free electron (m_e^*) and of the free hole (m_h^*).

$$\frac{1}{\mu_{eh}} = \frac{1}{m_e^*} + \frac{1}{m_h^*} \tag{10.1}$$

The energy and the radius of the electron-hole orbit with the lowest energy of an exciton, $E_{e1h1,b}$ and $a_{e1h1,b}$, respectively, can be estimated by taking into account μ_{eh}, the relative dielectric constant of the photovoltaic absorber (ε_r) and the ionization energy and the Bohr radius of the hydrogen atom ($E_{ion}^H = 13.56\,\text{eV}$, $a_0 = 0.051\,\text{nm}$). The radius $a_{e1h1,b}$ is called the exciton radius of the bulk semiconductor.

$$E_{e1h1,b} = \frac{\mu_{eh}}{m_e} \cdot \frac{1}{\varepsilon_r^2} \cdot E_{ion}^H \tag{10.2}$$

$$a_{e1h1,b} = \frac{m_e}{\mu_{eh}} \cdot \varepsilon_r \cdot a_0 \tag{10.3}$$

Effective electron and hole masses have been measured practically for all inorganic crystalline semiconductors. For PbS, for example, m_e^* and m_h^* are about $0.2 \cdot m_e$ (Barton, 1971) and $0.1 \cdot m_e$ (Cuff *et al.*, 1964), where m_e is the mass of the free electron ($9.1 \cdot 10^{-31}\,\text{kg}$). The exciton radius is therefore 18 nm for crystalline PbS. The values of $E_{e1h1,b}$ are usually much lower than the thermal energy in inorganic photovoltaic absorbers at room temperature due to the low effective mass of the exciton and due to the large relative dielectric constant. Therefore, excitons can be observed in most semiconductor crystals only at low temperatures.

In bulk semiconductors, the dimension of the crystal with a volume V is much larger than the $a_{e1h1,b}$. The energy of the first excitonic transition is lower than the band gap by a few meV in bulk semiconductors and is not important for solar cells based on conventional semiconductor absorbers such as crystalline silicon (c-Si) or gallium arsenide (GaAs).

The radius of semiconductor nano-crystals (r_{QD}) is less than $a_{e1h1,b}$ and therefore limits the radius of the exciton in the nano-crystal. The wave function of an exciton in a semiconductor nano-crystal can be compared with standing waves vibrating between fixed points like a string on a guitar. The half wavelength of the first harmonic of the string is equal to the length of the string. The frequency of a wave is proportional to the reciprocal wavelength. The energy of a wave is proportional to its frequency. Therefore, the energy of the standing wave of a string in a guitar increases with decreasing length of the string. The same happens in a semiconductor nano-crystal. The energy of an exciton increases with decreasing r_{QD}. The exciton binding energy becomes much larger than the thermal energy for very low values of r_{QD} so that excitons dominate electronic properties in quantum dots (QDs). The increase of the energy of an exciton with decreasing r_{QD} is called quantum confinement. A QD is a semiconductor nano-crystal that shows such quantum confinement.

The quantum confinement has consequences for optical absorption of QDs. An electron-hole pair photo-generated in a QD has to form an exciton since the interaction between the electron and the hole cannot be avoided. Therefore, optical absorption in QDs is only possible if the photon energy is larger than the sum of the band gap of the bulk semiconductor from which the QD is made and the binding energy of the exciton. In other words, photons with an energy a little bit larger than the band gap of the bulk semiconductor cannot be absorbed in a QD due to the absence of available electronic states for free holes and for free electrons in the QD. Therefore, the absorption edge of semiconductor QDs shifts to the energy of the first excitonic transition in semiconductor QDs. As a result, the absorption edge or optical band gap depends on the size of semiconductor nano-crystals.

Figure 10.5 compares the exciton radius in a bulk semiconductor and in a QD (a), the extension of the corresponding wave functions (b) and optical absorption in a bulk semiconductor and in a QD (c).

The increase of E_g due to quantum confinement presents the opportunity to use semiconductors with very low band gaps such as PbS (E_g is 0.37 eV for PbS crystals (Scanlon, 1958)) in solar cells and to tune band gaps for optimum absorption.

For application in solar cells, QDs have to be embedded into a surrounding matrix consisting of another material. The relative dielectric

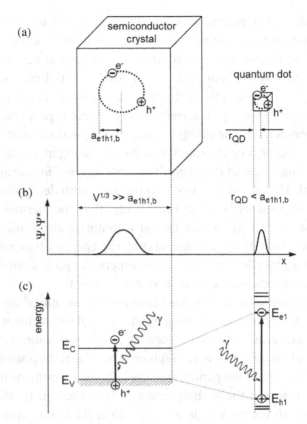

Figure 10.5. Schematic of a bulk semiconductor crystal (bulk) and a semiconductor QD (a), of the extension of an exciton in the bulk and in the QD (b) and of optical absorption in the bulk and in the QD (c). $a_{e1h1,b}$, r_{QD}, E_{h1} and E_{e1} denote the exciton radius in the bulk, the radius of the QD and the energies of the hole and of the electron for absorption in the first excitonic transition, respectively.

constant of the surrounding material (ε_{out}) is usually different than ε_r so that a correction with respect to the change of the Coulomb potential is needed. Both the quantum confinement effect and the correction term of the Coulomb energy depend not only on the size of the semiconductor nano-crystal but also on the shape, the given electron and hole wave functions, the band structure of the semiconductor crystal, the surface conditioning of QDs and the distance between neighboring QDs in a layer of QDs.

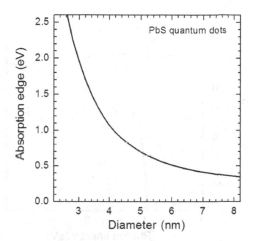

Figure 10.6. Dependence of the absorption edge on the diameter of PbS-QDs.

In a first approximation, the absorption edge of semiconductor QDs can be calculated as a function of r_{QD} (Brus, 1984):

$$E_{abs}(r_{QD}) \approx E_g + \frac{h^2}{8 \cdot r_{QD}^2 \cdot \mu_{eh}} - \left(\frac{1}{\varepsilon_{out}} + \frac{0.79}{\varepsilon_r}\right) \cdot \frac{q^2}{4\pi \cdot \varepsilon_0 \cdot r_{QD}}$$

$$(10.4)$$

Equation (10.4) gives the opportunity to calculate the absorption edge as a function of the diameter of semiconductor QDs. The dependence of the absorption edge of semiconductor QDs on the diameter of PbS-QDs is shown in Figure 10.6.

The absorption edge reaches a value of about 1 or 2 eV for a diameter of PbS-QDs of about 4 or 3 nm, respectively. This means that PbS-QDs with diameters between 3 and 4 nm are suitable as photovoltaic absorbers with respect to the sun spectrum.

The absorption coefficient of semiconductor QDs is of the same order as the absorption coefficient of the bulk semiconductor at larger photon energies where the absorption coefficient is usually somewhat higher in the region of the first excitonic transition and somewhat lower at photon energies a little bit above the first excitonic transition (see, for example, Moreels *et al.* (2009) for optical absorption in PbS-QDs). This is schematically demonstrated for PbS-QDs in Figure 10.7. The absorption coefficients of bulk PbS and PbS-QDs are equal at high photon energies.

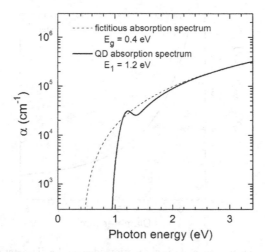

Figure 10.7. Absorption spectrum of a fictitious PbS-QD. The dotted line gives the approximate absorption spectrum of PbS.

High absorption coefficients above 10^4 cm^{-1} are reached close to the absorption edge of QDs. This makes semiconductor QD layers very attractive for solar cells with very thin absorber layer thicknesses.

PbS-QDs are of great interest due to the abundance of lead and sulfur in the earth's crust and due to the high absorption coefficient of PbS at photon energies above 1 eV.

10.2.2 *Solution-based fabrication of semiconductor quantum dots*

Colloidal semiconductor QDs are grown in solution at low temperatures between room temperature and 100–300°C within short time periods of between seconds and several minutes. Solutions with stabilized QDs can be stored for a long time and further processed by casting, layer-by-layer growth or printing. These technological advantages and the excellent absorption behavior make colloidal semiconductor QDs very interesting for future large-scale applications in solar cells.

Nearly perfect colloidal QDs of metal sulfides or metal selenides can be grown from organometallic precursors dissolved in organic compounds (see, for example, Brumer *et al.* (2005)). Incidentally, colloidal QDs have to be grown in an atmosphere free from oxygen and water in order to avoid oxidation of precursors.

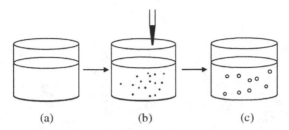

Figure 10.8. Fabrication steps of colloidal QDs consisting of preparation of separate organometallic and sulfur- or selenium-containing precursor solutions (a), fast nucleation of semiconductor nano-crystals during injection of the selenium or sulfur precursor solution into the hot organometallic precursor solution (b) and growth and stabilization of colloidal QDs at reduced temperature (c).

Preparation of colloidal QDs consists of three main steps (Figure 10.8). In the first step, two separate solutions with organometallic and sulfur or selenium precursors are prepared, for example, by dissolving lead (II) acetate or selenium powder in oleic acid (OA) and/or trioctylphosphin (TOP). In the second step, the solution with the sulfur or selenium precursors is injected into the hot solution with the organometallic precursor and fast nucleation of colloidal QDs takes place. Incidentally, instead of solutions with the sulfur or selenium precursors, H_2S or H_2Se can also be used for providing sulfur or selenium. During the third step, seeds grow to colloidal QDs with the desired diameter.

It is important to stabilize and passivate colloidal QDs in the suspension in order to avoid aggregation of colloidal QDs and continuation of chemical reactions at the surface of colloidal QDs. Stabilization and passivation of colloidal QDs is achieved with organic shells around colloidal QDs. The organic shell consists of, for example, TOP molecules (Figure 10.9(a)). Molecules stabilizing QDs are also called surfactant molecules.

Semiconductor QDs are extracted from the colloidal solution by centrifugation. A so-called stock solution contains stabilized colloidal QDs at high concentration. Layers of colloidal QDs can be cast directly from stock solutions. However, long surfactant molecules form barriers for the transfer of excitons and/or charge carriers between neighboring semiconductor QDs. Therefore, long surfactant molecules have to be exchanged by short surfactant molecules so that charge transport becomes possible in photovoltaic absorbers based on semiconductor QDs. For

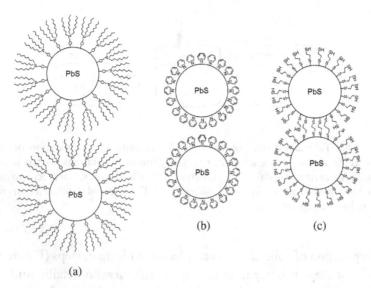

Figure 10.9. Examples for PbS-QDs with stabilizing surfactant molecules of TOP in stock solution (a), with short pyridine molecules as intermediate surfactants exchanged in solution (b) and with short ethane-dithiol molecules as linking surfactants exchanged on the deposited QD layer (c).

example, TOP molecules can be exchanged in solution by much smaller pyridine molecules (Figure 10.9(b)). Semiconductor QD absorber layers are mechanically stabilized by linking neighboring nano-crystals with very short so-called linker molecules, such as ethane-dithiol molecules (Luther *et al.*, 2008b) (Figure 10.9(c)).

10.2.3 *Colloidal quantum dot layers as nano-composite absorbers*

A photovoltaic absorber based on colloidal QDs is a nano-composite absorber consisting of semiconductor nano-crystals embedded in an organic matrix of linker molecules (Figure 10.10). The nano-composite absorber has a band gap given by the quantum confinement of the semi-conductor QDs and an absorption length of the order of several 100 nm.

The dielectric constant of a nano-composite absorber with colloidal QDs depends on the dielectric constants of the inorganic semiconductor and of the organic molecules, on the size of the semiconductor nano-crystals, on the interface between colloidal QDs and organic molecules and on the distance between neighboring semiconductor nano-crystals. The

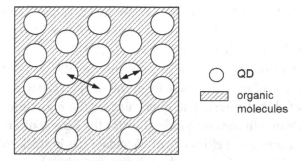

Figure 10.10. Schematic of a photovoltaic nano-composite absorber containing semiconductor QDs and organic linker molecules. The arrows mark the diameter of a given colloidal QD and the distance between two given colloidal QDs.

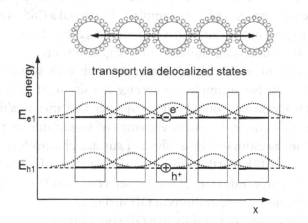

Figure 10.11. Chain of semiconductor QDs with ideally overlapping wave functions of free electrons and holes. The arrow marks charge transport in mini bands.

calculation of the dielectric function of nano-composites as an effective medium with colloidal QDs remains challenging.

10.2.4 *Charge transport and separation in quantum dot solar cells*

The wave functions of electrons, holes and excitons can partially overlap between neighboring QDs. The overlap of wave functions is also called coupling. Figure 10.11 shows an idealized picture of a chain of QDs. Ideally coupled wave functions can form so-called mini bands in which free electrons and holes may be transported at high mobility. However, numerous sources of disorder (fluctuations of distances, diameters and shapes of QDs, local variations of surface coverage with surfactant molecules, ...) make a band-like transport of free charge carriers in colloidal QD layers unlikely (Guyot-Sionnest, 2012).

The surface of QDs plays a decisive role for the charge transport in photovoltaic absorber layers made from colloidal QDs (see also task T10.4 at the end of this chapter). The quenching of the photoluminescence of colloidal QDs in solutions is an indication for the density of surface defects. The photoluminescence efficiency of colloidal QDs in stock solutions is of the order of 10% (see, for example, Talapin *et al.* (2001)), i.e. only one out of ten QDs has no recombination active defects. The photoluminescence efficiency of colloidal QDs has been increased to about 70% by growing a very thin layer of a semiconductor with a larger band gap around the semiconductor nano-crystal, for example, CdS around a CdSe-QD forming a passivating type I hetero-junction (Talapin *et al.*, 2001). However, this kind of surface passivation is not suitable for charge transport.

The density of surface defects increases with each surface treatment of colloidal QDs. For example, an average density of surface hole traps at CdSe-QDs increased from about 2 traps per QD after washing to 7 or 12 traps per QD after subsequent exchange of surfactants by pyridine or benzenedithiol as estimated by analysis of surface photovoltage transients (Zillner *et al.*, 2012). Furthermore, the distribution of surface states is very broad at semiconductor QDs (Fengler *et al.*, 2013) due to the large variability of defect configurations at QD surfaces.

An exciton dissociates rapidly in a QD containing surface states due to trapping at defect states. Charge transport in QD layers can be considered as a superposition of tunneling between defect states and neighboring QDs depending on the density and distribution of surface states at colloidal QDs. Figure 10.12 shows a chain of QDs with surface defects. Hopping between localized states dominates the charge transport. The density of defect states in colloidal QD layers is related to the volume occupied by one QD (about $100 \, nm^3$) and is therefore of the order of 10^{19}–$10^{20} \, cm^{-3}$, where only a small fraction of these defects are usually recombination active.

Electron and hole mobilities in QD layers can be of the order of 0.9 and $0.2 \, cm^2/(Vs)$, respectively, as shown with field-effect transistors with channels consisting of a PbSe-QD layers (Talapin and Murray, 2005). The electron and hole mobilities in colloidal QD layers are significantly

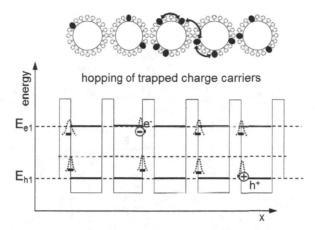

Figure 10.12. Chain of semiconductor QDs with surface states and localized wave functions of trapped electrons and holes. The arrows mark hopping events of a charge carrier between neighboring surface states. Incidentally, excitation of trapped charge carriers into delocalized states is also possible (trap-limited transport). Further, trapped charge carriers can lead to local variations in E_{h1} and E_{e1} (energetic disorder).

lower than for un-doped a-Si:H in which disorder and localized defect states also play a dominant role (see Chapter 9). The diffusion length of photo-generated charge carriers (L_{diff}) can be calculated by using Equation (9.16). In semiconductor QD layers, L_{diff} is of the order of 10 nm, which is much shorter than the absorption length. Therefore, drift is important for transport of photo-generated charge carriers in solar cells based on QDs.

The drift length of photo-generated charge carriers in colloidal QD layers (L_{drift}) can be estimated by using Equation (9.17) and by making a reasonable assumption about a built-in electric field. A value of L_{drift} of the order of 100 nm is obtained for an electric field of about 10^4 V/cm. This demonstrates that the realization of very efficient solar cells with QD absorber layers is challenging.

A QD-based solar cell with drift-controlled transport of photo-generated charge carriers was demonstrated for Schottky contacts with PbSe-QD layers (Luther *et al.*, 2008a). The external quantum efficiency of Schottky contact solar cells with PbSe-QD layers was increased to 60% by optimizing the layer structure for optical absorption (Law *et al.*, 2008). The importance of depletion regions for solar cells with PbS-QDs has been demonstrated based on a TiO_2/PbS-QD depletion region (Pattantyus-Abraham *et al.*, 2010).

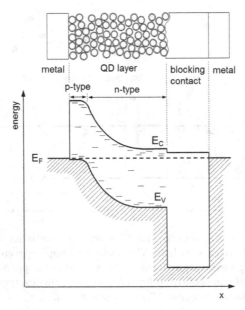

Figure 10.13. Idealized band diagram of a solar cell with a semiconductor QD absorber placed between a very thin highly p-type doped layer and a depleted n-type doped layer on top of a wide band gap buffer layer. The conduction and valence band edges (E_C and E_V) correspond to the E_{h1} and E_{e1} states in the QDs, respectively.

Colloidal QD layers can be p-type doped by adding silver atoms into PbS-QDs (Liu *et al.*, 2012). This can help to adjust regions of built-in electric fields within a pin concept (Chapter 9). The concept of depleted hetero-junction colloidal QD solar cells (Pattantyus-Abraham *et al.*, 2010) combined with a charge-selective pin structure seems most promising for reaching high energy conversion efficiencies with semiconductor QD absorbers. Figure 10.13 shows an idealized band diagram of a solar cell with a thin p-type doped QD layer and a moderately n-type doped QD layer deposited on a wide-gap semiconductor accepting electrons from the n-type doped QD layer. The valence and conduction band edges of the QD absorber layer are obtained by averaging over the distribution of diameters of QDs. The solar energy conversion efficiency of solar cells with a PbS-QD absorber has been increased to more than 6% by increasing the width of the depletion region in a junction between a highly doped p-type layer of PbS-QDs and a moderately n-type doped PbS-QD layer (Liu *et al.*, 2012).

The values of V_{OC} are quite low in comparison to the band gap of colloidal QD absorber layers. For example, values of V_{OC} of 0.51 and 0.24 V were obtained for E_g of about 1.3 eV (PbS-QDs, depleted, on TiO$_2$ (Pattantyus-Abraham *et al.*, 2010)) and 1.1 eV (PbSe-QDs, Schottky barrier (Luther *et al.*, 2008a)), respectively. The value of V_{OC} has been increased to 0.45 V for E_g of 1.03 eV in the case of ternary PbS$_x$Se$_{1-x}$ QDs (Schottky barrier (Ma *et al.*, 2009)).

10.3 Organic Solar Cells

10.3.1 *Organic semiconductors with conjugated π-electron systems*

Different kinds of hybridization of electrons in s- and p-orbitals are possible for the formation of chemical bonds between carbon atoms in organic molecules. Carbon bonds can exist in single, double and triple bonds. Double bonds are formed between carbon atoms in the case of sp^2 hybridization. A double bond with sp^2 hybridization is called a conjugated bond. The sp^2 hybridization is the origin for electrical conductivity and related phenomena in organic polymers (Heeger *et al.*, 1988).

The chemical C=C bonds are formed by the σ-orbitals in the xy-plane (Figure 10.14). The binding energy of the σ-orbitals is of the order of 4 eV. The p$_z$-orbitals along the z-axis do not participate in the chemical bond. The p$_z$-orbitals are quite large so that p$_z$-orbitals can overlap between neighboring carbon atoms with a double bond.

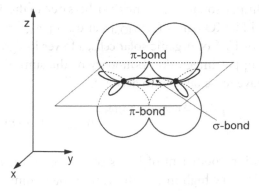

Figure 10.14. Configuration of a conjugated bond.

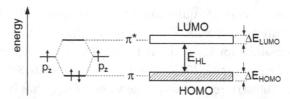

Figure 10.15. Formation of the HOMO and LUMO bands due to overlapping p_z orbitals in conjugated bonds and HOMO–LUMO gap.

The overlap of the p_z-orbitals leads to the formation of weaker bonding and of anti-bonding states. These bonds are also called π-bonds and electrons participating in π-bonds are called π-electrons. Bonding and anti-bonding states of overlapping p_z-orbitals form the so-called HOMO (highest occupied molecular orbital) and LUMO (lowest unoccupied molecular orbital) bands in molecules with a conjugated π-electron system (Figure 10.15). Electrons and missing electrons can be transferred in the LUMO and HOMO bands similarly to the transport of free electrons and holes in the conduction and valence bands of an inorganic semiconductor. Therefore, organic semiconductors can be formed from organic molecules with a conjugated π-electron system.

The difference between the energies at the highest value of the HOMO and the lowest value of the LUMO band is called the HOMO–LUMO gap (Figure 10.15) of an organic semiconductor (E_{HL}). The values of E_{HL} range between 1.5 and 2.5 eV for numerous organic semiconductors, i.e. related organic semiconductors are well suited as photovoltaic absorbers.

The maximum short-circuit current density (I_{SC}^{max}) of organic solar cells is reduced in comparison to inorganic absorbers due to the limited width of the HOMO and LUMO bands (ΔE_{HOMO} and ΔE_{LUMO}, respectively). For the calculation of I_{SC}^{max} of organic solar cells, a lower integration boundary of E_{HL} and an upper integration boundary of the sum of E_{HL}, ΔE_{HOMO} and ΔE_{LUMO} have to be taken into account.

$$I_{SC}^{max} = q \cdot \int_{E_{HL}}^{E_{HL}+\Delta E_{HOMO}+\Delta E_{LUMO}} \Phi_{sun}(h\nu) \cdot d(h\nu) \qquad (10.5)$$

The absorption coefficient of layers consisting of organic conjugated molecules can be very high in the absorption maximum. As an example, Figure 10.16 shows the configuration of a CuPc (copper phthalocyanine)

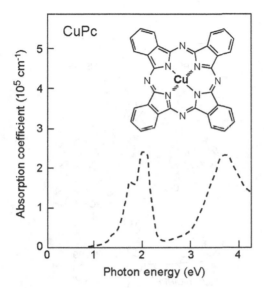

Figure 10.16. Configuration and absorption spectrum of CuPc (copper phthalocyanine) in the range of the air mass (AM)1.5 spectrum. Data points from Schechtman and Spicer (1970).

molecule and the absorption spectrum of a layer of CuPc molecules. Absorption sets in at about 1 eV and the absorption coefficient reaches a value of about $2.4 \cdot 10^5$ cm^{-1} in the maximum at photon energies around 2 eV for a CuPc layer (Schechtman and Spicer, 1970). This means that organic photovoltaic absorbers with a thickness of only 100–200 nm are already sufficient for complete absorption of light with photon energies around the absorption maximum.

There is practically an unlimited variability for the design of conjugated organic molecules. A first classification makes a difference between small conjugated organic molecules or oligomers and large conjugated organic molecules or conjugated polymers. For example, phthalocyanines such as CuPc or ZnPc and porphyrins belong to the class of oligomers. Oligomers can be packed in dense layers in which p_z-orbitals of neighboring molecules overlap so that charge transport is possible across the complete layer of the organic semiconductor.

The family of poly-(1,4-phenylene-vinylene) (PPV) played an important role for the development of organic polymer solar cells (see, for example, Brabec *et al.* (2001a)). PPV consists of a phenyl ring and two

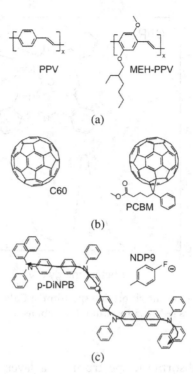

Figure 10.17. Examples of hole-conducting conjugated polymers of the PPV (poly-vinylene-phenylene) family (a), of C60 and PCBM acceptor molecules (b) and of a p-type doped organic semiconductor based on p-DiNPB molecules and NDP9 ionic dopants (c).

carbon atoms with a double bond. The semiconducting polymer is formed by repeating the basic unit (Figure 10.17).

Organic molecules can be modified by replacing, for example, a hydrogen atom by a side group. In this way, the structure or diameter of molecules can be changed and properties needed for technological processing such as solubility in certain organic solvents can be modified. MEH-PPV (see, for example, Yu *et al.* (1995)) is given as an example in Figure 10.17(a). Other families of conjugated polymers used for organic solar cells belong, for example, to polythiophenes (see, for example, Günes *et al.* (2007)).

The C60 molecule (Kroto *et al.*, 1985) plays a decisive role for organic solar cells (Figure 10.17(b)). A C60 molecule contains 60 carbon atoms arranged in a football-like sphere and belongs to the group of fullerenes. C60 molecules are stable and strong acceptor molecules (Haddon *et al.*, 1986), and can accept several electrons. C60 molecules are so stable that they can be deposited by thermal evaporation. The phenyl-C61-butyric acid methyl ester (PCBM) molecule (Hummelen *et al.*, 1995) is a soluble derivative of a C60 molecule, which can be deposited by casting or spin coating from solutions.

Organic semiconductors can be n-type or p-type doped. For example, p-type doping of organic semiconductors can be achieved by adding some small electron acceptor molecules. The small electron acceptor molecules capture electrons from conjugated organic molecules. The charge at the small immobile anions is compensated by mobile positive charge or holes in the π-electron system of the conjugated organic molecule. As an example, a p-DiNPB molecule p-type doped with a NDP9 molecule (Schüppel *et al.*, 2010) is shown in Figure 10.17(c).

Organic semiconductors can be n-type doped by adding electron donors, for example, metal ions or electron donor molecules. Organic semiconductors can be also highly doped. The most prominent highly p-type doped organic semiconductor is PEDOT:PSS in which poly(3,4-ethylendioxythiophene) is doped with poly-stryrenesulfonate (Groenendaal *et al.*, 2000).

The mobility of charge carriers and therefore the diffusion coefficient can vary over many orders of magnitude in organic semiconductors depending on ordering and architecture of molecules (Coropceanu *et al.*, 2007). For example, disordered chains in layers of conjugated polymers result in a hole mobility between 10^{-7} and 10^{-4} cm^2/(Vs) (see, for example, Lebedev *et al.* (1997)). Electron and hole mobilities of 0.4 and 0.8 cm^2/(Vs), respectively, were shown in pure organic crystals such as anthracene (Kepler and Hoestery, 1974). A high intra-chain hole mobility of 600 cm^2/(Vs) was found for local rigid conjugated ladder elements (Prins *et al.*, 2006). For comparison, the electron mobility in a graphene layer can be as high as $2 \cdot 10^5$ cm^2/(Vs) (Bolotin *et al.*, 2008). These examples give an impression about the importance of molecular design for optimization of charge transport in organic solar cells.

Organic molecules have the advantage that all chemical bonds are saturated. Therefore, organic semiconductors do not have electronic

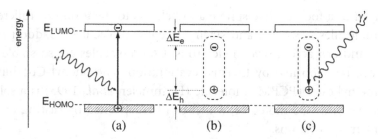

Figure 10.18. Schematic of light absorption in organic semiconductors (a), exciton formation (b) and radiative recombination of an exciton (c).

states caused by unsaturated dangling bonds or surface states in the HOMO–LUMO gap. This opens broad opportunities for the design of interfaces with organic semiconductors.

10.3.2 *Excitons in organic semiconductors*

After photo-generation of an electron-hole pair in an organic semiconductor, an exciton (see, for example, Scholes and Rumbles (2006)) is formed within an extremely short period of time. The energy of the electron and of the hole in the exciton is significantly lower than E_{HL} due to the low dielectric constant of organic semiconductors (Figure 10.18).

The exciton binding energy in organic semiconductors can be estimated using Equation (10.2) and taking into account a relative dielectric constant between 2 and 4 in organic semiconductors. Effective masses of electrons and holes of the order of the free electron mass can be assumed in organic semiconductors. The resulting exciton binding energy is therefore of the order of 0.4–0.8 eV in organic semiconductors.

The exciton binding energy corresponds to a loss in the maximum chemical potential of photo-generated electrons and holes and cannot be regained after dissociation of the exciton. Therefore, the high binding energy of excitons is an important fundamental loss mechanism limiting V_{OC} in organic solar cells.

The Bohr radius of the electron-hole orbit at the lowest energy of the exciton can be estimated using Equation (10.3). As a result, excitons are extended only over several atoms in organic semiconductors. This strong localization causes low diffusion constants and very short diffusion lengths of excitons in organic semiconductors.

The lifetime of excitons is of the order of 1 ns in organic semi-conductors (see, for example, Markov *et al.* (2005)). The diffusion length of excitons is of the order of 5 nm in PPV derivatives (Markov *et al.*, 2005) or 14 nm in ladder-type poly(p-phenylen) (Haugeneder *et al.*, 1999), i.e. the exciton diffusion length is much shorter than the absorption length in organic photovoltaic absorbers. Therefore, small aggregates of organic semiconductors and local dissociation of excitons at the boundaries are required for an organic nano-composite absorber.

10.3.3 *Donor–acceptor hetero-junction and organic nano-composite absorber*

Excitons are neutral quasi-particles. They have to dissociate within the photovoltaic absorber before photo-generated electrons and holes can be separated at charge-selective contacts. Organic molecules have the ability to attract or accept electrons from other molecules (electron-acceptor molecule) or to give away or donate an electron to another molecule (electron-donor molecule). The ability of molecules to accept or to donate electrons is used to dissociate excitons at molecular donor–acceptor hetero-junctions (see, for example, Yu *et al.* (1995)).

Excitons diffuse to a donor–acceptor hetero-junction. If the attractive force of an acceptor molecule to an electron that is part of an exciton is strong enough, then the exciton forms an intermediate state from which the electron can be transferred into the acceptor molecule (Sariciftci *et al.*, 1992) (dissociation of the exciton, see Figure 10.19). The lifetime of the charge-separated configuration is of the order of milliseconds at a temperature

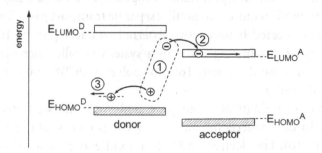

Figure 10.19. Principle of charge separation at a donor–acceptor hetero-junction including the steps of dissociation of an exciton (1), electron transfer into the LUMO band of the acceptor molecules (2) and polaronic hole transport in the donor molecules (3).

of 80 K, i.e. several orders of magnitude longer than the exciton lifetime (Smilowitz *et al.*, 1993).

The first donor–acceptor junction has been demonstrated for potential applications in solar cells with a copper phthalocyanine/perylene tetracarboxylic derivative double layer by Tang (1986). The ultra-fast photoinduced electron transfer from a conducting polymer into C60 (Sariciftci *et al.*, 1992; Brabec *et al.*, 2001b), and its application for solar cells has been shown by Sariciftci *et al.* (1993). C60 molecules or derivatives of C60 molecules are applied as acceptor molecules in all kinds of efficient organic solar cells.

The diffusion length of excitons is much shorter than the absorption length in organic semiconductors. Therefore, the donor–acceptor heterojunction is dispersed or folded in organic photovoltaic absorbers (blended). Domains of donor and acceptor molecules interpenetrate each other in organic photovoltaic absorbers on a scale of the order of 1–10 nm, i.e. organic photovoltaic absorbers are nano-composites, which are also called organic blends. The local dimensions of domains of donor or acceptor molecules can vary depending on deposition technology and post-treatment. The optimization of the dimensions of domains in donor–acceptor nano-composites is very important for reaching high quantum efficiencies in organic solar cells.

From the point of view of charge transport, organic photovoltaic absorbers can be treated as nano-composite semiconductors, the valence and conduction bands of which are formed by the HOMO band of donor molecules and by the LUMO band of acceptor molecules, respectively. However, in a donor–acceptor nano-composite, a certain density of each of the components is required for getting separate transport paths for electrons and holes connected between two external contacts (Figure 10.20). The connectivity of conducting units in a system is called percolation. The percolation threshold is about 16% for polymer/PCBM nano-composites (Hotta *et al.*, 1987).

The diffusion length of photo-generated and locally separated electrons and holes in donor–acceptor nano-composites can be obtained by using Equation (9.16). The electron and hole mobilities depend sensitively on the composition of donor–acceptor nano-composites and amount, for example, to the order of 10^{-5}–10^{-4} cm^2/(Vs) for blended polyfluorene/PCBM

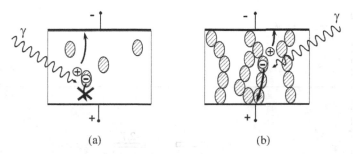

(a)   (b)

Figure 10.20. Schematic donor–acceptor nano-composite with light absorption, exciton dissociation and local charge separation without percolation (a) and with percolation (b) of conduction paths for photo-generated charge carriers.

layers (Pacios *et al.*, 2003). The lifetimes of charge-separated states are of the order of ms at low temperature (Smilowitz *et al.*, 1993). However, photocurrents can decrease within only 1–10 ns, for example, in MEH-PPV/C60 nano-composites (Lee *et al.*, 1993). Therefore, the resulting diffusion length of photo-generated electrons and holes can be of the order of 100 nm or even much less depending on disorder and/or the density of defects in donor–acceptor nano-composites.

Donor–acceptor nano-composites have two band gaps. The transport gap ($E_{HL}^{transport}$) is the difference between the lowest energy of the LUMO band of the acceptor molecules ($E_{LUMO}^{acceptor}$) and the highest energy of the HOMO band of the donor molecules (E_{HOMO}^{donor}).

$$E_{HL}^{transport} = E_{LUMO}^{acceptor} - E_{HOMO}^{donor} \qquad (10.6)$$

The absorption gap of a donor–acceptor nano-composite ($E_{HL}^{absorption}$) is related to the lowest of the HOMO–LUMO gaps of the acceptor or donor molecules, respectively.

$$E_{HL}^{absorption} = E_{LUMO}^{donor} - E_{HOMO}^{donor} \qquad (10.7')$$

or

$$E_{HL}^{absorption} = E_{LUMO}^{acceptor} - E_{HOMO}^{acceptor} \qquad (10.7'')$$

The absorption and transport band gaps of organic photovoltaic nano-composite absorbers limit I_{SC} and V_{OC}, respectively, of organic solar cells.

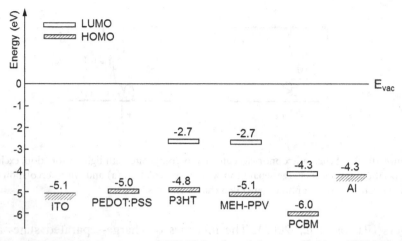

Figure 10.21. Ionization energy of HOMO bands and electron affinities of LUMO bands for P3HT, MEH-PPV, PEDOT:PSS and PCBM and work functions of Al and ITO.

10.3.4 *Charge-selective and ohmic contacts in organic solar cells*

The potential energy of electrons decreases and the potential energy of holes increases at the charge-selective electron contact and *vice versa* at the charge-selective hole contact. The change of the potential energy at charge-selective contacts can be realized, for example, with a pn-junction where minority photo-generated charge carriers become majority charge carriers.

The ionization energy of the HOMO band of donor molecules and the electron affinity of the LUMO band of acceptor molecules are important for charge-selective and ohmic contacts in organic solar cells from the material point of view of losses in potential energy. As an example, Figure 10.21 gives the energies of HOMO and LUMO bands of P3HT and MEH-PPV as donor molecules and of PCBM as an acceptor molecule. The work functions of the indium tin oxide (ITO) and Al contacts are given as well. PEDOT:PSS is used as a highly doped organic semiconductor at the ohmic contact with ITO.

The ionization energies of the HOMO bands of P3HT and MEH-PPV are well aligned with the ionization energy of highly doped PEDOT:PSS so that holes can be transferred from P3HT or MEH-PPV into PEDOT:PSS without additional losses in potential energy. Therefore, highly p-type doped PEDOT:PSS is a hole-collecting contact for P3HT. The ionization

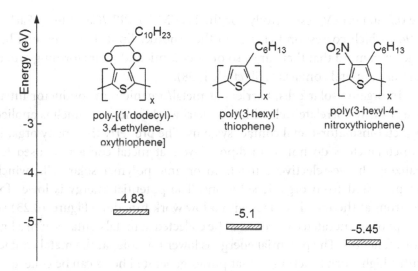

Figure 10.22. Ionization energy of HOMO bands for polythiophenes with different side groups (values taken from Scharber *et al.* (2006)).

energy of PEDOT:PSS is well aligned with the work function of ITO. Therefore ITO and PEDOT:PSS form an ohmic contact (see Chapter 5). The electron affinity of PCBM is well aligned with the work function of Al so that electrons can be transferred from PCBM into Al without significant losses in potential energy.

The ionization energies of HOMO bands can be engineered over a relatively wide range by varying side groups at a conjugated polymer. Figure 10.22 shows examples for polythiophenes with different side groups and corresponding ionization energies of the HOMO bands changing between −4.83 and −5.45 eV for the given examples (Scharber *et al.*, 2006). The values of V_{OC} increase with increasing ionization energy of the HOMO band (Scharber *et al.*, 2006), i.e. potential energy losses at contacts can be minimized by molecular engineering.

Donor and acceptor molecules electrically compensate each other in organic photovoltaic nano-composite absorbers and as such a donor–acceptor nano-composite absorber can be treated like an un-doped or intrinsic semiconductor. Furthermore, metal/organic semiconductor interfaces can often be treated as contacts close to the Schottky limit (Equation (5.5)) since all chemical bonds are saturated in organic molecules. It has been shown, for example, that the barriers for hole and electron injection are 0.2 and 0.1 eV, respectively, for the ITO/MEH-PPV/Ca system (Parker, 1994), which correspond closely to the Schottky limit. However, one has to keep in mind that there are also organic semiconductors forming dipole barriers at metal contacts (Hill *et al.*, 1998).

In organic solar cells, barriers at metal/organic semiconductor interfaces usually correlate well with the metal work function, which is applied in metal-insulator-metal contact systems. The property that many organic semiconductors do not form dipole layers at metal contacts is used for realizing charge-selective contacts in organic polymer solar cells, which are processed from organic solutions. The potential energy is lower for electrons at the metal contact with a low work function (Figure 10.23) so that photo-generated electrons can be collected at metal contacts with a low work function. The potential energy is lower for holes at the metal contact with a high work function so that photo-generated holes can be collected.

Work functions or potential energies of free charge carriers can also be engineered by doping of organic semiconductors. Interface dipoles can usually be neglected at charge-selective organic semiconductor/organic semiconductor hetero-junctions since chemical bonds are saturated in organic molecules and since the nature of the chemical bonds is similar in all organic semiconductors. Therefore, offsets of HOMO and LUMO

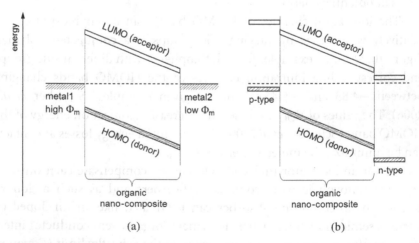

Figure 10.23. Band diagrams of donor–acceptor nano-composite absorbers contacted between two metals with different work functions (Φ_m) (a) and between p-type and n-type doped organic semiconductors (b).

bands at organic semiconductor/organic semiconductor hetero-junctions follow from the differences of the ionization energies and electron affinities, respectively. Figure 10.23(b) shows a schematic band diagram of a pin organic solar cell. In this type of solar cell, a donor–acceptor nano-composite absorber is sandwiched between a p-type (for example, p-DiNPB doped with NDP9) and an n-type (n-type doped C60) doped organic semiconductor. The pin concept is usually applied for organic solar cells based on small organic molecules, which can be deposited by thermal evaporation (see, for example, Schüppel *et al.*, (2010)).

Additional layers of organic molecules can reduce recombination at ohmic contacts by reducing diffusion of excitons to ohmic contacts. For example, a layer of bathophenanthroline (Bphen) molecules blocks diffusion of excitons from a C60 layer to an Al contact (Figure 10.24) but enables transfer of electrons from C60 to the ohmic contact (see, for example, Schüppel *et al.* (2010)).

The open-circuit voltage of organic solar cells operated under illumination of AM1.5 is given approximately by the difference of $E_{LUMO}^{acceptor}$ and E_{HOMO}^{donor} and an energy of the order of about 0.3 eV (Scharber *et al.*, 2006).

$$q \cdot V_{OC} \approx E_{LUMO}^{acceptor} - E_{HOMO}^{donor} - 0.3\,eV \tag{10.8}$$

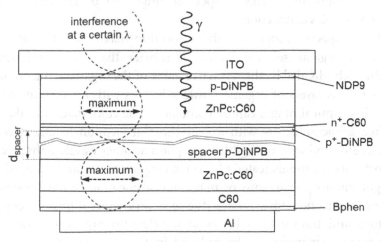

Figure 10.24. Schematic cross section of an organic pin–pin solar cell with a p-DiNPB spacer for optimization of absorption. The layer thicknesses are adjusted so that the maximum light intensity is reached in the donor–acceptor nano-composite absorber layers.

The energy of 0.3 eV in Equation (10.8) is related to the difference between the quasi-Fermi-levels of electrons and holes and the edges of the LUMO and HOMO bands of acceptor and donor molecules, respectively, in the donor–acceptor nano-composite absorber. The value of 0.3 eV is quite similar, for example, to the difference of the band gaps of c-Si (see Chapter 7) and GaAs (see Chapter 8) and V_{OC} in world record c-Si or GaAs solar cells, respectively, under illumination at AM1.5.

10.3.5 *Optimization routes of organic solar cells*

Organic semiconductors offer some unique opportunities for the optimization of solar cells. The absorption gap of organic semiconductors and the energies of the HOMO bands can be tuned over a relatively wide range. In this way, the discrepancy between the absorption and transport gaps can be minimized so that, in turn, V_{OC} can be maximized (see, for example, Wilke *et al.* (2012)) at maximum I_{SC}. However, it is not really clear how far losses due to exciton binding energy and due to energy needed for dissociation of excitons can be minimized.

Organic semiconductors can be highly doped so that ohmic contacts can be formed between layers of organic semiconductors with different absorption gaps. Therefore, stacked multi-junction organic solar cells absorbing light in a broader spectral range can be realized by using integrated series connection.

In the spectral range near the absorption maximum, the absorption length of organic semiconductors is shorter than the wavelength of exciting light divided by the refractive index of the organic semiconductor. Furthermore, layers of organic semiconductors with absorption maxima in distinct optical ranges can be combined. The combination of different organic semiconductors with a very short absorption length at certain wavelengths and with negligible absorption at other wavelengths gives the opportunity for sophisticated photon management leading to a maximum of light intensity in regions of maximum absorption in layer systems of organic semiconductors. Losses due to a very short diffusion length of excitons and transmission losses at wavelengths not very close to the absorption maximum can be reduced in this way. Figure 10.24 shows a schematic cross section of an optimized organic solar cell including a transparent spacer for optimum photon management between two stacked organic solar cells in pin-spacer-pin configuration (Drechsel *et al.*, 2005). The energy conversion efficiency can exceed 12% for such an organic solar cell at an overall thickness of the ZnPc:C60 nano-composite absorber less than 100 nm.

The mobility of electrons and holes usually increases with increasing temperature in organic nano-composite absorbers. This effect can even lead to an increase of the energy conversion efficiency with increasing temperature — in contrast to conventional solar cells based on inorganic semiconductors for which V_{OC} and the efficiency generally decrease with increasing temperature. Therefore, the energy conversion efficiency of organic solar cells can be optimized for an expected average temperature range at which an organic solar cell is operated.

10.4 Dye-Sensitized Solar Cells

10.4.1 *Light absorption in dye molecules*

The absorption of a photon by a dye molecule leads to the excitation of an electron from the ground state into an excited state of the dye molecule (Figure 10.25). The absorption spectra of dye molecules are usually rather narrow. There is a huge variability of natural and synthetic dye molecules absorbing light in different spectral regions, which are very interesting for photovoltaic solar energy conversion.

Dye molecules strongly absorb light in their absorption maximum. The absorption of dye molecules is described by the molar extinction coefficient (ε_{mol}). The molar extinction coefficient is defined by the decadic decrease of the intensity of light transmitted through a dye solution, in contrast to the absorption coefficient, which is defined by the exponential decrease of the intensity of transmitted light through a layer. The number of photons

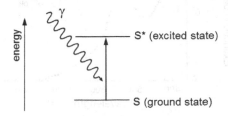

Figure 10.25. Photon absorption by a dye molecule.

passing a kuvette with transparent windows and a thickness d_K of the solution is given by the following equation (reflection losses are considered in N_{ph0}):

$$N_{ph}(\lambda) = N_{ph0}(\lambda) \cdot 10^{-\varepsilon_{mol} \cdot d_K \cdot C_{dye}} \qquad (10.9)$$

The concentration (C_{dye}) is given in the unit mol/l or M. Therefore the molar extinction coefficient is given in the unit of $M^{-1}cm^{-1}$. For example, 90% of photons are absorbed by the dye molecules if the product of the molar extinction coefficient, the thickness of the kuvette and the concentration of dye molecules in the solution is equal to 1.

Ruthenium-based dye molecules played a very important role in the development of dye-sensitized solar cells. As an example, Figure 10.26 shows the configuration and the spectrum of the molar extinction coefficient of the so-called N3 (N stands for Nazeruddin who developed the given dye molecule and a whole family of related dyes (Nazeeruddin *et al.*, 1993)) molecules dissolved in ethanol. The N3 dye molecule contains one Ru atom which is surrounded by two bi-pyridin, and two thiocyanate moieties. Two carboxylic groups are connected with each bi-pyridin moiety. The geometric size of such a molecule is of the order of 1 nm. The N3 dye molecule has a relatively broad extinction spectrum and a large molar extinction coefficient of about $14000\,M^{-1}cm^{-1}$ in the maximum at 532 nm. Absorption of N3 dye molecules sets in at wavelengths around 750 nm.

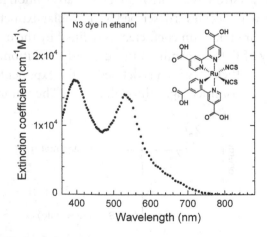

Figure 10.26. Configuration and extinction spectrum of N3 dye molecules in ethanol.

A second absorption maximum appears for N3 dye molecules at about 400 nm.

A density of about $4 \cdot 10^{16}$ N3 molecules/cm^3 is necessary for the absorption of 90% of incident photons with energies close to the absorption maximum. The density of about $4 \cdot 10^{16}$ N3 molecules/cm^3 would correspond to about 500 mono-layers or to an absorption length of about 250 nm of densely packed N3 dye molecules. Similarly high or even higher molar extinction coefficients are obtained for many other dye molecules absorbing light in a range between about 800 and 400 nm.

From the point of view of the absorption gap and of the absorption coefficient, dye molecules are very good absorbers for photovoltaic solar energy conversion. However, electrons photo-excited in dye molecules are not mobile in densely packed layers of dye molecules and so thick layers of dye molecules cannot be applied for solar cells.

10.4.2 *Dye sensitization of a sintered network of TiO$_2$ nanoparticles*

For application in solar cells, electrons photo-excited in dye molecules have to be transferred from excited dye molecules into a material in which electrons are mobile. Metal oxides with large band gaps are suitable materials for the transport of electrons. Dye molecules can be adsorbed on the surface of metal oxides with large band gaps.

The transfer of mobile charge carriers into an insulator is called injection. Ultra-fast electron injection from excited dye molecules into semiconductor electrodes has been shown (Eichberger and Willig, 1990).

Electrons photo-excited in a dye molecule can be injected into the conduction band of a metal oxide if the energy of the excited electron is equal to or larger than the conduction band edge of the metal oxide. The transfer of electrons photo-excited in a dye molecule at a photon energy lower than the band gap of a metal oxide is called sensitization of the metal oxide. Sensitization means that a metal oxide with a large band gap becomes sensitive in terms of a photocurrent induced by photons with energy less than the band gap of the metal oxide.

The anatase phase of TiO$_2$ is usually applied for electron transport in dye-sensitized solar cells (O'Regan and Grätzel, 1991). Figure 10.27 depicts the unit cell of anatase and the absorption spectrum of an anatase layer deposited by a sol-gel process. Titanium is sixfold coordinated in TiO$_2$.

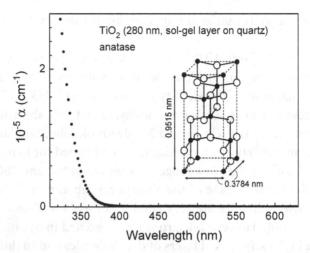

Figure 10.27.   Unit cell of TiO₂ (anatase) and absorption spectrum of a sol-gel TiO₂ layer.

The unit cell of anatase is quite large due to the distorted positions of four oxygen atoms from a plane in the central TiO_6-octahedron of the unit cell (for details see Diebold (2003) and references therein). The band gap of anatase varies between 3.2 and 3.4 eV depending on preparation. Optical absorption sets in for anatase layers or in anatase nanoparticles at wavelengths shorter than 370–380 nm.

Dye molecules can be adsorbed at the surface of anatase by linking with carboxylic groups resulting in a stable . . . -O-Ti-O-C-. . . surface bond configuration. Therefore carboxylic groups are usually attached to dye molecules used for dye-sensitized solar cells (Nazeeruddin *et al.*, 1993).

The injection of an excited electron from a dye molecule into the conduction band of TiO₂ (see also Figure 10.28) can be described by the following expression:

$$S^* \rightarrow S^+ + e^-_{TiO2} \tag{10.10}$$

S^* and S^+ denote the excited and charged states of a dye molecule. The mobility of electrons is about 1 cm²/(Vs) in TiO₂ crystals (Tang *et al.*, 1994), i.e. injected electrons can be transported well in TiO₂.

The dielectric constant of TiO₂ is larger than the dielectric constant of the surrounding ambience. Therefore, the potential energy of an electron injected into a TiO₂ nanoparticle increases towards the surface of the

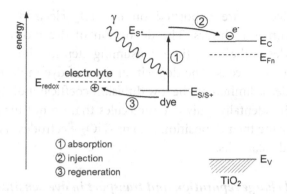

Figure 10.28. Elementary steps of photon absorption, electron injection from the excited dye molecule into the conduction band of TiO_2 and of regeneration of the dye molecule.

TiO_2 nanoparticle. This so-called dielectric screening (Keldysh, 1979) is important for low surface recombination at TiO_2 nanoparticles.

The diameter of a dye-sensitized TiO_2 nanoparticle should not be too small in order to avoid tunneling recombination of an electron injected into the TiO_2 nanoparticle with a charged dye molecule at the surface. The optimum diameter of TiO_2 nanoparticles is of the order of 25 nm for dye-sensitized solar cells.

The internal surface area of dye-sensitized solar cells has to be folded by 500–1000 times for optimum light absorption by adsorbed dyes. Large internal surface areas can be reached with porous or nanoporous structures. TiO_2 nanoparticles can be well prepared in large quantity (production of white pigments) and sintered to networks of interconnected nanoparticles at moderate temperatures of about 450°C.

A nanoporous electrode of interconnected TiO_2 nanoparticles can be produced, for example, by screen-printing of a paste containing TiO_2 nanoparticles on a transparent contact such as SnO_2:F followed by firing in air (O'Regan and Grätzel, 1991). TiO_2 nanoparticles are mixed with an organic binder to form a printable paste. Organic molecules of the binder are burned to CO_2 and sintering necks are formed between neighboring TiO_2 nanoparticles during firing in air for about 30 min. The internal surface area of a nanoporous TiO_2 electrode (np-TiO_2) is about 1000 times larger than the geometric area for a thickness of the np-TiO_2 electrode of about 30 μm.

Dye molecules are deposited on np-TiO$_2$ electrodes from a dye solution within 6–12 hours. The deposition of dye molecules into np-TiO$_2$ electrodes is the most time-consuming step for the preparation of dye-sensitized solar cells. The deposition time of dye molecules into np-TiO$_2$ electrodes is limited by the low diffusion coefficient of dye molecules in solution. Incidentally, only dye molecules that do not form clusters or aggregates during their deposition into np-TiO$_2$ electrodes can be used for dye-sensitized solar cells.

10.4.3 *Local charge separation and transport in dye-sensitized solar cells*

Local charge separation takes place in dye-sensitized solar cells due to the injection of excited electrons from dye molecules into TiO$_2$. The remaining charged dye molecules have to be neutralized or regenerated for subsequent absorption events. An electric contact covering the whole internal surface area of the dyed np-TiO$_2$ electrode is required for regeneration of charged dye molecules. Related electric contacts can be realized with electrolytes completely filling the pores between interconnected TiO$_2$ nanoparticles (O'Regan and Grätzel, 1991). The positive charge is transferred from the charged dye molecule into the electrolyte during the elementary step of regeneration (Figure 10.28).

An electrolyte is characterized by the redox potential (E$_{redox}$) of the charge-transporting ionic species. The redox potential of an electrolyte is equivalent to the Fermi-energy of a semiconductor. The density of ions in electrolytes is huge in comparison to the density of photo-generated charge carriers in photovoltaic absorbers. Therefore, the redox potential is fixed in dye-sensitized solar cells. The difference between the Fermi-energy of injected electrons in the interconnected TiO$_2$ nanoparticles (E$_{Fn}$) and the redox potential of the electrolyte corresponds to the Fermi-level splitting in illuminated bulk semiconductors.

The np-TiO$_2$ electrode with adsorbed dye molecules surrounded by a redox electrolyte can be considered as a complex photovoltaic nano-composite absorber (Figure 10.29). The maximum energy of the dye in the ground state and the minimum energy of the dye in the excited state correspond to the absorption gap of the nano-composite absorber. The redox potential of the electrolyte and the Fermi-energy of the electrons in

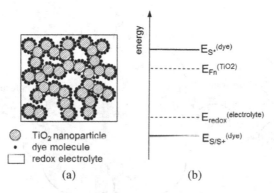

Figure 10.29. Nano-composite absorber of a dye-sensitized solar cell consisting of interconnected TiO_2 nanoparticles, dye molecules and electrolyte (a) and energy scheme of the nano-composite absorber in a dye-sensitized solar cell.

the np-TiO_2 correspond to the Fermi-energy of the majority and minority charge carriers, respectively, in the nano-composite absorber.

The regeneration of charged dye molecules requires a redox reaction during which an electron is transferred from an ionic species in the electrolyte to the charged dye molecule. The extraction of an electron from an ionic species in the electrolyte increases the oxidation state of the ionic species in the electrolyte, i.e. the ionic species is further oxidized (Figure 10.30). The transfer of an electron from an ionic species in the electrolyte to the charged dye molecule is equivalent to the transfer of a hole from the charged molecule into the electrolyte.

The iodide (I^-/I_3^-) redox electrolyte played a decisive role in the development of dye-sensitized solar cells (O'Regan and Grätzel, 1991). The first great advantage of the I^-/I_3^- electrolyte is that the negatively charged ionic species are repulsed from the negatively charged TiO_2 nanoparticles. In this way, the recombination of electrons from TiO_2 nanoparticles into the electrolyte is strongly suppressed. The second great advantage of the I^-/I_3^- electrolyte is that a catalytic activation is required for electron transfer at conducting electrodes. This is the prerequisite for realizing a very simple charge-selective contact with the np-TiO_2/dye/electrolyte nano-composite absorber and a SnO_2:F electrode into which only electrons can be transferred. The regeneration of charged dye molecules can be described by the following expression:

$$2S^+ + 3I^- \rightarrow 2S + I_3^- \tag{10.11}$$

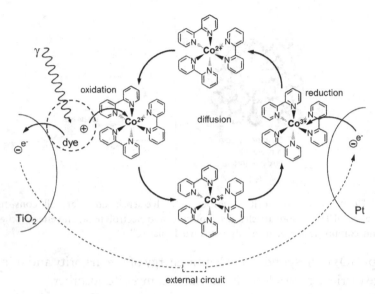

Figure 10.30. Example for charge transport with a hypothetic ionic Co species in a redox electrolyte of a dye-sensitized solar cell.

Equation (10.11) shows that two charged dye molecules and three I^- ions are involved in one elementary regeneration step. This complex redox reaction limits the regeneration rate.

An alternative redox reaction is based on ionic molecular complexes with a Co^{2+}/Co^{3+} redox couple (Nusbaumer *et al.*, 2001, 2003). Figure 10.30 illustrates the charge transport in an electrolyte via hypothetic OR-Co^{2+}/OR-Co^{3+} species.

$$S^+ + OR - Co^{2+} \rightarrow S + OR - Co^{3+} \qquad (10.12)$$

In Equation (10.12), OR$-$ denotes the organic rest of the ionic species containing Co^{2+} or Co^{3+}.

The concentration of ionic species with higher oxidation state increases in the nano-composite absorber of the dye-sensitized solar cell due to regeneration of charged dye molecules. Therefore, the ionic species with higher oxidation state diffuse into the bulk electrolyte towards a platinized counter electrode. The platinized electrode is connected with the ohmic contact collecting photo-generated electrons from np-TiO_2. Electrons flow from np-TiO_2 through an external load to the Pt nanoparticles where the

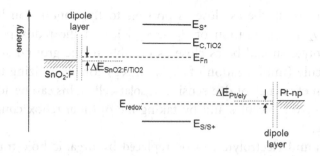

Figure 10.31. Energy diagram of a dye-sensitized solar cell under open-circuit voltage.

ionic species are reduced to a lower oxidation state as shown in Equations (10.13) and (10.14) for the reduction of I_3^- and $OR - Co^{3+}$, respectively.

$$2e_{Pt}^- + I_3^- \rightarrow 3I^- \tag{10.13}$$

$$e_{Pt}^- + OR - Co^{3+} \rightarrow OR - Co^{2+} \tag{10.14}$$

The charge-selective contact of dye-sensitized solar cells is formed at the interface where electrons, which are minority charge carriers in the nano-composite absorber, become majority charge carriers. At the same time, charge transfer from the redox electrolyte is blocked at the charge-selective contact. The charge-selective contact is at the np-TiO_2/SnO_2:F interface in dye-sensitized solar cells (Figure 10.31).

Permanent polarization takes place at hetero-contacts and metal–electrolyte contacts due to the formation of interface dipole layers. Interface dipoles introduce additional drops of the electrostatic potential reducing the potential at external leads.

The difference between E_{Fn} and E_{redox} in the nano-composite absorber of a dye-sensitized solar cell is reduced at the contacts by the voltage drop across the dipole at the SnO_2:F/np-TiO_2 interface ($\Delta E_{SnO2:F/TiO2}$) and by the dipole at the electrolyte/Pt interface ($\Delta E_{Pt/ely}$). The value of V_{OC} of a dye-sensitized solar cell becomes:

$$q \cdot V_{OC} = E_{Fn} - E_{redox} - \Delta E_{SnO2:F/TiO2} - \Delta E_{Pt/ely} \tag{10.15}$$

Dipoles at SnO_2:F/np-TiO_2 and at electrolyte/Pt interfaces should be minimized for getting a maximum photovoltage at the external leads of a dye-sensitized solar cell. Furthermore, the energy of the excited state of the dye should be as close as possible to the conduction band edge of np-TiO$_2$ (minimization of E_{S*}-$E_{C,TiO2}$), and the redox potential of the electrolyte should be as close as possible to the ground state of the dye molecule (minimization of E_{redox}-$E_{S/S+}$) for maximizing the energy conversion efficiency of dye-sensitized solar cells. This can be achieved by designing dye molecules and by taking the optimal redox couple in the electrolyte.

The liquid electrolyte can be replaced by organic hole-transporting materials such as spiro-OMeTAD (Bach *et al.*, 1998). TiO$_2$ can be also sensitized with the organic-inorganic hybrid material CH$_3$NH$_3$PbI$_3$ (Kim *et al.*, 2012).

10.4.4 *Passivation and co-sensitization of dye-sensitized solar cells*

Electrons injected into the conduction band of np-TiO$_2$ can recombine by back transfer to charged dye molecules or by reducing species in the electrolyte (Figure 10.32). Recombination of electrons in dye-sensitized solar cells can be considered as local surface recombination leading to an effective electron lifetime in nano-composite absorbers of dye-sensitized solar cells. Passivation strategies of TiO$_2$ nanoparticles play an important role for high-efficiency dye-sensitized solar cells.

The recombination rates to charged dye molecules (R_{S+}) and to species in the electrolyte (R_{ely}) are proportional to the density of free electrons in np-TiO$_2$ (n_{TiO2}). The density of available species in the electrolyte is much higher than n_{TiO2} while the density of charged dye molecules is of the order of n_{TiO2} and depends on the regeneration rate (R_{reg}). The rate constants are not well known for np-TiO$_2$/dye/electrolyte systems (Peter, 2009). The

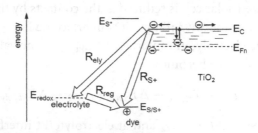

Figure 10.32. Trapping and transport of electrons in TiO$_2$ and recombination of electrons with charged dye molecules or ions or molecules in the electrolyte.

point is that the rate constants depend in a complex way on species in the electrolyte and their concentrations, on geometry, moieties and side groups of dye molecules as well as on bond configuration of dye molecules at TiO_2.

Elementary recombination events of electrons and electron diffusion are retarded by fast trapping of electrons within ps into exponentially distributed trap states near the conduction band edge of np- TiO_2. Exponentially distributed trap states are caused by disorder in the relatively large unit cell of anatase in TiO_2 nanoparticles. The energy and the density of trap states characterizing the exponential distribution are of the order of 100 meV and 10^{19}–10^{20} cm^{-3}, respectively (Dittrich, 2000).

De-trapping times are longer for electrons trapped in deeper trap states. The average de-trapping time decreases with increasing density of free electrons in np-TiO_2 (n_{TiO2}) since more of the deepest traps are occupied when E_{Fn} shifts towards E_C. The effective electron lifetime (τ_n) decreases and the effective electron diffusion coefficient increases with increasing n_{TiO2} so that their product and hence the effective electron diffusion length remain constant (Fisher *et al.*, 2000). The intensity dependent effective electron lifetime can be described by the change of the density of trapped electrons ($n_{t,TiO2}$) and the change of n_{TiO2} (Bisquert and Vikhrenko, 2004).

$$\tau_n = \tau_0 \cdot \left(1 + \frac{\partial n_{t,TiO2}}{\partial n_{TiO2}}\right) \qquad (10.16)$$

High values of I_{SC} are obtained for efficient dye-sensitized solar cells, for example, 21 mA/cm^2 for a solar cell with a Ru-based dye (Han *et al.*, 2012). A high value of I_{SC} gives evidence that the effective electron diffusion length is about three times larger than the thickness of the nano-composite absorber in efficient dye-sensitized solar cells. The value of n_{TiO2} can be estimated by using Equation (3.9) if assuming a reasonable lifetime of 1–10 ms and an absorption length of 30 μm in a nano-composite absorber of a dye-sensitized solar cell. The resulting value of n_{TiO2} is of the order of 10^{16}–10^{17} cm^{-3}.

The open-circuit voltage in dye-sensitized solar cells is related to the change of the Fermi-level of electrons in TiO_2. A value of V_{OC} of 0.743 V (Han *et al.*, 2012) corresponds to a change of the density of free electrons under illumination by more than 12 orders of magnitude. The density of free electrons in the dark ($n_{0,TiO2}$) is equilibrated by the redox potential of the electrolyte.

The value of $n_{0,TiO2}$ can be reduced and therefore V_{OC} can be increased if the difference between E_C and E_{redox} is increased. For a given electrolyte, an increase of the difference between E_C and E_{redox} can be achieved if the electron affinity of np-TiO$_2$ is reduced, for example, by implementing a surface dipole layer where the surface dipole layer should be dense enough to avoid penetration of screening ions from the electrolyte to the TiO$_2$ surface. Dipole layers can be realized by diade molecules consisting of a donor and an acceptor moiety linked with a bridge with a conjugated π-electron system (see, for example, Macor *et al.* (2012)). The acceptor moiety of the diade points to the TiO$_2$ surface. An additional advantage of a dense passivating diade layer is that the distance between the TiO$_2$ surface and the electrolyte is increased so that the recombination rate of electrons from np-TiO$_2$ into the electrolyte is reduced.

The spectral absorption range and therefore I_{SC} of a dye-sensitized solar cell can be increased by co-sensitization with dye molecules having complementary extinction spectra. Figure 10.33 shows an example of

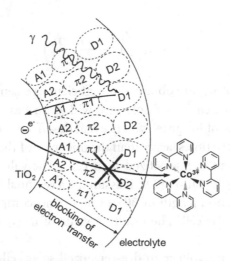

Figure 10.33. Principle of passivation and co-sensitization of TiO$_2$ nanoparticles in dye-sensitized solar cells with densely packed diade molecules containing small acceptor moieties (A1 and A2, pointed to the TiO$_2$ surface), π conjugated bridges (π1 and π2) and large light-absorbing donor moieties (D1 and D2, pointed to the electrolyte).

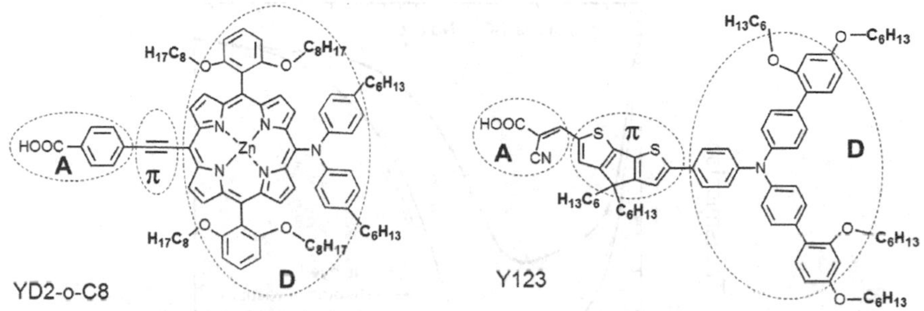

Figure 10.34. Diade dye molecules with acceptor, π-conjugated bridge and donor moieties used in co-sensitized solar cells with Co^{2+}/Co^{3+}-based redox electrolyte by Yella *et al.* (2011). The molecules contain long C_6H_{13} and/or C_8H_{17} side groups for additional passivation.

a well-passivated TiO_2 surface with diade molecules also providing co-sensitization by light-absorbing donor moieties.

Figure 10.34 shows the configuration of two diade molecules which have the functions of co-sensitization, TiO_2 surface passivation and minimization of $n_{0,TiO2}$ (Yella *et al.*, 2011). The molecules contain long chain side groups filling additional space between neighboring adsorbed molecules. The application of related diade molecules allowed the use of a Co^{2+}/Co^{3+}-based electrolyte giving the opportunity for further increase of the difference between E_C and E_{redox}. A value of V_{OC} as high as 0.935 V and a solar energy conversion efficiency of 12.3% were reached for a dye-sensitized solar cell with the given diade molecules (Yella *et al.*, 2011). However, the values of V_{OC} still remain quite low in comparison to the onset of optical absorption in dye-sensitized solar cells. A decrease of dye aggregation and additional passivation is reached by enveloping porphyrin molecules with alkoxyl and/or alkyl chains (Wang *et al.*, 2012).

10.4.5 *About the inherent stability of dye-sensitized solar cells*

Anatase is an efficient photocatalyst, which can be used, for example, for the oxidation of organic molecules or for water splitting (Fujishima and Honda, 1972). Free holes in the valence band of anatase contribute to the decomposition of organic molecules, which is useful for water cleaning (see, for example, Vautier *et al.* (2001)) but leads to degradation of dye-sensitized solar cells. Therefore, the illumination of dye-sensitized anatase nanoparticles with ultraviolet light must be avoided, i.e. optical filters

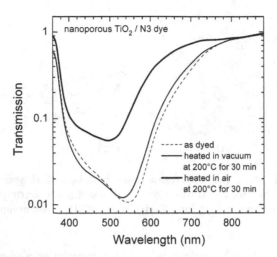

Figure 10.35. Transmission spectra of N3 dye molecules deposited on nanoporous TiO$_2$ before (dashed line) and after heating in vacuum (thin solid line) or air (thick solid line) at 200°C for 30 min.

transmitting visible light but absorbing ultraviolet light have to be placed in front of a dye-sensitized solar cell.

Oxygen or water molecules can be activated at titania surfaces so that oxidation of organic molecules becomes possible. This effect is used, for example, for cleaning of TiO$_2$ surfaces during firing in air at 450°C. The degradation of N3 dye molecules in air already starts at much lower temperatures and is demonstrated in Figure 10.35 for heating in air at 200°C for 30 min. Therefore, oxygen and water molecules have to be avoided in electrolytes of dye-sensitized solar cells.

Free holes and reactive surface defects can be thermally generated in nanoporous anatase. An example of decomposition of N3 dye molecules by less than 10% is given in Figure 10.35 after heating in vacuum at 200°C for 30 min. The band gap of anatase is about 3.2 eV. Therefore, the generation rate of thermally generated charge carriers at room temperature is reduced by about ten orders of magnitude in comparison to 200°C if considering the density of intrinsic charge carriers (Equation (2.32)). This shows that degradation can be neglected for dye molecules deposited on anatase and stored at room temperature in vacuum or in ambience free from oxygen or water molecules.

Free radicals leading to a degradation of dye molecules in dye-sensitized solar cells may be formed during photo-excitation of dye molecules. The number of excitation and neutralization cycles which have to sustain a dye molecule during the lifetime of a dye-sensitized solar cell can be estimated if taking into account a lifetime of the solar cell of 25 years and a density of dye molecules of about $5 \cdot 10^{16}$ cm^{-2} corresponding to about 500 mono-layers of dye molecules. The integrated flux of photons absorbed at AM1.5 is about $1 \cdot 10^{17}$ cm^{-2}s^{-1} for a dye-sensitized solar cell. Therefore, each dye molecule in a dye-sensitized solar cell can absorb about one or two photons within one second. It has been demonstrated that dye molecules sustain more than 10^9 cycles of excitation and neutralization, which is more than enough for the lifetime of a dye-sensitized solar cell. The stability of dye-sensitized solar cells has been tested under thermal aging at AM1.5 at 60°C and no degradation has been observed after 1000 h (Wang *et al.*, 2005).

The formation of free radicals in dye-sensitized solar cells is suppressed due to ultra-fast transfer of excited electrons from the dye molecule into the conduction band of TiO$_2$ (Eichberger and Willig, 1990). This provides an inherent stability of excited dye molecules. Therefore, degradation of dye-sensitized solar cells is related to components such as the sealing of contacts, which can be technologically optimized.

10.4.6 *Solar cells sensitized with methyl-ammonium lead iodide*

Methyl-ammonium lead iodide (CH$_3$NH$_3$PbI$_3$) is an ionic photovoltaic absorber with a band gap of 1.5 eV and a perovskite structure (Figure 10.36). The perovskite structure is defined by an octahedron of negatively charged iodine ions at the corners and a positively charged lead ion in the

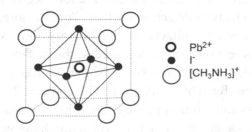

Figure 10.36. Perovskite structure of CH$_3$NH$_3$PbI$_3$.

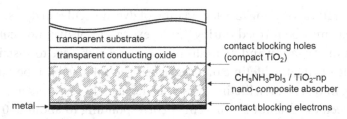

Figure 10.37. Schematic cross section of a solar cell based on a thin layer of nanoporous TiO$_2$ sensitized with methyl-ammonium lead iodide (CH$_3$NH$_3$PbI$_3$) and containing contact layers for blocking electrons (back contact) and holes (front contact).

center of the octahedron. The octahedron is stabilized in space by positively charged methyl-ammonia ions.

CH$_3$NH$_3$PbI$_3$ seems an intrinsically p-type doped semiconductor due to an excess of negative ions.

The highest solar energy conversion efficiency with a nano-composite absorber has been reached with a solid-state solar cell sensitized with CH$_3$NH$_3$PbI$_3$ (Burschka *et al.*, 2013). The absorption coefficient of CH$_3$NH$_3$PbI$_3$ is high so that a thin layer of a CH$_3$NH$_3$PbI$_3$/TiO$_2$ nano-composite absorber with a thickness of only 300–400 nm is sufficient for reaching I$_{SC}$ as high as 20 mA/cm^2 (Burschka *et al.*, 2013). Figure 10.37 shows a schematic cross section of a solar cell based on a thin layer of nanoporous TiO$_2$ sensitized with CH$_3$NH$_3$PbI$_3$.

The open-circuit voltage of a solid-state solar cell sensitized with CH$_3$NH$_3$PbI$_3$ can be as high as 0.993 V (Burschka *et al.*, 2013). A high V$_{OC}$ means that recombination losses caused by Shockley–Read–Hall and surface recombination (see Chapter 3) are very low. Thin layers of CH$_3$NH$_3$PbI$_3$ have a sharp onset of the surface photovoltage around the band gap (Supasai *et al.*, 2013) indicating a low degree of disorder and therefore a low density of defects in the bulk. Furthermore, recombination at the charge-selective contacts should be very low. A thin compact TiO$_2$ layer (thickness of 30 nm) deposited at the surface of a transparent conducting oxide (TCO) layer (see Chapter 9) has been applied for electron collection (Burschka *et al.*, 2013). This contact efficiently blocks the transfer and recombination of holes. Organic hole conductors are usually applied as charge-selective contacts for collecting holes in solar cells with CH$_3$NH$_3$PbI$_3$.

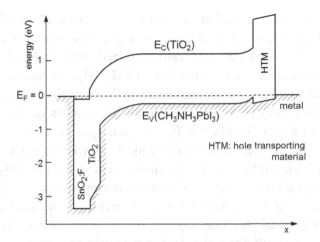

Figure 10.38. Idealized band diagram of a solar cell based on a thin layer of nanoporous TiO$_2$ sensitized with methyl-ammonium lead iodide (CH$_3$NH$_3$PbI$_3$).

Figure 10.38 shows an idealized band diagram of a solid-state solar cell sensitized with CH$_3$NH$_3$PbI$_3$. Similar to dye-sensitized solar cells, the conduction band of the nano-composite absorber of a solid-state solar cell sensitized with CH$_3$NH$_3$PbI$_3$ is formed by the conduction band of the sintered network of TiO$_2$ nanoparticles. The valence band of the solid-state nano-composite absorber is formed by the valence band of CH$_3$NH$_3$PbI$_3$.

The conduction band offsets (see Chapter 8) at the dispersed hetero-junctions between TiO$_2$ and CH$_3$NH$_3$PbI$_3$ cannot be large since V$_{OC}$ is rather large for solid-state solar cells sensitized with CH$_3$NH$_3$PbI$_3$. Therefore the absorption and transport gaps are very similar in solid-state solar cells sensitized with CH$_3$NH$_3$PbI$_3$, in contrast to organic solar cells.

The interface between CH$_3$NH$_3$PbI$_3$ and the hole conductor is assumed to be in accumulation so that a back surface field is formed for photo-generated electrons. This assumption is supported by the fact that a quasi-intrinsic accumulation layer is formed at the contact between CH$_3$NH$_3$PbI$_3$ and lead iodide (PbI$_2$) arising during moderate heating of CH$_3$NH$_3$PbI$_3$ (Supasai *et al.*, 2013). Furthermore, it is assumed that defect states can be neglected at the interface between CH$_3$NH$_3$PbI$_3$ and the hole conductor. Surface photovoltage measurements showed that charge separation caused by excitation from electronic states at the surface of CH$_3$NH$_3$PbI$_3$ is reduced after the formation of a CH$_3$NH$_3$PbI$_3$/PbI$_2$ interface (Supasai *et al.*, 2013). The formation of intrinsically passivated interfaces with $CH_3NH_3PbI_3$ seems important for reaching high solar energy conversion efficiencies with solid-state solar cells sensitized with $CH_3NH_3PbI_3$.

Solid-state solar cells sensitized with $CH_3NH_3PbI_3$ have some strong technological advantages in comparison to dye-sensitized solar cells. Very thin compact TiO_2 and thin nanoporous TiO_2 layers can be deposited homogeneously on TCO layers. The thin nanoporous TiO_2 layer is infiltrated with PbI_2, a yellowish precursor for the formation of $CH_3NH_3PbI_3$ (Burschka *et al.*, 2013). The PbI_2 precursor is converted into $CH_3NH_3PbI_3$ during dipping into a solution of methyl-ammonium iodide (CH_3NH_3I) within a time of the order of 1 min (Burschka *et al.*, 2013), which is much shorter than the time required for sensibilization of nanoporous TiO_2 electrodes with dye molecules (an overnight procedure).

10.5 Light Concentration by Nano-Photonic Concepts for Solar Cells

10.5.1 *Optical confinement for solar cells*

Scattering changes the direction of light. Light can be homogeneously scattered at nanoparticles in a homogeneous medium (Figure 10.39(a)). If the nanoparticle is placed at the boundary between two media with different

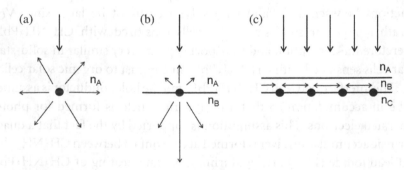

Figure 10.39. Homogeneous light scattering at a nanoparticle in a homogeneous medium (a), preferential forward scattering of light at a nanoparticle placed at the boundary between two materials with different refractive index (b) and trapping in a thin layer of light scattered at an array of nanoparticles displaced at the boundary between the thin layer and a material (c). n_A, n_B and n_C denote the refractive indices in the different media ($n_B > n_A, n_C$).

refractive indices, the light is scattered preferentially into the material with the higher refractive index (Figure 10.39(b)). If scattered light meets lateral periodic structures then constructive or destructive interference may occur depending on wavelengths of light, distances, heights and shapes of structures and optical constants of scattering and wave-guiding materials. Light of a given wavelength can be trapped in thin films by combining scattering and constructive interference (Figure 10.39(c)) (Atwater and Polman, 2010).

For trapping, light can be scattered at nano-structures by interaction with free charge carriers in metal nanoparticles (Basch *et al.*, 2012) or by Mie scattering in dielectric nanoparticles (Spinelli *et al.*, 2012). In an ideal case, scattered light is trapped and propagates only in the photovoltaic absorber. For example, the reflectivity of a silicon wafer has been reduced to 1.3% over a broad spectral range by forming an array of silicon nanoparticles coated with Si_3N_4 on top of the wafer (Spinelli *et al.*, 2012).

Figure 10.40 shows an example for interference of two lateral concentric wave sources. Related wave sources may arise, for example, by scattering of incoming light at two neighboring metal nanoparticles. Constructive interference appears when a maximum of one wave meets the maximum of the other wave. The intensity is increased in the regions of constructive interference but disappears in regions of destructive interference. There is no need to deposit absorber material in regions with destructive interference, i.e. the application of nano-photonic structures can help to drastically reduce the amount of absorber materials for solar cells.

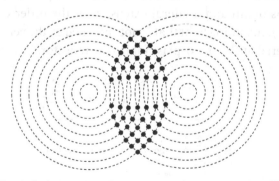

Figure 10.40. Example for two interfering concentric wave sources with a constructive interference pattern (black dots).

Furthermore, the extension of regions with constructive interference can be much smaller than the wavelength of the light, i.e. light can be concentrated in nano-structures so that the density of photo-generated charge carriers can be strongly increased in comparison to bulk photovoltaic absorbers. This effect gives the opportunity of increasing the Fermi-level splitting in photovoltaic absorbers and therefore also increasing V_{OC}.

Interference patterns depend on the geometry, the wavelength and other parameters of a structure at which interference takes place. A given interference pattern is also known as a mode. Localized and propagating modes are distinguished.

The planning of nano-photonic structures demands sophisticated simulation tools which take into account the Maxwell equations, the complete geometry of the system and all relevant material parameters and processes (see, for example, Ferry *et al.* (2011a)). The realization of theoretically optimized nano-photonic structures in solar cells requires sophisticated technologies for duplicating nano-photonic structures on large areas. Typical lateral structure sizes of nano-photonic structures are of the order of 50–500 nm. For comparison, a spatial resolution of 10 nm can be reached by soft imprinting over a range of more than 150 mm (Polman and Atwater, 2012).

10.5.2 *Enhancement of optical absorption in very thin absorbers*

The photon flux can be confined within a very thin absorber layer on top of a nano-photonic rear reflector. The excitation of plasmons in a well-conducting metal such as silver is directed by a strip or by a flat dot with height of tens of nm and a lateral extension of the order of 100–500 nm (Figure 10.41). Incoming light that is not absorbed by the very thin absorber layer is scattered at the nano-plasmonic rear reflector. The distribution of

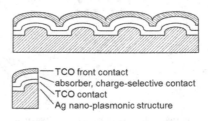

Figure 10.41.   Cross section of a solar cell with a possible nano-plasmonic rear reflector.

the photon flux in the very thin absorber layer follows from scattering at plasmons, Mie scattering and interference.

The structure shown in Figure 10.41 is very similar to the structure of thin-film solar cells but with one important difference: conventional thin-film solar cells do not have a nano-photonic rear reflector. The feasibility of a solar cell with an ultra-thin a-Si:H solar cell and a nano-photonic rear reflector has been demonstrated by the Polman group (Ferry *et al.*, 2011b). An energy conversion efficiency of 9.6% has been reached for an optimized structure and an intrinsic a-Si:H layer with a thickness of only 90 nm where I_{SC} was even higher for the solar cell with the ultra-thin a-Si:H absorber than for the present record cell based on a standard a-Si:H absorber (Benagli *et al.*, 2009). In addition, degradation of a-Si:H thin-film solar cells caused by the Staebler–Wronski effect (Staebler and Wronski, 1977) is strongly reduced in solar cells with an ultra-thin a-Si:H absorber due to the increased electric field in the un-doped a-Si:H layer.

Nano-photonic rear reflectors offer the opportunity of realizing solar cells with ultra-thin absorbers without folding of the internal surface area. For example, the optimal local thickness of the absorber layer in a solar cell based on $ZnO/In_2S_3:Cu/CuSCN$ is of the order of 30 nm (Belaidi *et al.*, 2008).

Wang *et al.* showed theoretically that about 90% of the incident light can be absorbed by a related solar cell in superstrate configuration with an absorber layer thickness of only 5 nm if taking into account a refractive index of 3.5 for the substrate (Wang *et al.*, 2013). However, a proof of concept with, for example, a QD absorber remains challenging.

10.5.3 *Nanowire array solar cells*

The amount of absorber material can be reduced in solar cells in which the nano-photonic structure is formed by a nano-composite consisting of light-absorbing nanowires and a dielectric matrix (see, for example, Kupec *et al.* (2010)). A photocurrent of more than 90% of the theoretical maximum can be achieved for solar cells based on an array of nanowires with a length of 2 μm, a diameter around 200 nm and a geometric FF of about 0.2, as shown by theoretical analysis (Kupec *et al.*, 2010).

The catalytic growth of semiconductor nanowires is possible by using liquid metal/semiconductor interfaces (Figure 10.42). In this

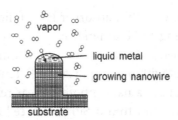

Figure 10.42. Schematic of catalytic growth of a semiconductor nanowire underneath a metallic droplet.

vapor–liquid–solid (VLS) deposition technique, first, an array of metallic nanoparticles can be deposited on a substrate by using nano-imprint lithography (Mårtensson *et al.*, 2004). The array of the metal nanoparticles defines the distance between the growing semiconductor nanowires. In the second step, the crystal growth is activated by increasing the temperature above the melting threshold of the metal nanoparticle. In a third step, atoms from the ambience are dissolved in the liquid metal droplet and diffuse to the liquid metal/semiconductor interface. The crystal growth is catalytically activated at the liquid metal/semiconductor interface so that incoming atoms are incorporated into the growing semiconductor nanowire. Sidewalls have to be passivated during the growth by sophisticated measures. Sidewalls of InP nanowires can be passivated by adding traces of HCl to the reaction gas (Wallentin *et al.*, 2013). The temperature is reduced to terminate the growth of semiconductor nanowires.

There are strong advantages for the catalytic growth of semiconductor nanowires based on $A^{III}B^{V}$ semiconductors. First, epitaxial growth of perfect semiconductor nanowires is possible. Second, doping of semiconductor nanowires can be performed with the usual dopants so that charge-selective and ohmic contacts can be easily formed. Third, strain can better relax in nanowires so that semiconductors with larger differences in the lattice constants can be grown on top of each other.

The space between semiconductor nanowires has to be filled with a dielectric such as SiO_2. Obviously, the density of interface states at the semiconductor/SiO_2 interface will be high so that the surface recombination velocity will be high as well. There is not enough space in semiconductor nanowires for additional measures of surface passivation. Therefore, the density of photo-generated free charge carriers has to be minimized by

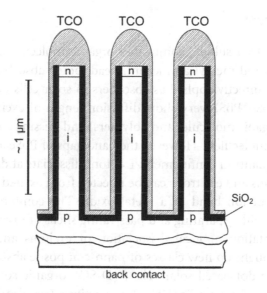

Figure 10.43. Schematic cross section of a solar cell based on an InP nanowire array.

implementing a drift field for free charge carriers. This has been realized in InP nanowire array solar cells (Wallentin *et al.*, 2013). Figure 10.43 shows a schematic cross section of a nano-composite solar cell based on an InP nanowire array with a drift field in an un-doped region of InP. The drift length is about 1.3 μm in an InP nanowire. The drift time of photo-generated charge carriers is of the order of 10–100 ps, which is much shorter than the non-radiative recombination lifetime (see also task T10.5 at the end of this chapter). The implementation of a drift field as a measure for surface passivation of semiconductor nanowires is not possible for conventional solar cells and opens the opportunity of reaching very high energy conversion efficiencies with nano-structured solar cells.

Despite the fact that the InP nanowires cover only 12% of the surface area of the solar cell, a value of I_{SC} (24.6 mA/cm^2) close to that of the planar record InP solar cell (29.5 mA/cm^2) has been achieved (Wallentin *et al.*, 2013). Further values of V_{OC} close to and even larger (0.906 V) than that of the planar record InP solar cell (0.878 V) have been reached with solar cells based on InP nanowire arrays (Wallentin *et al.*, 2013). This shows that nanowire structures are well suited for solar cells from the point of view of optimization of I_{SC} and reducing recombination losses.

10.6 Summary

Lead selenide, lead sulfide, conjugated organic molecules and polymers, dye molecules and methyl-ammonium lead iodide absorb light very well but cannot be directly applied as absorbers in solar cells due to very low band gaps (PbSe, PbS), very short diffusion length of excitons (layers of conjugated organic molecules and polymers) and missing charge transport (layers of dye molecules). However, the band gaps of PbSe and PbS can be increased by quantum confinement. Excitons dissociate at donor–acceptor hetero-junctions and electrons can be injected from excited dye molecules into the conduction band of a metal oxide. The combination of PbSe and PbS-QDs with separating and passivating shells of organic molecules, the implementation of local dissociation of excitons and local charge separation brought up new classes of nano-composite absorbers for solar cells: quantum dot-based solar cells (QD-SC), organic solar cells (OSC) and dye-sensitized solar cells (DSSC), respectively (see also Table 10.1).

Charge-selective contacts are realized differently in solar cells with nano-composite absorbers. QDs and conjugated organic molecules can be doped (pn- or pin-structures for charge separation). Electrons and holes are extracted at metal contacts with low or high work functions, respectively,

Table 10.1. Functionality and materials of nano-composite absorbers in QD-based solar cells (QD-SCs), organic solar cells (OSCs), dye-sensitized solar cells (DSSCs), solar cells with very thin absorber (vtaSCs) and in nanowire solar cells (nwSCs).

Type of solar cell	Functionality of the nano-composite	Materials
QD-SC	tunable band gap	$PbSe_xS_{1-x}$ QD core, passivating organic molecule shell
OSC	local dissociation of excitons	conjugated donor and acceptor molecules and polymers, C60 and PCBM as acceptor molecules
DSSC	local charge separation	dye molecules, nanoporous TiO_2, redox electrolyte
vtaSC	light confinement in a very thin absorber layer	structured metal or refraction array below the absorber layer
nwSC	light confinement in semiconductor nanowires	structured array of semiconductor nanowires

in organic polymer solar cells (metal-insulator-metal structures for charge separation). Charge-selective hetero-junctions can be applied in QD-SCs as well. In DSSCs electron transfer into the electrolyte and hole transfer into the TCO contact are blocked so that the charge-selective contacts are formed in DSSCs by the TCO/nano-composite absorber and electrolyte/nano-composite absorber contacts.

Scattering and interference of light at nanoparticles, nanostripes or nanowires are combined in photonic nano-composites with materials having different refractive indexes. Light can be confined in very thin layers or in nanowires of nanowire arrays by exploring nano-photonic concepts (see also Table 10.1). The thickness of the absorber layer is much thinner than the absorption length in solar cells with very thin absorbers (vtaSCs), for example, less than 100 nm in a-Si:H vtaSCs. Semiconductor nanowires such as InP nanowires are a part of the photonic nano-composite in nanowire solar cells (nwSCs). Therefore, solar cells with high energy conversion efficiency can be produced whereas the amount of absorber material can be strongly reduced. QD-SCs, OSCs and DSSCs can be combined with nano-photonic concepts as well.

Table 10.2 summarizes the values of I_{SC}, V_{OC}, FF and of the solar energy conversion efficiency for some kinds of record solar cells related to QD-SCs, OSCs, DSSCs, vtaSCs and nwSCs. Related material combinations are given as well.

High solar energy conversion efficiency (13.8%) has been reached with InP-nw solar cells for which the role of surface recombination has been strongly reduced (Wallentin *et al.*, 2013). For this reason, a high potential for further improvement of the solar energy conversion efficiency can be expected for nwSCs.

The solar energy conversion efficiency of vtaSCs based on a-Si:H (9.6% at present (Ferry *et al.*, 2011b) has potential for further increase and for implementation of nano-photonic structures in large-scale production. Limitations are given by the increase of shunts with decreasing thickness of un-doped a-Si:H towards 30 nm, similarly to ZnO/In_2S_3:Cu/CuSCN-based solar cells (Belaidi *et al.*, 2008).

Dye-sensitized solar cells play a decisive role in the development of the concept of nano-composite solar cells. Even at the beginning of the 1990s, DSSCs showed a relatively large energy conversion efficiency of 7.1–7.9%

Table 10.2. Values of short-circuit current density (I_{SC}), open-circuit voltage (V_{OC}), fill factor (FF) and solar energy conversion efficiency (η) for different types of nano-composite solar cells. For all DSSCs the electrode material is nanoporous TiO_2.

Type of solar cell	Material combinations	I_{SC} (mA/cm^2)	V_{OC} (V)	FF	η (%)	Reference
QD-SC	TiO_2 / PbS-QDs	16.2	0.51	0.58	5.1	(a)
QD-SC	pn-PbS-QDs	23.3	0.52	0.50	6.1	(b)
OSC	small molecules	6.091	2.674	0.735	12.0	(c)
OSC	polymer	17.45	0.754	0.7	9.2	(d)
DSSC	Co^{2+}/Co^{3+}, YD2-o-C8/Y123	17.75	0.94	0.76	12.9	(e)
DSSC	I^-/I_3^-, Ru-BD / Y1	20.88	0.743	0.727	11.28	(f)
DSSC	OMeTAD, Ru-dye	8.27	0.848	0.71	5.0	(g)
SSC	TiO_2 / $CH_3NH_3PbI_3$	20.0	0.993	0.73	15.0	(h)
vtaSC	a-Si:H, 90 nm	16.94	0.864	0.66	9.6	(i)
nwSC	InP-nw	24.6	0.779	0.724	13.8	(j)

References: (a) Pattantyus-Abraham *et al.* (2010), (b) Liu *et al.* (2012), (c) Pfeiffer (2013), Heliatek (2013), (d) He *et al.* (2012), (e) Grätzel (2012), (f) Han *et al.* (2012), (g) Wang *et al.* (2010), (h) Burschka *et al.* (2013), (i) Ferry *et al.* (2011b) and (j) Wallentin *et al.* (2013).

(O'Regan and Grätzel, 1991) and an optimum was soon reached for Ru-based dye molecules and iodide electrolyte. The optimization potential for this material combination seems quite narrow despite the fact that an increase of the energy conversion efficiency to 11.28% has been reached by co-sensitization (Han *et al.*, 2012). The value of V_{OC} has been increased to 0.94 V by replacing the iodide electrolyte by a cobalt-based electrolyte and an energy conversion efficiency of 12.9% has been reached by co-sensitization (Grätzel, 2012). However, a broad implementation of DSSCs is challenging for cost reasons and its limited outdoor application due to the liquid electrolyte. For example, large-scale DSSC modules degraded by more than 10% after only half a year of outdoor operation (Toyoda *et al.*, 2004).

The replacement of the liquid electrolyte by OMeTAD led to a drastic decrease of the energy conversion efficiency in DSSCs based on Ru dyes. An energy conversion efficiency of 5% has been reached (Wang *et al.*, 2010).

The replacement of the dye by $CH_3NH_3PbI_3$ perovskite allowed a reduction of the thickness of the nano-composite absorber to about 0.3–0.4 μm and to increase the energy conversion efficiency to almost 15%. This solar cell is a solid-state semiconductor-sensitized solar cell, which seems to have a great potential for further improvement for solution-based production on a large scale and with broad application.

Solar energy conversion efficiencies above 10% have been reached in several laboratories for OSC and commercialization of OSCs has already started for niche markets. A potential for solar energy conversion efficiencies above 15% is expected for OSC.

Low values of V_{OC} and FF are limiting the solar energy conversion efficiency of QD-SCs (Pattantyus-Abraham *et al.*, 2010), research on which just started some years ago. The replacement of the charge-selective hetero-junction with TiO_2 by a pn-junction led to a strong increase of I_{SC} (Liu *et al.*, 2012). However, innovative passivation concepts are required for further improvement of QD-SCs.

Nano-composite absorbers can be produced from earth-abundant elements, for example, from lead and sulfur (PbS-QDs). If assuming that stable PV modules can be fabricated from PbS-QD absorbers, only 10% of the yearly produced amount of Pb and S will be sufficient for electricity production for mankind (see also task T10.5 at the end of this chapter). This means that in terms of a material extraction cost index and annual electricity potential index (Wadia *et al.*, 2009), the strategic potential of PbS and $CH_3NH_3PbI_3$ is larger than the strategic potential of c-Si photovoltaics by five to ten orders of magnitude(!) Similar is true for vtaSCs based on a-Si:H or for OSCs.

The application of design rules for nano-composite absorbers and the consequent implementation of nano-photonic concepts allow for a drastic reduction in absorber material and processing temperatures. This leads to a reduction of the energy payback time and therefore to an improvement of the sustainability of photovoltaic solar energy production. However, competitive long-term stability of PV modules operated outdoors is a prerequisite for sustainability as well.

10.7 Tasks

T10.1: Solar cell with ultra-thin absorber

Estimate the factor by which the surface area of an absorber with very low diffusion or drift length has to be increased (roughness factor). The values of the absorption length, of the mobility and lifetime of photo-generated charge carriers are 2 μm and 1 cm^2/(Vs) and 1 ns, respectively.

T10.2: Surface atoms at colloidal QDs

Compare the number of surface atoms with the number of bulk atoms of colloidal PbS-QDs with diameters of 2, 4 and 6 nm. The density of PbS is 7.6 g/cm^2. The bond length is 0.297 nm in PbS.

T10.3: QD absorber layer

Ascertain I_{SC}^{max} for absorber layers with colloidal PbS-QDs with diameters of 2, 4 and 6 nm. The effective electron and hole masses, the relative dielectric constant of PbS, the relative dielectric constant of the shell and the band gap of PbS are about 0.2 and 0.1 times the free electron mass, 14, 3 and 0.37 eV, respectively. Discuss the importance of accurate measurements of effective masses and uncertainties of dielectric properties in QD absorbers.

T10.4: Strategic potential of PbS-QD solar cells and solar cells sensitized with $CH_3NH_3PbI_3$

Estimate the amount of lead which would be needed to completely support the electricity consumption of mankind produced only with solar cells based on PbS-QDs as compared with solar cells sensitized only with $CH_3NH_3PbI_3$. Make reasonable assumptions for the effective thickness of the absorber layer, for the electricity demand and for the efficiency of related solar cells. The density of PbS is 7.6 g/cm^3. Compare your estimated amount with the annual production of lead ($5.2 \cdot 10^6$ t/a in 2012 (Guberman, 2013)).

T10.5: Absorption and transport gaps in nano-composite absorbers

What role do the different absorption and transport gaps play for the maximum energy conversion efficiency of nano-composite solar cells? Compare with the Shockley–Queisser limit.

T10.6: InP-nw solar cell

The implementation of a drift field in pin InP-nw solar cells gives the opportunity of reducing the role of surface recombination despite poor surface passivation. Estimate the effective lifetime in an InP-nw solar cell and compare it with the drift time. Make a reasonable assumption about the surface recombination velocity. The drift mobility of photo-generated holes is 250 cm^2/(Vs) in InP.

A
Solutions to Tasks

A.1 Solutions to Chapter 1

T1.1: Power installed and energy produced

In Germany and in the south of Spain, the average power of the sun is 0.12 and 0.20 kW/m². The time of one year corresponds to 8760 h. With regard to Equation (1.11), one can write:

$$W_{el} = \frac{\langle P_{sun} \rangle}{P_{sun}(AM1.5)} \cdot t \cdot P_{inst} \tag{T1.1}$$

$$W_{el} = \frac{0.12(0.2)\text{kW/m}^2}{1\,\text{kW/m}^2} \cdot 8760\,\text{h} \cdot 1000\,\text{kW} \tag{T1.1'}$$

In total, $1.05 \cdot 10^6$ kWh and $1.75 \cdot 10^6$ kWh can be produced in Germany and in the south of Spain, respectively, during the first year of operation of a PV power plant with 1 MW$_p$.

T1.2: Degradation of PV modules and sustainability

(a) The relative decrease of the solar energy conversion efficiency can be expressed by a power law over the difference between 1 and the annual degradation rate and with the time as the power coefficient. The relative accumulated energy produced considers the integral of the power law. Figures T1.1 and T1.2 show the plots of the dependence of the relative decrease of the solar energy conversion efficiency and of the accumulated energy produced, respectively, over up to 250 years for k = 0.5, 1.0, 2.0 and 5.0%/year. For a degradation rate of 0.5%/year, a PV power plant still has about half of its initial efficiency after 138 years. For a degradation rate of 5%/year, a PV power plant has about half of its initial efficiency after about

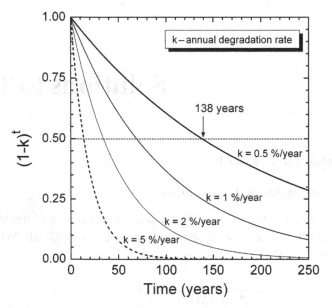

Figure T1.1. Relative decrease of the solar energy conversion efficiency as a function of time for annual degradation rates of k = 0.5, 1.0, 2.0 and 5.0%/year.

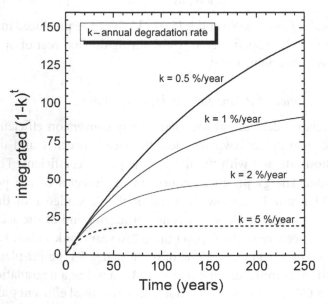

Figure T1.2. Relative increase of the accumulated energy produced over a long time period with PV modules having annual degradation rates of k = 0.5, 1.0, 2.0 and 5.0%/year.

13 years and the total energy which can be produced with this power plant is much lower than for PV power plants with lower annual degradation rates. Low degradation rates are very important for sustainable PV energy production.

(b) The value of power since installation (P_{inst}) decreases with increasing time due to the degradation rate (k_{deg}). The installed power, the time interval of one year and the number of years are assigned to P_{inst}^0, t_0 and N, respectively. It is assumed that 1000 kWh of electric energy are produced during the first year. Furthermore, it is assumed, unlike in T1.2(a), that the PV modules do not degrade during the first year and that the reduction of the installed power is considered after each year. The remaining power of the PV power plant after 100 years is therefore:

$$P_{inst}^{N=100} = P_{inst}^0 \cdot (1 - k_{deg})^{99} \tag{T1.2}$$

The remaining power of the PV power plant after 100 years is $0.61, 0.37$ and $0.22\,kW_p$ for degradation rates of $0.5, 1$ and 1.5%/year, i.e. the PV power plant is still producing energy, and for the degradation rate of 0.5%/year the remaining power is even more than 60% of the initial installation. This example shows the importance of building integration of PV systems.

The electric energy produced can be calculated by using the following equation:

$$W_{el} = \frac{\langle P_{sun} \rangle}{P_{sun}(AM1.5)} \cdot P_{inst}^0 \cdot t_0 \cdot \sum_{n=0}^{N-1} (1 - k_{deg})^n \tag{T1.3}$$

The maximum electric energy produced in 100 years with a PV power plant with degradation rates $0.5, 1$ and 1.5%/year amount to $81593, 64901$ and $52629\,MWh$, respectively. The energy payback time is related to the first year, i.e. energy payback factors of 50 and more can be easily reached with PV solar energy conversion.

T1.3: Photon energy in the maximum of blackbody radiation at T_s

The first derivative of Equation (1.5) has to be set to 0 for ascertaining the maximum of the sun spectrum. The following equation for the non-trivial solution is obtained:

$$exp\left(\frac{h\upsilon}{k_B \cdot T_S}\right) \cdot \left(3 - \frac{h\upsilon}{k_B \cdot T_S}\right) - 3 = 0 \qquad (T1.4)$$

This equation has a solution at:

$$h\upsilon|_{max} = 2.82 \cdot k_B \cdot T_s = 1.39\,\text{eV} \qquad (T1.5)$$

The sun spectrum has its maximum at 1.39 eV.

T1.4: Wavelength of light in the maximum of blackbody radiation at T_s

Equation (1.5) has to be transformed by replacing υ by c/λ and by replacing $d(h\upsilon)$ by:

$$d(h\upsilon) = -\frac{hc}{\lambda^2} \cdot d\lambda \qquad (T1.6)$$

The spectrum of a blackbody with the temperature of the sun as a function of wavelength is obtained:

$$\frac{dJ_S(\lambda)}{d(\lambda)} = \frac{8\pi \cdot c^2 \cdot h}{\lambda^5} \cdot \frac{1}{exp\left(\frac{hc}{\lambda \cdot k_B \cdot T_S}\right) - 1} \qquad (T1.7)$$

The first derivative of Equation (T1.7) has to be set to 0 for obtaining the wavelength in the maximum of the sun spectrum. The following equation for the non-trivial solution is obtained.

$$exp\left(\frac{hc}{\lambda \cdot k_B \cdot T_s}\right) \cdot \left(5 - \frac{hc}{\lambda \cdot k_B \cdot T_s}\right) - 5 = 0 \qquad (T1.8)$$

Equation (T1.8) has a solution at:

$$\lambda|_{max} = \frac{hc}{4.96 \cdot k_B \cdot T_s} = 502\,\text{nm} \qquad (T1.9)$$

The sun spectrum has its maximum at 502 nm.

T1.5: Maximum temperature of a solar cell under operation

(a) A solar cell is cooled only by thermal emission if convection is absent. The solar cell receives power from the solar radiation via the front surface and from the homogeneous blackbody radiation of the surrounding environment (at temperature T_0) via the front and back surfaces (sidewalls neglected). A part of the power received from sun is converted into electric power, which does not contribute to heating the solar cell. The solar cell is heated to the temperature T_{SC} and emits blackbody radiation via the front and back surfaces (sidewalls neglected). The area of the solar cell, the energy conversion efficiency and the Stefan–Boltzmann constant are denoted by A, η and σ_S, respectively. The balance of energy fluxes gives the following equation:

$$A \cdot P_{sun} + 2 \cdot A \cdot \sigma_s \cdot T_0^4 = \eta \cdot A \cdot P_{sun} + 2 \cdot A \cdot \sigma_s \cdot T_{SC}^4 \qquad \text{(T1.10)}$$

The temperature of the solar cell can be calculated after transforming Equation (T1.10):

$$T_{SC} = \left[T_0^4 + \frac{1 - \eta}{2 \cdot \sigma_s} \cdot P_{sun} \right]^{1/4} \qquad \text{(T1.11)}$$

The maximum temperature on earth is 300 K, the effective temperature on a satellite is approximated by 10 K (thermal radiation received from earth is neglected). Then the temperature of the solar cell is 356, 350 and 340 K for energy conversion efficiencies of 10, 20 and 40%, respectively, i.e. solar cells can reach about 70°C under operation on earth at AM1.5. The temperature of solar cells on a satellite is 298, 289 and 270 K for energy conversion efficiencies of 10, 20 and 40%, respectively, i.e. the temperature conditions are comparable to those on earth.

(b) In Equation (T1.11), the number 2 in the denominator has to be replaced by 1 for the rooftop or façade integration, with ideal thermal isolation at the backside of the PV module, and by 4 and 8 for free-standing PV modules with finned backsides. Equation (T1.11) remains as is for a free-standing PV module without finned backside. Figure T1.3 shows the temperature of a solar cell under illumination at $1\,\text{kW/m}^2$ as a function of the solar energy conversion efficiency between 0 and 44% for the four different cases of radiative cooling. For a complete rooftop or façade integration, the temperature of solar cells can increase to temperatures larger than 120°C (efficiency of 20%). The temperature of an illuminated solar cell or PV module can be strongly reduced for free-standing PV modules and even more for free-standing PV modules with finned backsides. As a remark, for radiative cooling the design of the finned area has to be realized in such a way that fins are not oriented face-to-face.

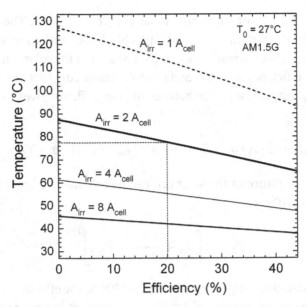

Figure T1.3. Efficiency dependence of the maximum temperature of solar cells illuminated at AM1.5G under different conditions of radiative cooling, taking into account radiating areas equal to the illuminated area (complete rooftop or façade integration with ideal thermal isolation), to the doubled illuminated area (free-standing PV module with un-finned backside) and to the illuminated area increased by the factors of 4 and 8 (finned backsides). Convection cooling is not considered.

As a consequence, the volume occupied by the solar cell with the fins would strongly increase. The strong increase of the volume can only be avoided if implementing convection cooling.

T1.6: Air mass

The angle of incidence (θ) is measured normal to the surface. The air mass is the ratio between the path of light and the thickness of the earth's atmosphere, i.e. the reciprocal of the cosine of θ:

$$AM(\theta) = \frac{1}{cos\theta} \qquad (T1.12)$$

At high angles of incidence, i.e. during the morning and evening hours, the air mass is overestimated due to the curvature of the earth and a correction term has to be implemented into Equation (T1.12).

T1.7: Tolerable series and parallel resistances

The combination of Equations (1.19) and (1.29′) and (1.29″) gives the expressions for calculating the tolerable series and parallel resistances.

$$R_{s,tol} < \frac{1}{100} \cdot \frac{\frac{k_B \cdot T}{q} \cdot ln\left(\frac{I_{SC}}{I_0} + 1\right)}{I_{SC}} \tag{T1.13′}$$

$$R_{p,tol} > 100 \cdot \frac{\frac{k_B \cdot T}{q} \cdot ln\left(\frac{I_{SC}}{I_0} + 1\right)}{I_{SC}} \tag{T1.13″}$$

Room temperature and a linear response of I_{SC} are assumed. The tolerable series and parallel resistances of the solar cell with $I_0 = 10^{-19}$ A/cm^2 and $I_{SC} = 0.025$ A/cm^2 are equal to 0.4 and 4000 Ωcm^2, respectively. The tolerable series and parallel resistances are 1 mΩcm^2 and 10 Ωcm^2, respectively, under operation at a concentration factor of 400.

T1.8: Current–voltage characteristics

The current–voltage characteristics are simulated by using the following equation:

$$I = I_{SC} - I_0 \cdot \left[exp\left(\frac{q \cdot U}{k_B \cdot T}\right) - 1\right] - \frac{U}{R_p} \tag{T1.14}$$

The current is multiplied by the potential for getting the power–voltage characteristics. The values of V_{OC} and FF are 1.017 V and 88.4% ($I_{SC} = 100$ mA/cm^2, $R_P = 100$ kΩcm^2), 0.957 V and 87.8% ($I_{SC} = 10$ mA/cm^2, $R_P = 100$ kΩcm^2), 1.017 V and 87.5% ($I_{SC} = 100$ mA/cm^2, $R_P = 1$ kΩcm^2) and 0.956 V and 80% ($I_{SC} = 10$ mA/cm^2, $R_P = 1$ kΩcm^2).

A.2 Solutions to Chapter 2

T2.1: I_{SC} density

The I_{SC} density can be calculated by taking into account that photons with energy below E_g and above $E_g + \Delta E$ ($E_g = 1.6$ eV, $\Delta E = 0.3$ eV) do not contribute to photocurrent generation.

$$I_{SC} = q \cdot \int_{E_g}^{\infty} \Phi_{sun}(h\nu) \cdot d(h\nu) - q \cdot \int_{E_g + \Delta E}^{\infty} \Phi_{sun}(h\nu) \cdot d(h\nu) \tag{T2.1}$$

The values of I_{SC} are about 24 and 15 mA/cm^2 for solar cells with band gaps of 1.6 and 1.9 eV, respectively. Therefore, I_{SC} is about 9 mA/cm^2 for a solar cell with a band gap of 1.6 eV and a width of the absorption band of 0.3 eV.

T2.2: Temperature-dependent I_{SC} density

The I_{SC} density increases by about 1% if the temperature of a c-Si solar cell increases from 25 to 60°C.

T2.3: Lambertian limit of light trapping

For Lambertian light scattering, the intensity of scattered light is distributed within a sphere as shown in Figure 2.13. The surface area of this sphere (A_0) corresponds to the total intensity of scattered light. The area of the sphere segment around the critical angle for total reflection (A_{loss}) corresponds to the intensity of the light lost due to scattering into a cone of angles less than the critical angle for total reflection (θ_C). The maximum intensity of scattered light is denoted by 2·R (R is the radius of the sphere).

The increase of the optical path is proportional to the ratio between the total intensity of scattered light and the intensity lost by scattering at angles less than θ_C. With respect to the geometric situation shown in Figure T2.1 it can be written:

$$X_{path} = \frac{A_0}{A_{loss}} = \frac{4\pi \cdot R^2}{2\pi \cdot R \cdot h} = \frac{2R}{h} \tag{T2.2}$$

$$tan\theta_C = \frac{h}{a} = \frac{a}{2R - h} \tag{T2.3}$$

Equation (T2.3) can be transformed to:

$$\frac{2R}{h} = \frac{1}{tan^2\theta_C} + 1 = \frac{1}{sin^2\theta_C} \tag{T2.4}$$

The limit of Lambertian light trapping is obtained by using Equations (T2.4), (T2.2), the definition of θ_C and the fact that scattered light passes through the absorber for a second time before total internal reflection.

$$X_{path} = 2 \cdot \left(\frac{n_S}{n_{AR}}\right)^2 \tag{T2.5}$$

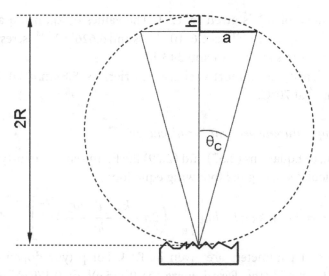

Figure T2.1. Geometric situation for Lambertian light scattering. The maximum intensity of scattered light is denoted by 2R.

T2.4: Temperature-dependent effective density of states

Equation (2.29) is used to calculate the effective density of states at the conduction band edge. The values of k_B, m_e and h are $1.38 \cdot 10^{-23}$ J/K, $9.1 \cdot 10^{-31}$ kg and $6.626 \cdot 10^{-34}$ Js, respectively. The temperatures are 273.14 and 343.14 K.

$$N_C = 2 \cdot \left(2\pi \cdot k_B \cdot T \cdot \frac{m_e^*}{h^2} \right)^{3/2} \tag{T2.6}$$

The values of N_C are $2.2 \cdot 10^{19}$ cm^{-3} at 0°C and $3.1 \cdot 10^{19}$ cm^{-3} at 70°C.

T2.5: Temperature-dependent density of minority charge carriers

With regard to Equations (2.31) and (2.29), the density of minority charge carriers (n_0) can be calculated using the following equation:

$$n_0 = \frac{32\pi}{p_0} \cdot \left(\frac{k_B \cdot T \cdot m_e}{h^2} \right)^3 \cdot \left(\frac{m_e^*}{m_e} \cdot \frac{m_h^*}{m_e} \right)^{3/2} \cdot exp\left(-q \cdot \frac{E_g(T)}{k_B \cdot T} \right) \tag{T2.7}$$

The value of m_e^* is given in T2.4. The values of k_B, m_e, q and h are $1.38 \cdot 10^{-23}$ J/K, $9.1 \cdot 10^{-31}$ kg, $1.6 \cdot 10^{-19}$ As and $6.626 \cdot 10^{-34}$ Js, respectively. The temperatures are 273.14 and 343.14 K.

The density of minority charge carriers is 8.8 cm^{-3} at 0°C and $6 \cdot 10^5$ cm^{-3} at 70°C.

T2.6: Temperature-dependent Fermi-energy

According to Equations (2.37) and (2.29) the Fermi-energy in p-type doped c-Si is calculated using the following equation:

$$E_F = E_V + k_B \cdot T \cdot \ln\left[\frac{2}{p_0} \cdot \left(2\pi \cdot \frac{k_B \cdot T \cdot m_h^*}{h^2}\right)^{3/2}\right] \tag{T2.8}$$

Values of parameters are given in T2.4. For p-type doped c-Si with $p_0 = 3 \cdot 10^{16}$ cm^{-3}, the Fermi-energy is 0.135 eV or 0.179 eV above the valence band edge at 0 and 70°C, respectively.

T2.7: Fermi-level splitting

According to Equations (2.41) and (2.31) the Fermi-level splitting is calculated using the following equation:

$$E_{Fn} - E_{Fp} = k_B \cdot T \cdot \ln\left(\frac{\left(\frac{n_i^2}{p_0} + \Delta n\right) \cdot (p_0 + \Delta p)}{n_i^2}\right) \tag{T2.9}$$

The value of n_i^2 is calculated from Equations (2.32) and (2.29) and amounts to about 10^{20} cm^{-6}. The Fermi-level splitting in p-type doped c-Si with $p_0 = 3 \cdot 10^{16}$ cm^{-3} is 0.627, 0.687 and 0.754 eV under illumination with $\Delta n = 10^{14}$, 10^{15} and 10^{16} cm^{-3}, respectively.

T2.8: Temperature-dependent Fermi-level splitting

The Fermi-level splitting is calculated by using Equation (T2.9). The reduced band gap (see task T2.2) has to be taken into account to calculate n_i^2 at 65°C. At 65°C the Fermi-level splitting is reduced by about 0.1 V in comparison to 25°C, i.e. the Fermi-level splitting is reduced by about 14%, which is higher than the increase of I_{SC} by about 1%.

A.3 Solutions to Chapter 3

T3.1: Minimum lifetime and light trapping

The thickness of the photovoltaic absorber should be larger than three times the absorption length (α^{-1}) divided by the factor of the increase of the optical path (X_{path}).

$$d_{abs} \geq \frac{3 \cdot \alpha^{-1}}{X_{path}} \tag{T3.1}$$

The value of X_{path} is equal to 1 for a single pass, 2 for an absorber with back reflector and of the order of 10 for an absorber with a Lambertian back reflector for light trapping (see Chapter 2). The corresponding values of d_{abs} are 300, 150 and 30 μm. The minimum lifetime condition is modified:

$$\tau_{min} = \frac{9 \cdot \alpha^{-2}}{X_{path}^2 \cdot D} \tag{T3.2}$$

The diffusion coefficient of photo-generated charge carriers (D) and α^{-1} are 15 cm^2/s and 100 μm, respectively, for a c-Si absorber. Values of τ_{min} of 60, 15 and 0.6 μs are obtained for a c-Si absorber with a single pass of light, with a back reflector and with a Lambertian back reflector for light trapping, respectively.

T3.2: Density of photo-generated charge carriers in a GaAs absorber

The density of photo-generated charge carriers can be estimated by combining Equations (3.9) and (T3.2):

$$\Delta n \approx \frac{3 \cdot \alpha^{-1}}{q \cdot X_{path} \cdot D} \cdot I_{SC}^{max} \tag{T3.3}$$

The value of I_{SC}^{max} for $E_g = 1.42$ eV (GaAs) is about 0.03 A/cm^2 under illumination at AM1.5 and X_{path} is equal to 2 for an absorber with a back reflector. The resulting densities of photo-generated charge carriers are about 10^{12} and $5 \cdot 10^{11}$ cm^{-3} for a GaAs absorber without and with a back reflector, respectively. For comparison, the density of photo-generated charge carriers is about three orders of magnitude higher in a c-Si absorber.

T3.3: Density of majority charge carriers in a c-Si absorber

The lifetime of minority charge carriers is limited by Auger recombination in c-Si. The density of majority charge carriers is obtained by using Equation (3.15). The maximum densities of majority charge carriers are $8.4 \cdot 10^{15}$ and $2.6 \cdot 10^{17}$ cm^{-3} for τ_{min} equal to 7 ms and 7 μs, respectively. The diffusion lengths of minority charge carriers are calculated with Equation (3.5) by using a diffusion constant of 15 cm^2 /s. The diffusion lengths of minority charge carriers are 3 mm and 95 μm for τ_{min} equal to 7 ms and 7 μs, respectively.

T3.4: Maximum limit of lifetime

A hypothetical lifetime of 157 s has to be related to densities of majority charge carriers and of defects in the bulk silicon crystal. Densities of majority charge carriers should be calculated for radiative and Auger recombination using Equations (3.12) and (3.15), respectively. For a lifetime of 157 s, densities of majority charge carriers of $2 \cdot 10^{12}$ and $6 \cdot 10^{13}$ cm^{-3} are obtained for radiative and Auger recombination, respectively. For comparison, the density of intrinsic charge carriers is about 10^{10} cm^{-3} in c-Si at room temperature, i.e. a related crystal is realistic from the point of view of doping. However, a lifetime of 157 s would require a density of defects in the bulk as low as $6 \cdot 10^5$ cm^{-3} (after Equation (3.21)), which is not realistic. It is interesting to note that radiative recombination limits the lifetime of minority charge carriers at very low densities of majority charge carriers.

T3.5: Intrinsic carrier lifetime

The intrinsic densities of charge carriers are about 10^{10} and 10^6 cm^{-3} for c-Si and GaAs, respectively (see Chapter 2). The lifetime is limited by radiative recombination for very low densities of free charge carriers. The intrinsic carrier lifetimes are about $3 \cdot 10^4$ s for c-Si and $5 \cdot 10^3$ s for GaAs.

T3.6: Lifetime at the solubility limit of doping

The lifetime is limited by Auger recombination at the solubility limit. With regard to Equation (3.15), the lifetime is as low as 50 ps at the solubility limit. A density of defects in the bulk of the order of $2 \cdot 10^{18}$ cm^{-3} can be tolerated at the solubility limit of doping (Equation (3.21)).

The diffusion coefficient is reduced at very high densities of free charge carriers. Therefore, the diffusion length of minority charge carriers is less than 100 nm at the solubility limit of doping.

T3.7: Total lifetime

The total lifetime is calculated using Equation (3.31) by taking into account Equations (3.12′) for the radiative recombination lifetime, (3.15) for the Auger recombination lifetime and (3.21′) for the Shockley–Read–Hall recombination lifetime.

$$\frac{1}{\tau_{total}} = B \cdot p_0 + C_A \cdot p_0^2 + \sigma_e \cdot v_{th} \cdot N_t + \frac{2 s_n}{d} \qquad \text{(T3.4)}$$

The total lifetime is 82 μs for the given c-Si absorber.

A.4 Solutions to Chapter 4

T4.1: Diffusion potential

The diffusion current density is calculated using Equation (4.12). The density of intrinsic charge carriers in Equation (4.12) can be obtained using Equation (2.32). Values of n_i^2 are about 10^{20} and 10^{12} cm^{-6} for c-Si and GaAs, respectively, at room temperature (according to Chapter 2). The diffusion potential of a pn-junction is 0.9 V for c-Si solar cells (a) and 1.38 V for GaAs solar cells (b).

T4.2: Space charge regions of a pn-junction

The thickness of the space charge regions are given by Equations (4.5). The sum of the potential drops across the space charge regions at the p-type and n-type doped sides of the pn-junction is equal to the diffusion potential:

$$U_D = U_n + U_p \qquad \text{(T4.1)}$$

The values of x_p and x_n are obtained by transforming Equations (4.5), (4.6) and (T4.1) to:

$$x_p = \sqrt{\frac{2 \cdot \varepsilon_r \cdot \varepsilon_0}{q \cdot N_A} \cdot \frac{U_D}{1 + \frac{N_A}{N_D}}} \qquad \text{(T4.2′)}$$

$$x_n = \sqrt{\frac{2 \cdot \varepsilon_r \cdot \varepsilon_0}{q \cdot N_D} \cdot \frac{U_D}{1 + \frac{N_D}{N_A}}} \qquad \text{(T4.2″)}$$

The diffusion potentials are calculated according to task T4.1. The relative dielectric constants of c-Si and GaAs are about 12 and 11, respectively. The values of x_p and x_n are about 350 and 0.35 nm, respectively, for a c-Si pn-junction with $N_A = 10^{16}$ and $N_D = 10^{19}\,cm^{-3}$ and about 12 and 120 nm, respectively, for a GaAs pn-junction with $N_A = 10^{18}$ and $N_D = 10^{17}\,cm^{-3}$.

T4.3: Maximum electric field at a pn-junction

The maximum electric field is reached at the pn-junction ($x = 0$). With respect to Equations (4.3) and (T4.2) the following equation is obtained for the maximum electric field at a pn-junction:

$$-\varepsilon_{max} = \sqrt{\frac{2 \cdot q \cdot N_A}{\varepsilon_r \cdot \varepsilon_0} \cdot \frac{U_D}{1 + \frac{N_A}{N_D}}} = \sqrt{\frac{2 \cdot q \cdot N_D}{\varepsilon_r \cdot \varepsilon_0} \cdot \frac{U_D}{1 + \frac{N_D}{N_A}}} \qquad (T4.3)$$

The maximum electric field is $-52\,kV/cm$ at a pn-junction with $N_A = 10^{16}$ and $N_D = 10^{19}\,cm^{-3}$ for c-Si and 200 kV/cm with $N_A = 10^{18}$ and $N_D = 10^{17}\,cm^{-3}$ for GaAs.

T4.4: Diode saturation current density of a semi-infinite pn-junction in the limit of Auger recombination

The diode saturation current density is calculated using Equation (4.34). The value of n_i^2 is given in task T4.1. The diffusion coefficients of electrons and holes can be approximated as 20 and 10 cm^2/s for c-Si. The diffusion lengths can be calculated according to Equation (3.5) where the maximum lifetimes depend on Auger recombination following from Equation (3.15). The minimum diode saturation current density of the semi-infinite pn-junction is calculated for a c-Si solar cell by:

$$I_0 = q \cdot n_i^2 \cdot \sqrt{C_A} \cdot (\sqrt{D_n} + \sqrt{D_p}) \qquad (T4.4)$$

It is interesting to remark that Equation (T4.4) does not depend on the densities of donors and acceptors. The Auger recombination rate constant (C_A) is about $2 \cdot 10^{-30}\,cm^6/s$ for c-Si (see Chapter 3).

The diode saturation current density of a semi-infinite pn-junction is $3.3 \cdot 10^{-13}\,A/cm^2$ for a c-Si solar cell in the Auger limit. Incidentally, this value is larger than the diode saturation current density of a record c-Si solar cell (Equation (1.22′)) and highlights the importance of additional passivation measures in order to reduce the geometry factor.

T4.5: Diode saturation current density of a semi-infinite pn-junction in the limit of radiative recombination

The diode saturation current density is calculated using Equation (4.34). The value of n_i^2 is given in task T4.1. The diffusion coefficients of electrons and holes can be approximated as 60 and $10 \, cm^2$ /s for GaAs. The diffusion lengths can be calculated according to Equation (3.5), where the maximum lifetimes depend on radiative recombination after Equation (3.12). The diode saturation current density of the semi-infinite pn-junction is calculated for a GaAs solar cell by:

$$I_0 = q \cdot n_i^2 \cdot \sqrt{B} \cdot \left(\sqrt{\frac{D_n}{N_A}} + \sqrt{\frac{D_p}{N_D}} \right) \tag{T4.5}$$

The radiative recombination rate constant is about $2 \cdot 10^{-10} \, cm^3$/s for GaAs (see Chapter 3).

The diode saturation current density of a semi-infinite pn-junction is $4 \cdot 10^{-20} \, A/cm^2$ for a GaAs solar cell with $N_A = 10^{18}$ and $N_D = 10^{17} \, cm^{-3}$. The obtained value of I_0 is lower than I_0 of a record GaAs solar cell and shows the optimization potential for the pn-junction.

T4.6: Temperature dependence of V_{OC} for c-Si solar cells

The general temperature dependence of V_{OC} is given by Equation (1.35) with $E_A = E_g$ (1.12 eV for c-Si at room temperature). The value of I_{00} in Equation (1.35) is obtained by combining Equations (2.32) and (T4.4):

$$V_{OC} = \frac{E_g}{q} - \frac{k_B}{q} \cdot \ln \left(\frac{q \cdot N_C \cdot N_V \cdot \sqrt{C_A} \cdot (\sqrt{D_n} + \sqrt{D_p})}{I_{SC}} \right) \cdot T \tag{T4.6}$$

The temperature dependence of V_{OC} is given by:

$$\frac{\Delta V_{OC}}{\Delta T} = -\frac{k_B}{q} \cdot \ln \left(\frac{q \cdot N_C \cdot N_V \cdot \sqrt{C_A} \cdot (\sqrt{D_n} + \sqrt{D_p})}{I_{SC}} \right) \tag{T4.7}$$

The values of I_{SC} at AM1.5, N_C, N_V, C_A, D_n and D_p are about $0.04 \, A/cm^2$, $10^{19} \, cm^{-3}$, $3 \cdot 10^{19} \, cm^{-3}$, $2 \cdot 10^{-30} \, cm^6$/s, $20 \, cm^2$ /s and $10 \, cm^2$/s, respectively. Therefore the temperature dependence of V_{OC} for a c-Si solar cell is about $1.4 \cdot 10^{-3}$ V/K.

T4.7: Diode saturation current density for SRH recombination at the pn-junction

The diode saturation current density for Shockley–Read–Hall (SRH) recombination is calculated using Equation (4.43) by taking into account the minimum SRH lifetime from Equation (3.21).

$$I_{0,SRH} = \frac{q}{2} \cdot n_i \cdot \delta_{SRH} \cdot v_{th} \cdot \sigma_{e(h)} \cdot N_t \qquad \text{(T4.8)}$$

The value of n_i is 10^{10} for c-Si, the thermal velocity (v_{th}), the capture cross section $(\sigma_{e(h)})$ and the characteristic length δ_{SRH} amount to about 10^7 cm/s, 10^{-15} cm^2 and 10^{-6} cm, respectively.

The diode saturation current density for SRH recombination in a c-Si solar cell is about 10^{-11} A/cm^2 for $N_t = 10^{12}$ cm^{-3} and 10^{-7} A/cm^2 for $N_t = 10^{16}$ cm^{-3}.

T4.8: Dependence of V_{OC} on SRH recombination at the pn-junction

According to the two-diode model given by Equation (4.45), taking $R_S = 0$ and $R_p \to \infty$ and setting the current to 0, the following quadratic equation is obtained:

$$0 = \left[exp \frac{q \cdot V_{OC}}{k_B \cdot T} \right]^2 + \frac{I_{0,SRH}}{I_0} \cdot \left[exp \frac{q \cdot V_{OC}}{k_B \cdot T} \right] - \frac{I_{SC} + I_0 + I_{0,SRH}}{I_0}$$

$$\text{(T4.9)}$$

The solution of Equation (T4.9) with positive sign has a physical meaning, and V_{OC} can be calculated:

$$V_{OC} = \frac{2 \cdot k_B \cdot T}{q} \cdot ln \left[\sqrt{ \left(\frac{I_{0,SRH}}{I_0} \right)^2 + \frac{I_{SC} + I_0 + I_{0,SRH}}{I_0} } - \frac{I_{0,SRH}}{2 \cdot I_0} \right]$$

$$\text{(T4.10)}$$

For very low $I_{0,SRH}$, Equation (T4.10) leads to the known solution for V_{OC} given by Equation (1.19). The behavior of Equation (T4.10) is shown in Figure T4.1 for $I_0 = 10^{-26}, 10^{-20}, 10^{-14}$ and 10^{-8} A/cm^2 ($I_{SC} = 0.04$ A/cm^2).

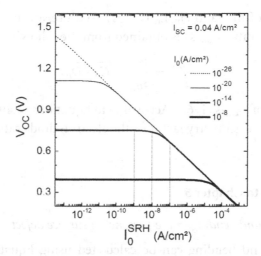

Figure T4.1. Dependence of V_{OC} on the diode saturation current density limited by SRH recombination at the pn-junction for different values of the diode saturation current density limited by bulk recombination ($I_0 = 10^{-26}$, 10^{-20}, 10^{-14} and 10^{-8} A/cm^2, dotted, thin solid, medium solid and thick solid lines, respectively) for a I_{SC} of 0.04 A/cm^2.

The plot shows that SRH recombination at the pn-junction limits V_{OC} completely for values of $I_{0,SRH}$ larger than the square root of I_0. On the other hand, the value of $I_{0,SRH}$ has to be about two orders of magnitude less than square root of I_0 in order to get rid of the influence of SRH on V_{OC}.

T4.9: Geometry factor of solar cells with a pn-junction

The geometry factor is calculated using Equation (4.49′). The functions of sinh (x) and cosh (x) are defined as:

$$\sinh(x) = \frac{1}{2} \cdot (e^x - e^{-x}) \qquad (T4.11')$$

$$\cosh(x) = \frac{1}{2} \cdot (e^x + e^{-x}) \qquad (T4.11'')$$

The argument of the sinh and cosh is the ratio between the thickness of the base and the diffusion length. With regard to Equations (3.5) and (3.21′), the argument of the sinh and cosh is obtained from the following equation:

$$\frac{H_p}{L_n} = H_p \cdot p_0 \cdot \sqrt{\frac{C_A}{D_n}} \qquad (T4.12)$$

The value of the ratio H_p/L_n is 0.047. With regard to Equations (3.5) and (3.21'), the ratio $s_{\infty n}/s_n$ is obtained from the following equation.

$$\frac{s_{\infty n}}{s_n} = \frac{p_0}{s_n} \cdot \sqrt{C_A \cdot D_n} \qquad (T4.13)$$

The value of $s_{\infty n}/s_n$ is 6.3. According to Equations (4.49') and (T4.11') and (T4.11''), the geometry factor of the electron diode saturation current density is 0.2.

A.5 Solutions to Chapter 5

T5.1: Surface band bending in the presence of surface defects

The surface band bending can be calculated using Equations (5.9) and (5.15), (5.16), (5.13') and (5.9').

$$\varphi_0 = \left[\frac{E_g - 2 \cdot (k_B \cdot T \cdot \ln(N_C/N_D))}{2 \cdot q} - \frac{\varepsilon_r \cdot \varepsilon_0}{4 \cdot q} \cdot \frac{N_d}{N_{st}^2} \cdot \left(\frac{E_g}{q}\right)^2\right]$$
$$- \sqrt{\left[\frac{E_g - 2 \cdot (k_B \cdot T \cdot \ln(N_C/N_D))}{2 \cdot q} - \frac{\varepsilon_r \cdot \varepsilon_0}{4 \cdot q} \cdot \frac{N_d}{N_{st}^2} \cdot \left(\frac{E_g}{q}\right)^2\right]^2 - \left[\frac{E_g - 2 \cdot (k_B \cdot T \cdot \ln(N_C/N_D))}{2 \cdot q}\right]^2} \qquad (T5.1)$$

For the semiconductor with $E_g = 1.1\,\text{eV}$ the surface band bending is 0.33 ($N_{st} = 10^{13}\,\text{cm}^{-2}$) or 0.025 V ($N_{st} = 10^{11}\,\text{cm}^{-2}$). The values of φ_0 amount to 0.605 ($N_{st} = 10^{13}\,\text{cm}^{-2}$) and 0.007 V ($N_{st} = 10^{11}\,\text{cm}^{-2}$) for the semiconductor with $E_g = 1.42\,\text{eV}$ (b).

T5.2: Fermi-level pinning

The change of the charge in surface states (ΔQ_{st}) has to be compared with the change of the charge in the space charge region (ΔQ_{st}) at the semiconductor surface by assuming that:

$$\Delta E_{FS} = q \cdot \Delta\varphi_0 \qquad (T5.2)$$

A shift of E_{FS} is impossible (Fermi-level pinning) if ΔQ_{st} cannot be compensated by ΔQ_{SC}, i.e.:

$$\Delta Q_{SC} < \Delta Q_{st} \tag{T5.3}$$

The value of ΔQ_{SC} can be calculated using Equation (5.9'):

$$\Delta Q_{sc} = \sqrt{\frac{2 \cdot \varepsilon_r \cdot \varepsilon_0}{q} \cdot N_D \cdot (\varphi_0 + \Delta\varphi_0)} - \sqrt{\frac{2 \cdot \varepsilon_r \cdot \varepsilon_0}{q} \cdot N_D \cdot \varphi_0} \tag{T5.4}$$

The value of ΔQ_{st} can be calculated using Equation (5.11) and by taking into account that $N_{st} = N_{st,a} = N_{st,d}$:

$$\Delta Q_{st} = 2 \cdot N_{st} \cdot \Delta E_{FS} \tag{T5.5}$$

The value of φ_0 can be set to the surface band bending in thermal equilibrium, which has been calculated in task T5.1 (a) for the conditions of this task. The values of ΔQ_{st} are about $2 \cdot 10^{10}$ cm^{-2} (a) and $2 \cdot 10^{12}$ cm^{-2} (b). The corresponding values of ΔQ_{SC} are about $7 \cdot 10^{10}$ cm^{-2} (a) and $3 \cdot 10^{10}$ cm^{-2} (b). Therefore, a shift of E_{FS} by 0.1 eV is possible for $N_{st} = 10^{11}$ cm^{-2} (a) but impossible for $N_{st} = 10^{13}$ cm^{-2} (b).

T5.3: Engineering of surface band bending with interface defects

The surface band bending for n-type doped semiconductors can be calculated from Equation (5.16). For p-type doped semiconductors, Equation (5.16) changes to:

$$\varphi_0 = -\frac{2 \cdot B + A}{2} + \sqrt{\left(\frac{2 \cdot B + A}{2}\right)^2 - B^2} \tag{T5.6}$$

To calculate parameter A, the value of N_D in Equation (5.15') is replaced by N_A for p-type doped semiconductors. The differences $E_C - E_{F0}$ (n-type) and $E_{F0} - E_V$ (p-type) are 0 for the given conditions. To calculate parameter B, the value of $N_{st,a}$ in Equation (5.15'') is replaced by $N_{st,d}$ for p-type doped semiconductors.

The values of φ_0 are (a) 0.249 V for n-type ($N_D = 10^{18}$ cm^{-3}) and -0.908 V for p-type ($N_A = 10^{19}$ cm^{-3}) and (b) 1.13 V for n-type ($N_D = 10^{18}$ cm^{-3}) and -0.22 V for p-type ($N_A = 10^{19}$ cm^{-3}) doped semiconductors.

A low band bending is desired for contacts with low tunneling resistance. A surface band bending of the order of 0.25 or −0.25 V can be easily reached at surfaces of moderately or highly doped n-type or p-type doped semiconductors, respectively, with high densities of surface defects where an excess of donor or acceptor defects by about three to five times is sufficient.

T5.4: Transmission probability of tunneling contacts with inorganic and organic semiconductors

Equation (5.26) is applied. For the organic semiconductors a value of m_e^* equal to m_e is assumed. The transmission probability is $9 \cdot 10^{-4}$ for the given organic and 0.027 for the given inorganic semiconductor.

T5.5: Recombination contact between highly doped semiconductors

The barrier heights for electron and hole tunneling ($\Phi_{B,e}$ and $\Phi_{B,h}$) are given by the drops of the potential across the space charge regions of the n-type and p-type doped regions ($\Phi_{B,e} = q \cdot U_n$ and $\Phi_{B,h} = q \cdot U_p$). The values of U_n and U_p follow from the condition of charge neutrality and from considering the band diagram (Figure 5.16).

$$U_n \cdot N_D^+ = U_p \cdot N_A^- \tag{T5.7}$$

$$U_n = \frac{E_{g2} - \Delta E_C}{q} - U_p \tag{T5.8}$$

The values of U_n and U_p are:

$$U_n = \frac{E_{g2} - \Delta E_C}{q} \cdot \frac{1}{1 + \frac{N_D^+}{N_A^-}} \tag{T5.9$'$}$$

$$U_p = \frac{E_{g2} - \Delta E_C}{q} \cdot \frac{1}{1 + \frac{N_A^-}{N_D^+}} \tag{T5.9$''$}$$

The values of U_n and U_p are 1.283 and 0.257 V, respectively. The electron and hole tunneling resistances follow from Equation (2.29). The recombination resistance of the degenerated pn-hetero-junction can be calculated:

$$R_{c,tunn} = R_{c,tunn,e} + R_{c,tunn,h} = \frac{3 \cdot \hbar}{8 \cdot q \cdot v_{th}} \cdot \sqrt{\frac{1}{N_D^+ \cdot m_e^* \cdot \varepsilon_r \cdot \varepsilon_0}}$$

$$\cdot \exp\left(\frac{8}{3} \cdot \sqrt{\frac{m_e^* \cdot \varepsilon_r \cdot \varepsilon_0}{N_D^+}} \cdot \frac{U_n}{\hbar}\right) + \frac{3 \cdot \hbar}{8 \cdot q \cdot v_{th}}$$

$$\cdot \sqrt{\frac{1}{N_A^+ \cdot m_h^* \cdot \varepsilon_r \cdot \varepsilon_0}} \cdot \exp\left(\frac{8}{3} \cdot \sqrt{\frac{m_h^* \cdot \varepsilon_r \cdot \varepsilon_0}{N_A^+}} \cdot \frac{U_p}{\hbar}\right)$$

$$\text{(T5.10)}$$

The values of $R_{c,tunn,e}$ and $R_{c,tunn,h}$ are about $6 \cdot 10^{-4}$ and $2 \cdot 10^{-7}$ Ωcm^2, respectively, i.e. tunneling of holes limits the given recombination contact. The tolerable series resistance of a triple-junction solar cell is about one order of magnitude larger than for a single-junction solar cell due to the increase of V_{OC} and the reduction of I_{SC}. A tolerable series resistance of the order of 0.2 Ωcm^2 is obtained for illumination at AM1.5 for a V_{OC} of about 2.4 V and an I_{SC} of about 0.012 mA/cm^2 for a multi-junction solar cell. The tolerable series resistance of the multi-junction solar cell becomes equal to the resistance of one tunneling contact at concentrated sunlight with a concentration factor of 300. For illumination at higher concentrated sunlight a reduction of the given N_A^- to a value closer to N_D^+ or an increase of N_D^+ will be beneficial.

T5.6: Schottky barriers for charge separation and ohmic contacts

A semiconductor at a Schottky barrier is depleted, i.e. photo-generated minority charge carriers are separated towards the metal–semiconductor contact. In the case of Fermi-level pinning, the surface recombination velocity at the metal–semiconductor contact is very high due to the high density of surface defects and so the contact behaves as an ohmic contact for photo-generated minority charge carriers. This may be a technological advantage since the formation of a pn-junction can be omitted in the solar cell. However, (i) the diffusion potential across a pn-junction can be varied over a wider range than the surface band bending of a Schottky barrier and (ii) the density of defects can be reduced to negligible values at a pn-homo-junction, which is not possible with a Schottky barrier. Therefore, high solar energy conversion efficiencies cannot be obtained with solar cells based on Schottky barriers.

A.6 Solutions to Chapter 6

T6.1: Shockley–Queisser limit

Equation (1.15) is used to calculate the solar energy conversion efficiency. The power of the sun is 0.1 W/cm^2 in the Shockley–Queisser limit. From Figure T6.1, the values of I_{SC} are 0.069, 0.034, 0.013 and 0.003 A/cm^2 for band gaps of 0.3, 1.3, 2.0 and 2.7 eV, respectively.

The values of V_{OC} are calculated by using Equations (1.19) and Equation (6.13) to ascertain the diode saturation current density. Equation (6.13) has to be written in the following way if the values of E_g are given in units of eV.

$$I_0 = \frac{2 \cdot q^3 \cdot \Omega_S \cdot k_B \cdot T_0}{h^3 \cdot c^2} \cdot (E_g[eV])^2 \cdot exp\left(-\frac{E_g}{k_B \cdot T_0}\right) \qquad \text{(T6.1)}$$

The pre-factor (I_0') in Equation (T6.1) amounts to:

$$I_0' = \frac{2 \cdot q^3 \cdot \Omega_S \cdot k_B \cdot T_0}{h^3 \cdot c^2} = 1630\frac{A}{cm^2} \qquad \text{(T6.2)}$$

The corresponding values of I_0 and V_{OC} are $1.5 \cdot 10^{-3}$ A/cm^2 and 0.1 V, $5.2 \cdot 10^{-19}$ A/cm^2 and 1.007 V, $2.3 \cdot 10^{-30}$ A/cm^2 and 1.68 V, $8 \cdot 10^{-42}$ A/cm^2 and 2.309 V for band gaps of 0.3, 1.3, 2.0 and 2.7 eV, respectively.

The FF are obtained from Figure T6.2 by extrapolating. The values of FF are 0.5, 0.89, 0.92 and 0.95 for band gaps of 0.3, 1.3, 2.0 and 2.7 eV, respectively.

The solar energy conversion efficiencies of solar cells in the Shockley–Queisser limit are about 3, 30.5, 20 and 6.6% for band gaps of 0.3, 1.3, 2.0 and 2.7 eV, respectively. Incidentally, these values are accurate within the error caused by extrapolation.

T6.2: Maximum efficiency of tandem solar cells

The solar energy conversion efficiency of tandem solar cells is calculated with the following equation.

$$\eta_{tandem} = \frac{I_{SC}^{bottom-single}}{2} \cdot \frac{FF_{bottom} \cdot V_{OC}^{bottom} + FF_{top} \cdot V_{OC}^{top}}{P_{sun}} \qquad (T6.3)$$

The values of $I_{SC}^{bottom-single}/2$ are obtained from Figure T6.1 and amount to 0.033, 0.0235 and 0.0135 A/cm^2 for the band gaps of the bottom cells of 0.5, 1.0 and 1.5 eV, respectively.

The band gaps of the top solar cells correspond to the band gaps of single solar cells with I_{SC} equal to $I_{SC}^{bottom-single}/2$. The values of the band gaps of the top cells are 1.35, 1.65 and 2.00 eV and of the bottom cells are 0.5, 1.0 and 1.5 eV, respectively.

The values of I_0 are calculated using Equations (T6.1) and (T6.2) and amount to $1.9 \cdot 10^{-6}$, $7.8 \cdot 10^{-20}$, $3.1 \cdot 10^{-14}$, $1.2 \cdot 10^{-24}$, $3 \cdot 10^{-23}$ and $2.3 \cdot 10^{-30}$ A/cm^2 for the band gaps of 0.5, 1.35, 1.0, 1.65, 1.5 and 2.0 eV, respectively.

Equation (1.19) is applied for calculating the values of V_{OC}, which are equal to 0.254, 1.055, 0.711, 1.335, 1.236 and 1.666 V for the band gaps of 0.5, 1.35, 1.0, 1.65, 1.5 and 2.0 eV, respectively.

The FF are found from Figure T6.2 and amount to 0.74, 0.89, 0.86, 0.88, 0.88 and 0.92 for the band gaps of 0.5, 1.35, 1.0, 1.65, 1.5 and 2.0 eV, respectively.

The solar energy conversion efficiencies of ideal tandem solar cells are about 37, 42 and 36% for bottom cells with a band gap of 0.5, 1.0 and 1.5 eV, respectively.

T6.3: Band gaps of absorbers in multi-junction solar cells

The I_{SC} of a multi-junction solar cell with five sub-cells is calculated using Equation (6.18).

$$I_{SC}^{k=5} = \frac{I_{SC}^{bottom-single}}{5} \qquad (T6.4)$$

The ideal top cell has a band gap corresponding to a single solar cell with I_{SC}^{top} equal to $I_{SC}^{k=5}$. The band gaps of the bottom and top cells have the indices E_{g1} and E_{g5}, respectively. The band gaps are obtained by using Equation (6.19).

$$(6-j) \cdot I_{SC}^{k=5}|_{j \in (1,5)} = q \cdot \int_{E_{g(j)}}^{\infty} \Phi_{sun}(h\upsilon) \cdot d(h\upsilon) \qquad (T6.5)$$

For the multi-junction solar cell with the bottom cell with a band gap of 0.5 eV the values of I_{SC}^{single} are 0.0132, 0.0264, 0.0396, 0.0528 and 0.066 A/cm^2 corresponding to band gaps of $E_{g5} = 2.05$ eV, $E_{g4} = 1.53$ eV, $E_{g3} = 1.18$ eV, $E_{g2} = 0.8$ eV and $E_{g1} = 0.5$ eV, respectively.

For the multi-junction solar cell with the bottom cell with a band gap of 1.0 eV the values of I_{SC}^{single} are 0.0094, 0.0188, 0.0282, 0.0376 and 0.047 A/cm^2 corresponding to band gaps of $E_{g5} = 2.17$ eV, $E_{g4} = 1.75$ eV, $E_{g3} = 1.47$ eV, $E_{g2} = 1.22$ eV and $E_{g1} = 1.0$ eV, respectively.

T6.4: Concentrator multi-junction solar cells

The I_{SC} of an ideal triple-junction solar cell is calculated using Equation (6.18).

$$I_{SC}^{k=3} = \frac{I_{SC}^{bottom-single}}{3} \qquad (T6.6)$$

The value of $I_{SC}^{bottom-single}$ is 0.06 A/cm^2 ($E_{g1} = 0.7$ eV). Therefore, the values of E_{g2} and E_{g3} are 1.17 and 1.75 eV corresponding to band gaps of single-junction solar cells with I_{SC} equal to 0.04 and 0.02 A/cm^2, respectively.

The solar energy conversion efficiency of the triple-junction solar cell is calculated from the following equation:

$$\eta_{triple} = \frac{I_{SC}^{Eg1-single}}{3}$$
$$\cdot \frac{FF_{Eg1}^{X} \cdot V_{OC}^{Eg1,X} + FF_{Eg2}^{X} \cdot V_{OC}^{Eg2,X} + FF_{Eg2}^{X} \cdot V_{OC}^{Eg2,X}}{P_{sun}(AM1.5)} \qquad (T6.7)$$

Incidentally, the concentration factor in the current density and in the power of the sunlight canceled out in Equation (T6.7). The procedures for finding I_0, V_{OC} and FF are identical to those in previous tasks. The values of η_{triple} are 53.5, 56, 57 and 58% for the concentration factors of 20, 200, 500 and 1000, respectively.

T6.5: Maximum concentration factor of sunlight

An absorber cannot emit photons at higher energy than it receives, i.e. the maximum temperature of an absorber cannot exceed the temperature of the sun. The power of sunlight multiplied with the maximum concentration factor of sunlight is balanced by the power of thermal radiation of the absorber heated to the maximum temperature of $T_S \approx 5800$ K.

$$X_{max} \cdot P_{sun} = \sigma_S \cdot T_s^4 \qquad (T6.8)$$

The energy flux received from the sun is the product of the solar constant (1356 W/m^2) and the maximum concentration factor (X_{max}). The Stefan–Boltzmann constant is $5.67 \cdot 10^{-8}$ W/(m^2 K^4). The maximum concentration factor of sunlight is about 46000.

T6.6: Thermodynamic limit of solar energy conversion

The maximum efficiency of thermal energy conversion is reached with the Carnot process (efficiency η_C):

$$\eta_C = 1 - \frac{T_0}{T_A} \qquad (T6.9)$$

The temperatures T_A and T_0 correspond to the temperature of the absorber or medium and to the ambient temperature, respectively. The absorber or medium is heated by the power of sunlight, which is proportional to the temperature at the surface of the sun (T_S) to the power of 4 (Stefan–Boltzmann law). The heated absorber emits thermal radiation, which is lost for the conversion into work. The emission losses of the absorber are proportional to the absorber temperature to the power of 4. The total efficiency of two subsequent processes is the product of the efficiencies of both processes. Therefore, the efficiency at the thermodynamic limit (η_{th}) is given by the following equation:

$$\eta_{th} = \left(1 - \left(\frac{T_A}{T_S}\right)^4\right) \cdot \left(1 - \frac{T_0}{T_A}\right) \qquad (T6.10)$$

Equation (6.26) has a maximum of 85% at a temperature of the absorber equal to 2478 K (for more details, see Würfel (2005)).

T6.7: Critical remark on solar cells with metal intermediate band

The photo-generation rate is increased due to absorption via the metal intermediate band (Figure T6.3).

Ideal absorption is assumed for maximum solar energy conversion efficiency. Ideal absorption must also be taken into account for the thermal generation rate. As a consequence, the diode saturation current density increases drastically if implementing a metal intermediate band due to the spectrum of thermal radiation. Therefore, the Fermi-level splitting is drastically reduced and the solar energy conversion efficiency of a solar cell with a metal intermediate band cannot exceed the efficiency of a solar cell with a band gap, which is equal to the higher of the differences between the valence or conduction band edges and the energy of the metal intermediate band. Incidentally, the nature of ideal charge-selective contacts has no influence on the Fermi-level splitting in an ideal PV absorber.

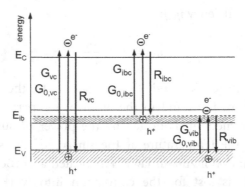

Figure T6.3. Generation and recombination processes in a PV absorber with a metal intermediate band meaning that there are occupied and unoccupied states available for excitation. G_{vc}, $G_{0,vc}$ G_{ibc}, $G_{0,ibc}$, G_{vib} and $G_{0,vib}$ denote photo-generation and thermal generation from the valence into the conduction band of the PV absorber, from the metal intermediate band into the conduction band of the absorber and from the valence band of the absorber into the intermediate band, respectively. Recombination takes place from the conduction band into the valence band of the absorber (R_{vc}), from the conduction band of the absorber into the metal intermediate band (R_{ibc}) and from the metal intermediate band into the conduction band of the absorber (R_{vib}).

A.7 Solutions to Chapter 7

T7.1: n-type doped base and p-type doped emitter

Equation (7.1′) is applied to calculate n_0 of the n-type doped base. The value of n_0 is $1 \cdot 10^{16}$ cm^{-3} if assuming a thickness of the base of 200 μm and a value of D_p which is 3.5 times smaller than 26 cm^2/s. The thickness of the p-type doped emitter is obtained by using the following equation:

$$d_{em} \leq \sqrt{\frac{D_n}{100 \cdot C_A \cdot p_0^2}} \qquad (T7.1)$$

The maximum useful density of acceptors is about the effective density of states at the valence band edge ($1 \cdot 10^{19}$ cm^{-3}). The value of d_{em} is about 200 nm.

T7.2: Silver required for c-Si-based photovoltaics

A bus bar collecting the current from a square with an area of a L^2 is considered for the estimation of the amount of silver. The length of the bus bar and its cross section are L and A_{cr}, respectively. The specific conductivity and the density of silver are denoted as σ_{Ag} and ρ_{Ag}, respectively. According to the definition of the $R_{s,tol}$ (Equation (1.29′)) and to the resistance (Equation (1.30)) the cross section of the bus bar can be expressed as:

$$A_{cr} = \frac{100}{\sigma_{Ag}} \cdot L \cdot \frac{I_{SC}}{V_{OC}} \qquad (T7.2)$$

Following the definition of the density, the mass of silver required for the collection of the current from an area L^2 (m_{Ag-L^2}) is calculated after equation:

$$m_{Ag-L^2} = 100 \cdot \frac{\rho_{Ag}}{\sigma_{Ag}} \cdot L^2 \cdot \frac{I_{SC}}{V_{OC}} \qquad (T7.3)$$

The mass of silver needed for 1 W of installed power is obtained from m_{Ag-L^2} by dividing Equation (T7.3) by the product of FF, I_{SC} and V_{OC}.

$$m_{Ag-W} = 100 \cdot \frac{\rho_{Ag}}{\sigma_{Ag}} \cdot \frac{L^2}{FF \cdot V_{OC}^2} \qquad (T7.4)$$

For an efficiency of 25% of a c-Si solar cell, the value of L^2 is about 40 cm^2 for 1W$_p$. As a result, a mass of about 170 mg of silver is required for a c-Si solar cell giving a power of 1 W$_p$. Therefore, silver is an important cost factor for c-Si solar cells. However, the amount of silver can be reduced if higher power losses caused by resistive heating are allowed for.

The electricity demand of mankind can be estimated if assuming 2 kW per capita, a number of $7 \cdot 10^9$ people and a performance of 25% (solar cells produce electricity only during day time). The amount of silver which would be required for complete electricity production with c-Si-based PV is about $9 \cdot 10^6$ t, i.e. about 400 times larger than the annual silver production. Therefore, silver has to be replaced in c-Si solar cells as a strategic issue for global electricity production with PV power plants.

T7.3. Boron-doped c-Si(100) wafers

Large silicon crystals can be doped very homogeneously with boron since boron has a segregation coefficient close to 1. Furthermore, boron is an acceptor in c-Si and minority charge carriers have a longer diffusion length in p-type than in n-type doped c-Si. Therefore, a base with a given thickness can be higher p-type than an n-type doped, which leads to a higher diffusion potential and therefore to a higher V_{OC}.

T7.4. Diffusion of phosphorus in c-Si

The diffusion coefficient of phosphorus in c-Si is about 10^{-15} cm^2/s at 900°C. The thickness of the emitter can be calculated by using Equation (7.25) and amount to 56 nm for a diffusion time of 1000 s.

T7.5. Diffusion of copper in c-Si

The diffusion from localized copper contaminations corresponds to diffusion from an exhaustible source and therefore Equation (7.26) has to be applied for calculating the spread of Cu. The diffusion coefficient of copper in c-Si is $D_i = 5 \cdot 10^{-5}$ cm^2/s at 900°C. A contamination of one mono-layer of Cu corresponds to a density of about $Q_{is} = 10^{15}$ cm^{-2}. The density of Cu should be lower than $N_{i,max} = 10^{12}$ cm^{-3} with respect to the minimum lifetime condition in c-Si absorbers and Shockley–Read–Hall recombination (see Chapter 3). The spread $x_{s,Cu}$ of the contamination can be defined as the depth at which $N_{i,max}$ is reached. The value of $x_{s,Cu}$ can be calculated by using the following equation:

$$x_{s,Cu} = \sqrt{4 \cdot D_i \cdot t \cdot ln\left(\frac{Q_{is}}{N_{i,max} \cdot \sqrt{\pi \cdot D_i \cdot t}}\right)} \qquad \text{(T7.5)}$$

The spreads of Cu are 1.4 mm, 4.2 mm and 1.2 cm after diffusion from a Cu contamination of only one mono-layer at 900°C for 10, 100 and 1000 s, respectively. This example demonstrates why any copper contamination has to be avoided for the production of c-Si solar cells, if the processes include conventional emitter diffusion and/or surface passivation by oxidation at high temperatures.

T7.6. V_{OC} of PERL and HIT c-Si solar cells

In PERL c-Si solar cells, a conventional pn-junction is used as a charge-selective contact. Very high values of V_{OC} can only be achieved for a base with a diffusion length of minority charge carriers much larger than the thickness of the base and simultaneously with extremely low surface recombination velocities (see Chapter 4). The overall surface recombination rate is still influenced by the ohmic contacts despite the back surface field due to the relatively low barrier for minority charge carriers and due to photo-generation in barrier regions of the back surface field of PERL c-Si solar cells.

In HIT c-Si solar cells, a-Si:H layers passivate the complete c-Si base and form charge-selective contacts with (i) additional barriers for minority charge carriers due to the band offsets at a-Si:H/c-Si hetero-junctions. In addition, (ii) the thin a-Si:H layers do not contribute to photo-generation. Therefore, the density of minority charge carriers is stronger reduced along the ohmic contacts for HIT than for PERL c-Si solar cells. Furthermore, (iii) the thickness of the base is only 98 μm in HIT c-Si solar cells while the thickness of the base is 260 μm in PERL c-Si solar cells. This improves the ratio between the thickness of the base and the diffusion length of minority charge carriers. Therefore, the geometry factor of the diode saturation current density can be stronger reduced for HIT than for PERL c-Si solar cells.

There is a disadvantage of HIT c-Si solar cells caused by absorption losses in the TCO front contact layer which limit I_{SC}.

T7.7. Potential of kerfless wafer transfer technologies

In conventional wafer technology, about half of a large c-Si crystal is transformed to slick during wire sawing and subsequent polishing of the wafers, i.e. half of the energy treated for the production of large c-Si crystals is wasted. Kerfless wafer transfer technologies avoid this waste in energy. In addition, the thickness of wafers can be strongly reduced to 25 μm by using exfoliation of c-Si absorbers. This means that the amount of c-Si can be reduced by up to 20 times in total. However, one has to keep in mind that the solar energy conversion efficiency of solar cells produced on exfoliated c-Si absorbers cannot reach the high efficiency of c-Si solar cells produced by advanced wafer-based technologies. This problem can be overcome by applying beam-induced wafering techniques, for example. It can be expected that kerfless wafer transfer technologies can lead to a reduction of the energy payback time of c-Si solar cells by more than 30%. However, one has to keep in mind that dozens or even hundreds of wafers can be cut by wire sawing from one large c-Si crystal at the same time. This parallel processing is not possible with kerfless wafer transfer technologies from one large c-Si crystal. Therefore, parallel processing of many large c-Si crystals with big machines will be required for high throughput in mass production of c-Si wafers by kerfless wafer transfer.

A.8 Solutions to Chapter 8

T8.1: Type I and type II semiconductor hetero-junction

The band gaps are taken from Figure 8.2 and the natural valence band offsets in relation to GaAs are taken from Figure 8.6. Then the natural conduction band offsets are calculated by using the following equation:

$$\Delta E_C = (E_{g2} - E_{g1}) - \Delta E_V \qquad (T8.1)$$

Natural GaP/InP, InN/AlAs and GaAs/InAs interfaces form type I, type II and type I hetero-junctions, respectively.

T8.2: *Influence of an interface dipole on the band offset*

The charge density at the interface dipole layer can be estimated from the reciprocal area of one unit cell.

$$\Delta Q_{pol} = \frac{0.1 \cdot q}{a_{lc}^2} \qquad (T8.2)$$

With respect to Equations (8.7) and (T8.2), the change of the band offset caused by the interface dipole layer can be estimated:

$$\Delta E_{dipol} = q \cdot \frac{0.1 \cdot q}{\varepsilon \cdot \varepsilon_0 \cdot a_{lc}} \qquad (T8.3)$$

The value of ΔE_{dipol} is about 0.27 eV if assuming a lattice constant of 0.57 nm. According to Figure 8.6 and to the band gaps of the given semiconductors, the natural valence band offset at the GaAs/InAs interface is about 0.06 meV and the natural conduction band offset is about 0.23 eV at the GaAs/AlAs interface. Therefore, the type of the hetero-junctions can change depending on the direction and the value of the interface dipole layer at the hetero-junction.

T8.3: *Tuning of band gaps in lattice-matched epitaxy*

According to Figure 8.3, all band gaps of $Ga_xIn_{1-x}P$ with $x > 0.07$ are larger than all band gaps of $Ga_xIn_{1-x}As$. Therefore, the maximum and minimum band gaps of $Ga_xIn_{1-x}As_yP_{1-y}$ are related to the band gaps of $Ga_xIn_{1-x}P$ and $Ga_xIn_{1-x}As$ with the $a_{lc} = 0.58$ nm. According to the lattice constants of the binary III–V semiconductors and to Vegard's law (Equation (8.8)) stoichiometry parameters of $x = 0.16$ and $x = 0.64$ are obtained for $Ga_xIn_{1-x}P$ and $Ga_xIn_{1-x}As$, respectively. Incidentally, the stoichiometry parameter of $Ga_{0.16}In_{0.84}P$ is larger than 0.07, i.e. the assumption made before was correct. According to the dependence of the band gap on the stoichiometry and bowing parameters (Equation (8.2)) the band gaps of $Ga_{0.16}In_{0.84}P$ and $Ga_{0.64}In_{0.36}As$ are 1.51 and 0.93 eV, respectively.

T8.4: *Absorber resistance of solar cells based on III–V semiconductors*

The thickness of the absorber layer is about 3 μm in GaAs solar cells and a density of majority charge carriers of 10^{17} cm^{-3} can be assumed. The resistance of the GaAs absorber can be calculated according to Equation (7.3) and according to the values of the mobility of holes and electrons.

$$R_{absorber}^{GaAs} = \frac{d_{abs}}{q \cdot p(n)_0 \cdot \mu_{p(e)}} \approx 10^{-5} - 10^{-6} \, \Omega cm^2 \qquad (T8.4)$$

A GaAs absorber with a thickness of 3 μm and a density of majority charge carriers of 10^{17} cm^{-3} has a resistance of 10^{-5}–10^{-6} Ωcm^2 . For comparison, the tolerable series resistance of a GaAs solar cell operated at AM1.5 is about 0.3 Ωcm^2 ($V_{OC} = 1.107$ V, $I_{SC} = 29.6$ mA/cm^2) (Kayes *et al.*, 2011). Therefore, the bulk resistance of solar cells based on III–V semiconductors is not critical for operation at highly concentrated sunlight.

T8.5: Epitaxial layers in tandem solar cells with III–V semiconductors

The layer system of the GaAs/Al$_{0.36}$Ga$_{0.64}$As tandem solar cell consists of 13 separate layers (layers 1–13). The epitaxy starts with the growth of a buffer layer (layer 1) for reducing the density of defects being present at the surface of the GaAs substrate. The buffer layer and the substrate are n-type doped. A barrier layer for holes (layer 2) is grown next. The n-type and p-type doped layers 3 and 4 absorb sunlight and form the pn-junction of the GaAs bottom cell. Layer 5 forms a barrier for electrons. The tunneling junction is formed by highly p-type and n-type doped GaAs layers (layers 6 and 7). A barrier layer for holes is deposited next (layer 8). The n-type and p-type doped layers 9 and 10 absorb sunlight and form the pn-junction of the Al$_{0.36}$Ga$_{0.64}$As top cell. Layers 11 and 12 are responsible for the transport of holes towards the ohmic front contact while layer 12 forms an additional barrier layer for electrons. Layer 13 is made from highly p-type doped GaAs and forms the ohmic contact with a metal layer system.

T8.6: Maximum efficiency of a lattice-matched triple-junction solar cell

The I_{SC} of the triple-junction solar cell is limited by the top cell since the band gap of the top cell is significantly larger than the optimum band gap of the top cell for the given bottom cell (see Chapter 6). The value of I_{SC} is obtained from Figure T6.1. The values of the diode saturation current densities of ideal solar cells are calculated for the top, middle and bottom cells using Equation (6.12) and the corresponding FF can be taken from

Figure T6.2 by extrapolating. The values of V_{OC} are calculated for the top, middle and bottom cells by using Equation (1.19).

$$\eta_{max}^{triple} = \frac{I_{SC}}{P_{sunlight}}$$
$$\cdot [FF_{top} \cdot V_{OC,top} + FF_{middle} \cdot V_{OC,middle} + FF_{bottom} \cdot V_{OC,bottom}] \tag{T8.5}$$

The maximum energy conversion efficiency of the lattice-matched triple-junction solar cell is 47%. For comparison, the maximum efficiency of a triple-junction solar cell with optimum band gaps is 50%.

T8.7: Lattice mismatch for lattice-matched epitaxy

First, the stoichiometry parameters are calculated by using Equation (8.2), the values of the corresponding band gaps of the binary III–V semiconductors for the Γ valleys (1.42, 0.42, 1.35 and 2.9 eV for GaAs, InAs, InP and GaP, respectively) and of the bowing parameters ($C = 0.477$ for $Ga_xIn_{1-x}As$ and $C = 0.65$ for $Ga_xIn_{1-x}P$ for $x < 0.64$).

$$x = \sqrt{\left[\frac{E_g(A) - E_g(B) - C}{2 \cdot C}\right]^2 + \frac{E_g(A_xB_{1-x}) - E_g(B)}{C}}$$
$$- \frac{E_g(A) - E_g(B) - C}{2 \cdot C} \tag{T8.6}$$

The stoichiometries $Ga_{0.97}In_{0.03}As$ and $Ga_{0.43}In_{0.57}P$ are calculated. According to the lattice constants of the binary III–V semiconductors and to Vegard's law (Equation (8.8)), the lattice constants of $a_{lc} = 0.56653$ nm and $a_{lc} = 0.56894$ nm are found for $Ga_{0.97}In_{0.03}As$ and $Ga_{0.43}In_{0.57}P$, respectively. For comparison, the lattice constant of germanium is 0.56575 nm. The lattice mismatch is defined by Equation (8.13). The lattice mismatch is 0.14% between Ge and $Ga_{0.97}In_{0.03}As$ and 0.42% between $Ga_{0.97}In_{0.03}As$ and $Ga_{0.43}In_{0.57}P$.

T8.8: Lattice mismatch for lattice-mismatched epitaxy

The stoichiometry parameters are calculated with regard to Equation (T8.6). The stoichiometries $Ga_{0.91}In_{0.09}As$ and $Ga_{0.39}In_{0.61}P$ are calculated for the triple-junction solar cell grown on Ge and $Ga_{0.7}In_{0.3}As$ for the inverted metamorphic triple-junction solar cell. The lattice constants 0.56905, 0.57747 and 0.57062 nm are found for $Ga_{0.97}In_{0.03}As$, $Ga_{0.7}In_{0.3}As$ and $Ga_{0.43}In_{0.57}P$, respectively. The lattice mismatch is 0.58% between Ge and $Ga_{0.91}In_{0.09}As$ and 0.28% between $Ga_{0.91}In_{0.09}As$ and $Ga_{0.39}In_{0.61}P$. For comparison, the lattice mismatch is 2.1% between $Ga_{0.7}In_{0.3}As$ and GaAs.

T8.9: Type II hetero-junctions for charge separation

Figure T8.1 shows the band diagram of a type II pn-hetero-junction. It is assumed that the semiconductor with index 1 is n-type (N_{D1}) and the semiconductor with index 2 p-type doped (N_{A2}). The diffusion potential can be expressed by the following equations:

$$q \cdot U_D = U_1 + U_2 = E_{g2} - (E_{C1} - E_F) - (E_F - E_{V2}) - \Delta E_C$$
$$\text{(T8.7')}$$

$$q \cdot U_D = E_{g1} - (E_{C1} - E_F) - (E_F - E_{V2}) - \Delta E_V \qquad \text{(T8.7'')}$$

Equations (T8.7') and (T8.7'') show that the maximum possible $q \cdot V_{OC}$ (ultimate efficiency; see Chapter 2) is given by the differences between E_{g2} and ΔE_C or between E_{g1} and ΔE_V, i.e. $q \cdot V_{OC}$ is lower than each of the band gaps. As consequence, the solar energy conversion efficiency of a solar cell with a type II hetero-junction as a charge-selective contact is less than the solar energy conversion efficiency of a solar cell with a pn-homo-junction. For avoiding losses caused by the band offsets, one of the band offsets has to be reduced to 0. In this case, the semiconductor with the larger band gap acts as a window layer and highly doped $p^{++}n^{++}$-tunneling homo-junctions will be required for integrated series connection in related multi-junction solar cells. However, it is practically impossible to reduce simultaneously one of the band offsets to 0 for a series of PV absorbers and defects at the

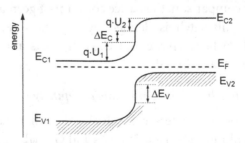

Figure T8.1. Band diagram of a type II pn-hetero-junction.

charge-selective hetero-junction to 0. Therefore, type II charge-selective hetero-junctions are not useful for multi-junction solar cells with very high solar energy conversion efficiency.

T8.10: Finger grid of concentrator solar cells

The emitter of multi-junction concentrator solar cells consists of a highly n-type doped window layer with a density of donors of about $5 \cdot 10^{18} \, \text{cm}^{-3}$ and a thickness of about 100 nm. The electron mobility of highly doped III–V semiconductors is of the order of $1000 \, \text{cm}^2 \, /\text{Vs}$. The emitter resistance given by Equation (7.11) should not be larger than the tolerable series resistance defined by Equation (1.29'). The specific resistance of the emitter is given by Equation (7.8). Therefore, the distance between grid fingers can be calculated using the following equation:

$$a \leq \sqrt{\frac{1}{100} \cdot \frac{V_{OC}}{I_{SC}} \cdot 12 \cdot d_{em} \cdot q \cdot n_0 \cdot \mu_{n+}} \qquad \text{(T8.8)}$$

The value of V_{OC} of a triple-junction solar cell illuminated with highly concentrated sunlight is about 3 V. The value of I_{SC} of a triple-junction solar cell illuminated with sunlight concentrated by a factor of 2000 is about $30 \, \text{A/cm}^2$. The resulting distance between grid fingers is about $100 \, \mu\text{m}$. Incidentally, the width of the grid fingers should be about $5 \, \mu\text{m}$ or even less for minimizing losses by shading. Photolithography is required for producing related finger grids for concentrator solar cells.

A.9 Solutions to Chapter 9

T9.1: Bulk and tunneling resistances at grain boundaries in a TCO layer

The length, the height and the width of the TCO layer are denoted by L, H and B, respectively. The bulk resistance of the TCO layer can be calculated using Equation (9.2).

$$R_{bulk} = \frac{1}{q \cdot n_0 \cdot \mu_n} \cdot \frac{L}{B \cdot H} \qquad \text{(T9.1)}$$

The tunneling resistance through the barrier of one depletion region can be calculated by using Equation (5.30) and by keeping in mind the effective mass of the tunneling electron. Furthermore, the tunneling resistance has to be doubled since tunneling occurs via defect states at the grain boundary. The tunneling resistance of the TCO layer is obtained by taking into account the number of barriers across the length and the cross section of tunneling barriers perpendicular to the direction of the current flow.

$$R_{tunn} = 2 \cdot R_{c,tunn,e} \cdot \frac{L}{d_g} \cdot \frac{1}{BH} \tag{T9.2}$$

$$R_{c,tunn,e} \approx \frac{\Omega cm^2}{1.26 \cdot 10^9} \cdot \sqrt{\frac{10^{19} cm^{-3}}{N_D^+}}$$

$$\cdot exp\left(79 \cdot \sqrt{\frac{10^{19} cm^{-3}}{N_D^+}} \cdot \sqrt{\frac{m_e^*}{m_e} \cdot \frac{\Phi_B}{eV}}\right) \tag{T9.3}$$

The specific bulk resistance is $6 \cdot 10^{-4}$ Ωcm. The specific tunneling resistance is given by:

$$\rho_{tunn} = \frac{2}{d_g} \cdot R_{c,tunn,e} \tag{T9.4}$$

The specific tunneling resistances are 0.01 Ωcm for $\varphi_0 = 1.2$ V (a) and $6 \cdot 10^{-5}$ Ωcm for $\varphi_0 = 0.3$ V (b), i.e. the specific tunneling resistance is larger than the specific bulk resistance of the TCO layer for $\varphi_0 = 1.2$ V but lower for $\varphi_0 = 0.3$ V. This example demonstrates the importance of the control of defects at grain boundaries in TCO layers.

T9.2: TCO contact stripe of thin-film solar cells

Values of 8 mA/cm^2 and 1.7 V can be assumed for I_{SC} and V_{OC}, respectively, of a pin–pin a-Si:H solar cell with respect to the world record a-Si:H solar cell (see Table 9.3). Transformation of Equations (9.3) and (9.4) and considering that the specific resistance is the reciprocal conductivity gives the following expression for the width of the TCO stripe:

$$L \leq \sqrt{\frac{1}{100 \cdot R_\square} \cdot \frac{V_{OC}}{I_{SC}}} \tag{T9.5}$$

The resulting value of L is 6.6 mm

T9.3: Optical losses in TCO layers

The AM1.5d spectrum can be found on the internet (see, for example, RReDc (2013)). Optical losses in the ZnO:Al are caused by reflection and free carrier absorption. The refractive index of ZnO is calculated with respect to the Sellmeier Equation (9.5) and the parameters given by Sun and Kwok (1999) for ZnO thin films. The reflection losses are considered in the transmission coefficient of the ZnO:Al layer, which is given by Equation (9.6) and the parameter Equations (9.7) and (9.8).

The transmission losses due to free carrier absorption are calculated by using the Lambert–Beer law with an absorption coefficient which is proportional to the wavelength to the power of 3 and which is proportional to n_0 expressed in the pre-factor α_0.

$$\alpha(\lambda, n_0) = \alpha_0(n_0) \cdot \left(\frac{\lambda}{\lambda_0}\right)^3 \qquad \text{(T9.6)}$$

The value of α_0 is about $200 \, \text{cm}^{-1}$ at a wavelength $\lambda_0 = 1 \, \mu\text{m}$ and for n_0 equal to $5 \cdot 10^{19} \, \text{cm}^{-3}$ (Weiher, 1966).

The value of I_{SC} is obtained by integrating the product of the photon flux and the amount of light transmitted through the ZnO:Al layer from 330 nm to the wavelength corresponding to the band gap of the PV absorber (λ_g). The thickness of the TCO layer is denoted by d_{TCO}.

$$I_{SC} = q \cdot \int_0^{\lambda_g} \Phi_{sun}(\lambda) \cdot T_{tr} \cdot [1 - \exp(-\alpha(\lambda, n_0) \cdot d_{TCO})] \cdot d\lambda$$

$$\text{(T9.7)}$$

Figure T9.1 shows a plot of I_{SC} as a function of λ_g for an ideal solar cell without losses and with optical losses due to reflection and free carrier absorption in the ZnO:Al surface layer.

The values of I_{SC} are 19.6 mA/cm² (a-Si:H, $\lambda_g = 730$ nm) and 42.3 mA/cm² (CuInSe$_2$, $\lambda_g = 1230$ nm) for AM1.5d without losses. Keep in mind that I_{SC} would be larger by a factor 1.1 for AM1.5G. The values of I_{SC} are reduced by about 14 and 15% if considering only reflection losses for a-Si:H and CuInSe$_2$, respectively, with a 0.5-μm thick ZnO surface layer. The reduction of I_{SC} increases to 19 (27) and 24 (44)% for a-Si:H and CuInSe$_2$, respectively, with a 0.5 (1.0)-μm thick ZnO:Al surface layer

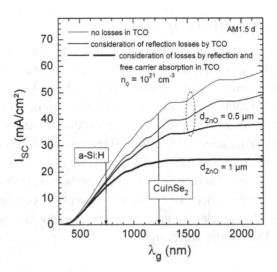

Figure T9.1. Dependence of I_{SC} on the wavelength corresponding to the band gap of a PV absorber (λ_g) without consideration of losses in the TCO (thinnest solid line), with consideration of optical losses due to reflection in a 0.5 μm thick TCO (ZnO) surface layer (thin solid line) and with consideration of optical losses due to reflection and free carrier absorption in a 0.5- and 1-μm thick ZnO:Al surface layer (thick and thickest solid lines, respectively). The values of λ_g are shown for a-Si:H and CuInSe$_2$ absorber layers.

(consideration of reflection and free carriers absorption). For comparison, I_{SC} is reduced by about 25 and 20% for the a-Si:H and CuIn$_{0.8}$Ga$_{0.2}$Se$_2$ record solar cells, respectively, in comparison to the maximum I_{SC} at the corresponding values of λ_g.

T9.4: Burstein shift in ZnO:Al

The density of states integrated from E_C to $E_C + \Delta E_g$ is equal to the density of free electrons in the degenerated semiconductor or metal oxide. Equation (2.16$'$) can be used while E_C is set to 0:

$$\Delta E_g = \left[\frac{cm^3}{6.7 \cdot 10^{21}} \cdot \left(\frac{m_e}{m_e^*} \right)^{3/2} \cdot N_D \right]^{2/3} eV \qquad (T9.8)$$

The resulting values of ΔE_g are 0.23 eV and 1.06 eV for densities of free electrons equal to 10^{20} and 10^{21} cm^{-3}, respectively.

T9.5: Open-circuit voltage in an a-Si:H solar cell

The densities of free electrons and holes in the n-type and p-type doped regions are of the order of 10^{18} cm^{-3}, which correspond to a built-in potential of about 1.4 V across the un-doped a-Si:H absorber layer (the densities of states at the conduction and valence band edges are about 10^{20} cm^{-2} and a band gap of 1.7 eV is assumed). The density of photo-generated charge carriers can be estimated by using Equation (3.8). The maximum I_{SC} of a solar cell with an a-Si:H absorber is about 16 mA/cm^2. A lifetime of photo-generated charge carriers of 10 ns and a thickness of the absorber layer of 1 μm are considered. Then the density of photo-generated charge carriers is about 10^{13} cm^{-3}, i.e. about five orders of magnitude less than the densities of the majority charge carriers in the p-type and n-type doped regions. A difference of five orders of magnitude in the densities of free charge carriers corresponds to a difference of 0.3 V in the chemical potentials. The value of V_{OC} can be estimated as the difference between the built-in potential and the differences of the chemical potentials for electrons and holes in the un-doped absorber and the p-type and n-type doped regions, respectively. Therefore V_{OC} is about 0.8 V for a-Si:H solar cells. The remaining built-in potential under illumination of about 0.6 V is sufficient for realizing a drift length of the order of the thickness of the un-doped a-Si:H absorber. Under light trapping, the thickness of the un-doped a-Si:H absorber layer can be reduced so that higher densities of photo-generated charge carriers can be obtained and a lower remaining built-in potential can be tolerated for sufficient drift. For comparison, V_{OC} is 0.886 V for the world record pin a-Si:H solar cell (Benagli *et al.*, 2009).

T9.6: Chalcopyrite tandem solar cell

The values of I_{SC}, V_{OC} and FF of the single-junction bottom ($E_{g,bottom} = 1.17$ eV, CuIn$_{1-x}$Ga$_x$Se$_2$ absorber) and top ($E_{g,top} = 1.67$ eV, CuGaSe$_2$ absorber) cells are 35.4 and 16.4 mA/cm^2, 0.74 and 0.795 V, and 77.5 and 69.2%, respectively. The value of I_{SC} of the tandem solar cell cannot be larger than I_{SC} of the single-junction top cell which is less than half of I_{SC} of the single-junction bottom cell. The values of V_{OC} and FF of the top cell remain unchanged in comparison to the single-junction cell with CuGaSe$_2$ absorber whereas the values of V_{OC} and FF are slightly reduced for the bottom cell in comparison to the single-junction cell with the $CuIn_{1-x}Ga_xSe_2$ absorber.

Under the assumption of ideal light transmission of the top cell at photon energies below $E_{g,top}$ and under the assumption of an ideal recombination contact between the top and bottom cells, the energy conversion efficiency of the tandem solar cell would be about 18%, which is significantly less than the efficiency of the world record chalcopyrite solar cell. The efficiency of the tandem solar cell would remain below 20% even if the I_{SC} of the single-junction top cell could be increased to half of I_{SC} of the single-junction bottom cell. The value of V_{OC} of the single-junction top cell should be larger than 1.0 V for getting an efficiency of the tandem cell above 21%. It is impossible to improve the world record of chalcopyrite solar cells without solving the V_{OC} problem of thin-film solar cells.

T9.7: Advantages of CdTe thin-film technology

First, CdTe exists in only one phase and sublimates in correct stoichiometry. Therefore, the stoichiometric CdTe phase forms in a relatively wide technological window. Second, CdTe has a large tolerance against incorporation of impurities from the targets and from atmosphere due to segregation. Third, $CdCl_2$ treatment at reduced temperature enables the formation of large grains. During this process the density of defects in the crystallites is reduced and additional segregation of impurity atoms towards grain boundaries takes place. Fourth, an intimate ohmic contact is formed at further reduced temperature by incorporation of acceptors into CdTe near the CdTe/contact layer interface.

T9.8: Strategic potential of CdTe, chalcopyrite and kesterite solar cells

The mass of an element (m_i) required for 1 W_p of installed power can be calculated by using the following equation:

$$\frac{m_i}{W_p} = \rho_{abs} \cdot \frac{M_i}{M_{abs}} \cdot \frac{d_{abs}}{P_{sun} \cdot \eta} \tag{T9.9}$$

The symbols $\rho_{abs}, M_i, M_{abs}, d_{abs}, P_{sun}$ and d_{abs} denote the density of the absorber material, the molar mass of the element i, the molar mass of the absorber material, the thickness of the absorber layer, the power of sunlight per cm^2 $(0.1\,W/cm^2)$ and the efficiency of the solar cells with the given absorber, respectively. Efficiencies of 12% for CdTe, 15% of chalcopyrite and 10% of kesterite PV modules are assumed. The densities of CdTe, chalcopyrite and kesterite absorbers are 5.85 and about 5.7 and 5 g/cm^3, respectively. Values of d_{abs} of 5 and 3 μm are considered for solar cells based on CdTe and chalcopyrite and kesterite absorbers, respectively. The molar masses of Te, In and Sn are 127.6, 114.8 and 118.7 g/mol, respectively. The molar masses of CdTe, $CuIn_{0.75}Ga_{0.25}Se_2$ and $Cu_2ZnSn(Se_xS_{1-x})_4$ absorbers are 240 g/mol, 324.4 g/mol and about 500 g/mol, respectively.

(a) The masses needed for one W_p are about 0.13 g/W_p (CdTe), 0.03 g/W_p ($CuIn_{0.75}Ga_{0.25}Se_2$) and 0.03 g/W_p ($Cu_2ZnSn(Se_xS_{1-x})_4$).

The installed power of PV power plants should be of the order of 10^{14} W_p if all electricity will be produced only by PV, if considering a population of $7 \cdot 10^9$ individuals and if taking into account the fact that solar cells do not produce electricity during the night.

(b) The required total masses of Te, In and Sn would be about 10^7 t (CdTe-based PV), $3 \cdot 10^6$ t (chalcopyrite-based PV) and $3 \cdot 10^6$ t (kesterite-based PV). The required amounts of Te, In and Sn would exceed the annual production of Te, In and Se by more than 10^5 (!), 10^3 (!) and 10 times, respectively, i.e. conventional CdTe and chalcopyrite thin-film solar cells are not suitable for global electricity production.

A.10 Solutions to Chapter 10

T10.1: Solar cell with ultra-thin absorber

The diffusion and drift lengths have to be compared (Equations (9.14) and (9.15)).

$$L = \sqrt{\frac{k_B \cdot T}{q} \cdot \mu \cdot \tau} \qquad (9.14)$$

$$L_{drift} = \mu \cdot \tau \cdot \varepsilon \qquad (9.15)$$

The diffusion length amounts to 50 nm. An electric field of about 10^4 V/cm can be assumed under illumination. Then the drift length is 100 nm. The effective thickness of the absorber layer should be larger than three times the absorption length of the absorber material. Therefore, the surface area should be increased by a factor of 60 if taking into account the drift length.

T10.2: Surface atoms at colloidal QDs

The number of atoms in a QD is obtained by using the definitions of the density and of the molar mass and by taking into account a sphere with a radius which is half of the diameter of the colloidal QD. The molar mass of PbS (M_{PbS}) is 239.3 g/mol. The Avogadro number (N_A) is $6.024 \cdot 10^{23}$ per mol. Lead and sulfur atoms are considered. The bond length is denoted by δ_{bond}. The number of bulk atoms of a QD (N_{bulk}) is given by the radius reduced by the bond length. The number of surface atoms of a QD (N_{surf}) is equal to the difference between N_{PbS-QD} and N_{bulk}.

$$N_{PbS-QD} = 2 \cdot \frac{\rho_{PbS}}{M_{PbS}} \cdot N_A \cdot \frac{4\pi}{3} \cdot r_{QD}^3 = 160 \cdot r_{QD}^3 [nm^3] \qquad (T10.1)$$

$$N_{bulk} = \frac{8\pi}{3} \cdot \frac{\rho_{PbS}}{M_{PbS}} \cdot N_A \cdot (r_{QD} - \delta_{bond})^3 \qquad (T10.2)$$

$$N_{surf} = N_{PbS-QD} - N_{bulk} \qquad (T10.3)$$

The numbers of bulk atoms are 55, 786 and 3149 for PbS-QDs with diameters of 2, 4 and 6 nm, respectively. At the same time, the numbers of surface atoms are 105, 494 and 1171 for PbS-QDs with diameters of 2, 4 and 6 nm, respectively.

T10.3: QD absorber layer

The effective mass of excitons and the absorption edge of the PbS-QD absorber can be calculated using Equations (10.1) and (10.4). The increase of the absorption gap due to quantum confinement in PbS-QDs can be expressed by the following equation:

$$\Delta E_{QD} = \frac{h^2}{8 \cdot r_{QD}^2 \cdot \mu_{e,h}} \approx 5.6 \cdot \frac{eV}{r_{PbS-QD}^2 [nm^2]} \qquad (T10.4)$$

Reduction of the energy by the Coulomb potential in PbS-QDs can be expressed by the following equation:

$$\Delta E_{Coul} = \left(\frac{1}{\varepsilon_{out}} + \frac{0.79}{\varepsilon_{in}} \right) \cdot \frac{q^2}{4\pi \cdot r_{QD} \cdot \varepsilon_0} \approx 0.56 \cdot \frac{eV}{r_{PbS-QD} [nm]}$$

$$(T10.5)$$

The band gaps are 5.4, 1.5 and 0.8 eV for PbS-QDs with diameters of 2, 4 and 6 nm, respectively, which correspond to a maximum I_{SC} of less than 1 mA/cm^2 (2 nm), 28 mA/cm^2 (4 nm) and 53 mA/cm^2 (6 nm). In literature, for example, a value of the absorption edge of 1.3 eV is given for PbS-QDs with a diameter of 3.5 nm. Differences are caused by variations of effective electron and hole masses, by polarization phenomena at interface dipole layers, penetration of solvents, disordered distribution of distances between QDs in a layer etc.

T10.4: Strategic potential of PbS-QD solar cells and solar cells sensitized with $CH_3NH_3PbI_3$

Equation (T9.7) is applied for calculating the mass of Pb (m_{Pb}) required for 1 W_p of installed power. A solar energy conversion efficiency of 10% and an effective thickness of the PbS-QD absorber in terms of compact PbS of 100 nm are assumed. The molar masses of Pb and PbS are 207 and 239 g/mol, respectively. Less than 7 mg of Pb is needed for one W_p. For a solar cell sensitized with $CH_3NH_3PbI_3$ a very similar value is obtained. The corresponding mass of Pb for complete electricity production for mankind would be about $7 \cdot 10^5$ t which is ten times less than the annual lead production. Therefore, solar cells based on PbS-QDs and solar cells sensitized with $CH_3NH_3PbI_3$ have tremendous strategic potential.

T10.5: Absorption and transport gaps in nano-composite absorbers

The fundamental limitation of the solar energy conversion efficiency of nano-composite solar cells cannot be described by the Shockley–Queisser limit due to the presence of an absorption gap limiting I_{SC} and a transport gap limiting V_{OC}. The difference between the absorption and transport gaps is generally needed in nano-composite solar cells as the driving force for local charge separation in an effective bulk semiconductor.

In nano-composite absorbers, the densities of mobile charge carriers in the dark are not dominated by thermal generation as in ideal PV absorbers. In nano-composite solar cells, the density of free charge carriers in the dark are dominated by electrical, chemical and electrochemical interac- tions at interfaces between materials transporting separately electrons or holes. The creation of a model describing fundamental limitations of the efficiency of nano-composite solar cells is still challenging.

T10.6: InP-nw solar cell

The effective lifetime related to surface recombination can be calculated following Equation (3.27). The density of defects at the interface between InP and SiO_2 seems to be quite large since SiO_2 does not fit well with InP (passivation of $A^{III}B^V$ semiconductors with oxides leads to non-stoichiometry and therefore to defect states). A surface recombination velocity of the order of 10^5–10^4 cm/s can be assumed, which gives a lifetime of photo-generated charge carriers of 0.1–1 ns. The electric field is extended over about 1 μm and the electrical field in the intrinsic InP layer is about 10^4 V/cm. Equation (9.15) can be used to calculate the drift time. The resulting drift time is about 0.04 ns, i.e. significantly shorter than the effective lifetime. Therefore, surface passivation does not limit the I_{SC} of InP-nw solar cells.

Bibliography

Aberle, A. G. (2000), Surface passivation of crystalline silicon solar cells: a review, *Prog. Photovolt.: Res. Appl.*, 8, pp. 473–487.

Aberle, A. G., Glunz, S., Warta, W. (1992), Impact of illumination level and poxide parameters on Shockley–Read–Hall recombination at the Si-SiO$_2$ interface, *J. Appl. Phys.*, 71, pp. 4422–4431.

Abou-Ras, D., Caballero, R., Kaufmann, C. A., Nichterwitz, M., Sakurai, K., Schorr, S., Unold, T. Schock, H. W. (2008), Impact of the Ga concentration on the microstructure of CuIn$_{1-x}$Ga$_x$Se$_2$, *Phys. Stat. Sol. (RRL)*, 3, pp. 135–137.

Adachi, S. (1989), Optical dispersion relations for GaP, GaAs, GaSb, InP, InAs, InSb, Al$_x$Ga$_{1-x}$As, and In$_{1-x}$Ga$_x$As$_y$P$_{1-y}$, *J. Appl. Phys.*, 66, pp. 6030–6040.

Adachi, S., Kimura, T., Suzuku, N. (1993), Optical properties of CdTe: experiment and modeling, *J. Appl. Phys.*, 74, pp. 3435–3441.

Albohn, J., Füssel, W., Sinh, N. D., Kliefoth, K., Fuhs, W. (2000), Capture cross section of defect states at the Si/SiO$_2$ interface, *J. Appl. Phys.*, 88, pp. 842–849.

Alferov, Zh. I., Andreev, V. M., Kagan, M. B., Protoasov, I. I., Trofim, V. G. (1971), Solar energy converters based on p-n Al$_{1-x}$Ga$_x$As-GaAs heterojunctions, *Sov. Phys. Semicond.*, 4, pp. 2047–2048, translated from (1970), *Fiz. Tekh. Poluprovodn.*, 4, pp. 2378–2379.

Alferov, Zh. I., Andreev, V. M., Korol'kov, V., I., Portnoi, E. L., Tret'yakov, D. N. (1969), Injection properties of n-Al$_{1-x}$Ga$_x$As-p-GaAs heterojunctions, *Sov. Phys. Semicond.*, 2, pp. 843–844, translated from (1968), *Fiz. Tekh. Poluprovodn.*, 2, pp. 1016–1017.

Alferov, Zh. I., Andreev, V. M., Portnoi, E. L., Trukan, M. K. (1970), AlAs/GaAs heterojunction injection lasers with low room-temperature threshold, *Sov. Phys. Semicond.*, 3, pp. 1107–1110, translated from (1969), *Fiz. Tekh. Poluprovodn.*, 3, pp. 1328–1332.

Algora, C., Ortiz, E., Rey-Stolle, I., Diaz, V., Pena, R., Andreev, V. M., Khvostikov, V. P., Rumyantsev, V. D. (2001), A GaAs solar cell with an efficiency of 26.2% at 1000 suns and 25% at 2000 suns, *IEEE Trans. Electr. Dev.*, 48, pp. 840–844.

465

Allred, A. L. (1961), Electronegativity values from thermochemical data, *J. Inorg. Nucl. Chem.*, 17, pp. 215–221.

Allsop, N., Schönmann, A., Belaidi, A., Muffler, H.-J., Mertesacker, B., Bohne, W., Strub, E., Röhrich, J., Lux-Steiner, M. C., Fischer, C.-H. (2006), Indium sulfide thin films deposited by the spray ion layer gas reaction technique, *Thin Solid Films*, 513, pp. 52–56.

Alperovich, V. L., Paulish, A. G., Terekhov, A. S. (1994), Domination of adatom-induced over defect-induced surface states on p-type GaAs(Cs,O) at room temperature, *Phys. Rev. B*, 50, pp. 5480–5483.

Anderson, P. W. (1958), Absence of diffusion in certain random lattices, *Phys. Rev.*, 109, pp. 1492–1515.

Andersson, B. A. (2000), Materials availability for large-scale thin-film photovoltaics, *Prog. Photovolt.: Res. Appl.*, 8, pp. 61–76.

Archer, J. (1960), Stain films on silicon, *J. Phys. Chem. Solids*, 14, pp. 104–110.

Ård, M. A., Granath, K., Stolt, L. (2000), Growth of Cu(In, Ga)Se$_2$ thin films by coevaporation using alkaline precursors, *Thin Solid Films*, 361–362, pp. 9–16.

Aspnes, D. E., Harbison, J. P., Studna, A. A., Florez, L. T. (1988), Reflectance-difference spectroscopy system for real-time measurements of crystal growth, *Appl. Phys. Lett.*, pp. 957–859.

Aspnes, D. E., Kelso, S. M., Logan, R. A., Bhat, R. (1986), Optical properties of Al$_x$Ga$_{1-x}$As, *J. Appl. Phys.*, 60, pp. 754–767.

Aspnes, D. E., Studna, A. A. (1983), Dielectric functions and optical parameters of Si, Ge, GaP, Gas, GaSb, InP, InAs, and InSb from 1.5 to 6 eV, *Phys. Rev. B*, 27, pp. 985–1009.

Atwater, H. A., Polman, A. (2010), Plasmonic for improved photovoltaic devices, *Nature Materials*, 9, pp. 205–213.

Auger, M. P. (1923), Sur les rayons β-secondaires produits dans un gas par des rayons X, *C. R. A. S.*, 177, pp. 169–171.

Bach, U., Lupo, D., Comte, P., Moser, J. E., Weissörtel, F., Salbeck, J., Spreitzer, H., Grätzel, M. (1998), Solid-state dye-sensitized mesoporous TiO$_2$ solar cells with high photon-to-electron conversion efficiencies, *Nature*, 395, pp. 583–585.

Bag, S., Gunawan, O., Gokman, T., Zhu, Y., Todorov, T. K., Mitzi, D. B. (2012), Low band gap liquid-processed CZTSe solar cell with 10.1 % efficiency, *Energy Environ. Sci.*, 5, pp. 7060–7065.

Bardeen, J. (1947), Surface states and rectification at a metal semiconductor contact, *Phys. Rev.*, 71, pp. 717–727.

Barton, C. F. (1971), Electron effective mass in PbS single crystals by infrared reflectivity measurements, *J. Appl. Phys.*, 42, pp. 445–450.

Basch, A., Beck, F. J., Soderstrom, T., Varlamov, S., Catchpole, K. R. (2012), Combined plasmonic and dielectric rear reflectors for enhanced photocurrent in solar cells, *Appl. Phys. Lett.*, 112, pp. 243903-1–243903-3.

Bean, K. E. (1978), Anisotropic etching of silicon, *IEEE Transactions*, ED-25, pp. 1185–1193.

Becker, P., De Bievre, P., Fujii, K., Glaeser, M., Inglis, B., Luebbig, H., Mana, G. (2007), Consideration on future redefinition of the kilogram, the mole and of other units, *Metrologia*, 44, pp. 1–14.

Bedair, S. M., Lamorte, M. F., Hauser, J. R. (1979), A two-junction cascade solar-cell structure, *Appl. Phys. Lett.*, 34, pp. 38–39.

Belaidi, A., Dittrich, T., Kieven, D., Tornow, J., Schwarzburg, K., Lux-Steiner, M. (2008), *Phys. Stat. Sol. (RRL)*, 2, pp.172–174.

Benagli, S., Borello, D., Vallat-Sauvain, E., Meier, J., Kroll, U., Hötzel, J., Spitznagel, J., Steinhauser, J., Castens, L., Djeridane, Y. (2009), High-efficiency amorphous silicon devices on LPCVD-ZNO TCO prepared in industrial KAI-M R&D reactor, *Proc. 24th European Photovoltaic Solar Energy Conference*, pp. 2293–2298.

Benick, J., Hoex, B., van de Sanden, M. C. M., Kessles, W. M. M., Schultz, O., Glunz, S. (2008), High efficiency n-type Si solar cells on Al_2O_3-passivated boron emitters, *Appl. Phys. Lett.*, 92, pp. 253504-1–253504-3.

Bentzen, A., Holt, A., Christensen, J. S., Svensson, B. G. (2006), High concentration in-diffusion of phosphorus in Si from spray-on source, *J. Appl. Phys.*, 99, pp. 064502-1–064502-8.

Bersch, E., Rangan, S., Bartynski, R. A. (2008), Band offsets of ultrathin high-k oxide films with Si, *Phys. Rev. B*, 78, pp. 085114-1–085114-10.

Bett, A. W., Dimroth, F., Guter, W., Hoheisel, R., Oliva, E., Philipps, S. P., Schöne, J., Siefer, G., Steiner, M., Wekkeli, A., Welser, E., Meusel, M., Köstler, W., Strobl, G. (2009), *24th European Photovoltaic Solar Energy Conference and Exhibition, 21–25 September, Hamburg (Germany)*, pp. 1–6.

Bhardwaj, A., Gupta, B. K., Raza, A., Sharma, A. K., Agnihotri, O. P. (1981), Fluorine-doped SnO_2 films for solar cell application, *Solar Cells*, 5, pp. 39–49.

Birkmire, R. W., Eser, E. (1997), Polycrystalline thin film solar cells: present status and future potential, *Annu. Rev. Mater. Sci.*, 27, pp. 625–653.

Bischoff, F. (1954), Verfahren zum Herstellen von reinstem kristallinem Germanium, Verbindungen von Elementen der III. und V. oder II. und VI. Gruppe des periodischen Systems und von oxydischem Halbleitermaterial, *Deutsches Patentamt*, Auslegeschrift 1 140 549, pp. 1–3.

Bisquert, J., Vikhrenko, V. S. (2004), Interpenetration of time constants measured by kinetic techniques in nanostructured semiconductor electrodes and dye-sensitized solar cells, *J. Phys. Chem. B*, 108, pp. 2313–2322.

Bodnar, I. V. (2008), Study of single crystals of the $CuIn_3Se$ ternary compound, *Semiconductors*, 42, pp. 1030–1033.

Brabec, C. J., Sariciftci, N. S., Hummelen, J. C. (2001a), Plastic solar cells, *Adv. Funct. Mater.*, 11, pp. 15–26.

Brabec, C. J., Zerza, G., Cerullo, G., De Silvestri, S., Luzzati, S., Hummelen, J. C., Sariciftci, S. (2001b), Tracing photoinduced electron transfer process in conjugated polymer/fullerene bulk heterojunctions in real time, *Chem. Phys. Lett.*, 340, pp. 232–236.

Braunstein, R., Moore, A. R., Herman, F. (1958), Intrinsic optical absorption in germanium-silicon alloys, *Phys. Rev.*, 109, pp. 695–710.

Brillson, L. J. (1978), Transition in Schottky barrier formation with chemical reactivity, *Phys. Rev. Lett.*, 40, pp. 260–263.

Britt, J., Ferekides, C. (1993), Thin-film CdS/CdTe solar cell with 15.8% efficiency, *Appl. Phys. Lett.*, 62, pp. 2851–2852.

Brodsky, M. H., Title, R. S., Weiser, K., Pettit, G. D. (1970), Structural, optical, and electrical properties of amorphous silicon films, *Phys. Rev. B*, 1, pp. 2632–2641.

Brooks, W. E. (2011), Silver, *U.S. Geological Survey, Mineral Commodity Summaries*, January, pp. 146–147.

Brus, L. (1984), Electron-electron and electron-hole interactions in small semi-conductor crystallites: the size dependence of the lowest excited state, *J. Chem. Phys.*, 80, pp. 4403–4409.

Burschka, J., Pellet, N., Moon, S. J., Baker, R. H., Gao, P., Nazeeruddin, M. K., Grätzel, M. (2013), Sequential deposition as route to high-performance perovskite-sensitized solar cells, *Nature*, 499, pp. 316–320.

Burstein, E. (1953), Anomalous optical absorption limit in InSb, *Phys. Rev. (Letters to the Editor)*, 93, pp. 632–633.

Capasso, F., Cho, A. Y., Mohammed, K., Foy, P. W. (1985), Doping interface dipoles: tunable heterojunction barrier heights and band-edge discontinuities by molecular beam epitaxy, *Appl. Phys. Lett.*, 46, pp. 664–666.

Caplan, P. J., Poindexter, E. H., Deal, B. E., Razouk, R. R. (1979), ESR centers, interface states, and oxide fixed charge in thermally oxidized silicon wafers, *J. Appl. Phys.*, 50, pp. 5847–5854.

Carlberg, T. (1986), Calculated solubilities of oxygen in liquid and solid silicon, *J. Electrochem. Soc.*, 139, pp. 1940–1942.

Carlin, J. F. (2013), Tin, *U.S. Geological Survey, Mineral Commodity Summaries*, January, pp. 170–171.

Carlson, D. E., Wronski, C. R. (1976), Amorphous silicon solar cell, *Appl. Phys. Lett.*, 28, pp. 671–673.

Casey, H. C., Miller, B. I., Pinkas, E. (1973), Variation of minority-carrier diffusion length with carrier concentration in GaAs liquid-phase epitaxial layers, *J. Appl. Phys.*, 44, pp. 1281–1287.

Chapin, D. M., Fuller, C. S., Pearson, G. L. (1954), A new silicon p-n junction photocell for converting solar radiation into electrical power, *J. Appl. Phys. (Letters to the Editor)*, 25, pp. 676–677.

Chelikowsky, J. R., Cohen, M. L. (1976), Nonlocal pseudopotential calculations for the electronic structure of eleven diamond and zinc-blende semiconductors, *Phys. Rev. B*, 14, pp. 556–582.

Chirilă, A., Reinhardt, P., Pianezzi, F., Blosech, P., Uhl, A. R., Fella, C., Kranz, L., Keller, D., Gretener, C., Hagendorfer, H., Jaeger, D., Erni, R., Nishiwaki, S., Buecheler, S., Tiwari, A. N. (2013), Potassium-induced surface modification of Cu(In,Ga)Se$_2$ thin films for high-efficiency solar cells, *Nature Materials*, 12, pp. 1107–1111.

Cho, A. Y., Arthur, J. R. (1975), Molecular beam epitaxy, *Prog. Solid State Chem.*, 10, pp. 157–191.

Cody, G. D., Tiedje, T., Abeles, B., Brooks, B., Goldstein, Y. (1981), Disorder and the optical-absorption edge of hydrogenated amorphous silicon, *Phys. Rev. Lett*, 47, pp. 1480–1483.

Cojacaru-Mirédin, O., Choi, P., Wuerz, R., Raabe, D. (2011), Atomic-scale characterization of the CdS/CuInSe2 interface in thin-film solar cells, *Appl. Phys. Lett.*, 98, pp. 103504-1–103505-3.

Contreras, M. A., Mansfield, L. M., Egaas, B., Li, J., Romero, M., Noufi, R., Rudiger-Voigt, E., Mannstadt, W. (2012), Wide bandgap Cu(In,Ga)Se$_2$ solar cells with improved energy conversion efficiency, *Prog. Photovolt.: Res. Appl.*, 20, pp. 843–850.

Coropceanu, V., Cornil, J., Da Silva Filho, D. A., Olivier, Y., Silbey, R., Brédas, J.-L. (2007), Charge transport in organic semiconductors, *Chem. Rev.*, 107, pp. 926–952.

Cowley, A. M., Sze, S. M. (1965), Surface states and barrier height of metal-semiconductor systems, *J. Appl. Phys.*, 36, pp. 3212–3220.

Cuff, K. F., Ellett, M. R., Kuglin, C. D., Williams, L. R. (1964), The band structure of PbTe, PbSe, and PbS, *Proc. 7th Int. Conf. de Physique des Semiconducteurs*, 7, pp. 677–684.

Czochralski, J. (1918), Ein neues Verfahren zur Messung der Kristallisations-geschwindigkeit der Metalle, *Z. Physik. Chemie*, 92, pp. 219–221.

Däweritz, L., Hey, R. (1990), Reconstruction and defect structure of vicinal GaAs(001) and (Al$_x$Ga$_{1-x}$As(001) surfaces during MBE growth, *Surface Science*, 236, pp. 15–22.

Deal, B. E., Grove, A. S. (1965), General relationship for the thermal oxidation of silicon, *J. Appl. Phys.*, 36, pp. 3770–3778.

Delbos, S. (2012), Kesterite thin films for photovoltaics: a review, *EPJ Photovoltaics*, 3, pp. 35004-p1–35004-p13.

de Nobel, D. (1959), Phase equilibria and semiconducting properties of CdTe, *Philips Res. Rep.*, 14, pp. 361–399.

Devlin, W. J., Wood, C. E. C., Stall, R., Eastman, L. F. (1980), A molybdenum source, gate and drain metallization system for GaAs MESFET layers grown by molecular beam epitaxy, *Solid-State Electron.*, 23, pp. 823–829.

Diebold, U. (2003), The surface science of titanium dioxide, *Surface Science Reports*, 48, pp. 53–229.

Dimmler, B., Dittrich, H., Menner, R., Schock, H. W. (1987), Performance and optimization of heterojunctions based on Cu(Ga,In)Se$_2$, *IEEE PVSC*, pp. 1454–1460.

Dimroth, F. (2006), Photovoltaic hydrogen generation process and device, *BRD Patent*, 2006/042650, pp. 1–24.

Dimorth, F., Grave, M., Beutel, P., Fiedeler, U., Karcher, C., Tibbits, T. N., Oliva, E., Siefer, G., Schachtner, M., Wekkeli, A., Bett, A. W., Krause, R., Piccin, M., Blanc, N., Drazek, C., Guiot, E., Ghyselen, B., Salvetat, T., Tauzin, A., Sgnamarcheix, T., Dobrich, A., Hannappel, T., Schwarzburg, K. (2014), Wafer bonded four-junction GaInP/GaAs/GaInAsP/GaInAs concentrator solar cells with 44.7% efficiency, *Prog. Photovolt: Res. Appl.*, 22, pp. 277–282.

Dimroth, F., Lanyi, P., Schubert, U., Bett, A. W. (2000), MOVPE grown Ga1-xInxAs solar cells for GaInP/GaInAs tandem applications, *J. Electronic Mats.*, 29, pp. 42–46.

Dingemans, G., Seguin, R., Engelhart, P., van de Sanden, M. C. M., Kessels, W. M. M. (2010), Silicon surface passivation by ultrathin Al_2O_3 films synthesized by thermal and plasma atomic layer deposition, *Phys. Stat. Sol. (RRL)*, 4, pp. 10–12.

Dingemans, G., van de Sanden, M. C. M., Kessels, W. M. M. (2011), Excellent Si surface passivation by low temperature SiO_2 using an ultrathin Al_2O_3 capping film, *Phys. Stat. Sol. (RRL)*, 5, pp. 22–24.

Dismukes, J. P., Ekstrom, L., Paff, R. J. (1964), Lattice parameter and density in germanium-silicon alloys, *J. Phys. Chem.*, 68, pp. 3021–3027.

Dittrich, T. (2000), Porous TiO_2: electron transport and application to dye sensitized injection solar cells, *Phys. Stat. Sol. (a)*, 182, pp. 447–455.

Dittrich, T., Belaidi, A., Ennaoui, A. (2011), Concepts of inorganic solid-state nanostructured solar cells, *Sol. Energy Mater. Sol. C.*, 95, pp. 1527–1536.

Dittrich, T., Bitzer, T., Rada, T., Timoshenko, V. Yu., Rappich, J. (2002), Non-radiative recombination at reconstructed Si surfaces, *Solid-State Electron.*, 46, pp. 1863–1872.

Dittrich, T., Burke, T., Koch, F., Rappich, J. (2001), Passivation of an anodic oxide/p-Si interface stimulated by electron injection, *J. Appl. Phys.*, 89, pp. 4636–4642.

Dixon, J. R., Ellis, J. M. (1961), Optical properties of n-type indium arsenide in the fundamental absorption edge region, *Phys. Rev.*, 123, pp. 1560–1566.

Drechsel, J., Männig, B., Kozlowski, F., Pfeiffer, M., Leo, K., Hoppe, H. (2005), Efficient organic solar cells based on a double p-i-n architecture using doped wide-gap transport layers, *Appl. Phys. Lett.*, 86, pp. 244102-1–244102-3.

Dullweber, T., Gatz, S., Hannebauer, H., Falcon, T., Hesse, R., Schmidt, J., Brendel, R. (2012), Towards 20% efficient large-area screen-printed rear-passivated silicon solar cells, *Prog. Photovolt.: Res. Appl.*, 20, pp. 630–638.

Eades, W. D., Swanson, R. M. (1985), Calculation of surface generation and recombination velocities at the Si-SiO_2 interface, *J. Appl. Phys.*, 58, pp. 4267–4276.

Eichberger, R., Willig, F. (1990), Ultra-fast electron injection from excited dye molecules into semiconductor electrodes, *Chem. Phys.*, 141, pp. 159–173.

Eisele, S. J., Röder, T. C., Köhler, J. R., Werner, J. H. (2009), 18.9% efficient full area laser doped silicon solar cells, *Appl. Phys. Lett.*, 95, pp. 133501-1–133501-3.

Engelhardt, P., Herder, N.-P., Grischke, R., Merkle, A., Meyer, R., Brendel, R. (2007), Laser structuring for back junction silicon solar cells, *Prog. Photovolt.: Res. Appl.*, 15, pp. 237–243.

Fa, Carl Zeiss (1939), Verfahren zur Erhöhung der Lichtdurchlässigkeit optischer Teile durch Erniedrigung des Brechungsexponenten an den Grenzflächen dieser optischen Teile, *Patentschrift*, 685767, pp. 1–3.

Fahey, P. M., Griffin, P. B., Plummer, J. D. (1989), Point defects and dopant diffusion in silicon, *Rev. Mod. Phys.*, 61, pp. 289–384.

Fengler, S., Zillner, E., Dittrich, T. (2013), Density of surface states at CdSe quantum dots by fitting temperature-dependent surface photovoltage transients with random walk simulations, *J. Phys. Chem. C*, 117, pp. 6462–6468.

Ferry, V. E., Polman, A., Atwater, H. A. (2011a), Modelling light trapping in nanostructured solar cells, *ACS Nano*, 5, pp. 10055–10064.

Ferry, V. E., Verschuuren, M. A., van Lare, M. C., Schropp, R. U. I., Atwater, H. A., Polman, A. (2011b), Optimized spatial correlations for broadband light trapping nanopatterns in high efficiency ultrathin film a-Si:H solar cells, *Nano Lett.*, 11, 4239–4245.

Fischer, C.-H., Allsop, N. A., Gledhill, S. E., Köhler, T., Krüger, M., Sáez-Araoz, R., Fu, Yanpeng, Schwieger, R., Richter, J., Wohlfahrt, P., Bartsch, P., Lichtenberg, N., Lux-Steiner, M. C. (2011), The spray-ILGAR®(ion layer gas reaction) method for the deposition of thin semiconductor layers: process and applications for thin film solar cell, *Sol. Energ. Mater. Sol. C.*, 95, pp. 1518–1526.

Fisher, A. C., Peter, L. M., Ponomarev, E. A., Walker, A. B., Wijayantha, K. G. U. (2000), Intensity dependence of the Bach reaction and transport of electrons in dye-sensitized nanocrystalline TiO_2 solarcells, *J. Phys. Chem. B*, 104, pp. 949–958.

Fraunhofer-Institut für Solare Energiesysteme ISE (2013), *Weltrekord-Solarzelle mit 44.7 Prozent Wirkungsgrad*, Presseinformation, 23 September 2013.

Fujishima, A., Honda, K. (1972), Electrochemical photolysis of water at a semiconductor surface, *Nature*, 238, pp. 37–38.

Gärtner, W. (1959), Depletion-layer photoeffects in semiconductors, *Phys. Rev.*, 116, pp. 84–87.

Gast, M., Köntges, M., Brendel, R. (2008), Lead-free on-laminate-laser-soldering: a new module assembling concept, *Prog. Photovolt.: Res. Appl.*, 16, pp. 151–157.

George, M. W. (2008), Tellurium, *U.S. Geological Survey, Mineral Commodity Summaries*, January 2008, pp. 170–171.

George, M. W. (2013), Tellurium, *U.S. Geological Survey, Mineral Commodity Summaries*, January, pp. 164–165.

Gloeckler, M. Sankin, I. Zhao, Z. (2013), CdTe solar cells at the threshold to 20% efficiency, *39th IEEE Photovoltaic Specialists Conference*, June 16–21, Tampa (Florida).

Goetzberger, A., Voß, B., Knobloch, J. (1997), *Sonnenenergie: Photovoltaik*, B. G. Teubner, Stuttgart.

Grant, R. W., Waldrop, J. R., Kraut, E. A. (1978), Observation of the orientation dependence of interface dipole energies in Ge-GaAs, *Phys. Rev. Lett.*, 40, pp. 656–659.

Grätzel, M. (2012), Status and progress in dye sensitized solar cells, *Proc. 27th European Photovoltaic Solar Energy Conf. and Exhibition* (EU-PVSEC), pp. 2158–2162.

Green, M. A., Emery, K., Hishikawa, Y., Warta, W., Dunlop, E. D. (2013), Solar cell efficiency tables (version 42), *Prog. Photovolt.: Res. Appl.*, 21, pp. 827–837.

Griffith, D. J. (2004), *Introduction into Quantum Mechanics*, 2nd ed., Prentice Hall, Singapore.

Grimm, A., Kieven, D., Lauermann, I., Lux-Steiner, M. Ch., Hergert, F., Schwieger, R., Klenk, R. (2012), Zn(O,S) layers for chalcopyrite solar cells sputtered from a single target, *EPJ Photovoltaics*, 3, pp. 30302-p1–30302-p4.

Grimsehl, E. (founded), Schallreuter, W. (continued), Haferkorn, H. (revised) (1988), *Grimsehl Lehrbuch der Physik, Band 3: Optik*, 19th ed., BSB B. G. Teubner Verlagsgesellschaft, Leipzig.

Grimsehl, E. (founded), Schallreuter, W. (continued), Lösche, A. (new ed.) (1990), *Grimsehl Lehrbuch der Physik, Band 4: Struktur der Materie*, 18th corrected ed., BSB B. G. Teubner Verlagsgesellschaft, Leipzig.

Groenendaal, L., Jonas, F., Freitag, D., Pielartzik, H., Reynolds, J. R. (2000), Poly(3,4-ethylenedioxythiophene) and its derivatives: past, present, and future, *Adv. Mater.*, 12, pp. 481–494.

Guberman, D. E. (2013), Lead, *U.S. Geological Survey, Mineral Commodity Summaries*, January, pp. 90–91.

Guillemoles, J. F. (2000), Stability of Cu(In,Ga)Se2 solar cells: a thermodynamic approach, *Thin Solid Films*, 361–362, pp. 338–345.

Günes, S., Neugebauer, H., Sariciftci, N. S. (2007), Conjugated polymer-based organic solar cells, *Chem. Rev.*, 107, pp. 1324–1338.

Guo, Q., Ford, G. M., Yang, W.-C., Walker, B. C., Stach, E. A., Hillhouse, H. W., Agrawal, R. (2010). Fabrication of 7.2% efficient CZTSSe solar cells using CZTS nanocrystals, *J. Am. Chem. Soc.*, 132, pp. 17384–17386.

Guo, Q., Kim, S. J., Kar, M., Shafarman, W. N., Birkmire, R. W., Stach, E. A., Agrwal, R., Hillhouse, H. W. (2008), Development of CuInSe2 nanocrystal and nanoring inks for low-cost solar cells, *Nano Lett.*, 8, pp. 2982–2987.

Gur, I., Fromer, N. A., Geier, M. L., Alivisatos, A. P. (2005), Air-stable all-inorganic nanocrystal solar cells produced from solution, *Science*, 310, pp. 462–465.

Gutsche, H. (1958), Method for producing highest-purity silicon for electric semiconductor devices, *US Patent*, 3,042,494, pp. 1–6.

Guyot-Sionnest, P. (2012), Electrical transport in colloidal quantum dot films, *J. Phys. Chem. Lett.*, 3, pp. 1169–1175.

Haase, F., Kajari-Schröder, S., Winter, R., Nese, M. Brendel, R. (2012), Back-contacted high-efficiency silicon solar cells — conversion efficiency dependence on cell thickness, *Photovoltaics International*, 18th ed., 4th quarter, pp. 50–57.

Haddon, R. C., Brus, L. E., Raghavachari, K. (1986), Electronic structure and bonding of icosahedral C60, *Chem. Phys. Lett.*, 125, pp. 459–464.

Haines, W. G., Bube, R. H. (1978), Effects of heat treatment on the optical and electrical properties of indium-tin oxide films, *J. Appl. Phys.*, 49, pp. 304–307.

Hall, R. N. (1952), Electron-hole recombination in germanium, *Phys. Rev.*, 87, pp. 387.

Hall, R. N. (1953), Segregation of impurities during the growth of germanium and silicon crystals, *J. Phys. Chem.*, 57, pp. 836–839.

Hammond, M. L. (2001), 2–Silicon epitaxy by chemical vapor deposition, in *Handbook of Thin Film Deposition Processes and Techniques*, 2nd ed., ed. Seshan, K., William Andrew Inc., Norwich, pp. 45–110.

Han, L., Islam, A., Chen. H., Malapaka, C., Chiranjeevi, B., Zhang, S., Yang, X., Yanagida, M. (2012), *Energy Environ. Sci.*, 5, pp. 6057–6060.

Han, L., Mandlik, P., Cherenack, K. H., Wagner, S. (2009), Amorphous silicon thin-film transistors with field-effect mobilities of $2 cm^2/(Vs)$ for electrons and $0.1 cm^2/(Vs)$ for holes, *Appl. Phys. Lett.*, 94, pp. 162105-1–162105-3.

Haug, F.-J., Rudmann, D., Zogg, H., Tiwari, A. N. (2003), Light soaking effects in $Cu(In,Ga)Se_2$ superstrate solar cell, *Thin Solid Films*, 431–432, pp. 431–435.

Haugeneder, A., Negas, M., Kallinger, C., Spirkl, W., Lemmer, U., Feldmann, J., Scherf, U., Harth, E., Gügel, A., Müllen, K. (1999), Exciton diffusion and dissociation in conjugated polymer/fullerene blends and heterostructures, *Phys. Rev. B*, 59, pp. 15346–15351.

He, Z., Zhong, C., Su, S., Xu, M., Wo, H., Cao, Y. (2012), Enhanced power-conversion efficiency in polymer solar cells using an inverted device structure, *Nature Photonics*, 6, pp. 591–595.

Heeger, A. J., Kivelson, S., Schrieffer, J. R., Su, W.-P. (1988), Solitons in conducting polymers, *Rev. Mod. Phys.*, 60, 781–850.

Heliatek (2013), Neuer Weltrekord für organische Solarzellen: Heliatek behauptet sich mit 12% Zelleffizienz als Technologieführer, *Pressemitteilung*, 13.1.2013, pp. 1–3.

Helms, C. R., Poindexter, E. H. (1994), The silicon-silicon-dioxide system: its microstructure and imperfections, *Rep. Prog. Phys.*, 57, pp. 791–852.

Herberholz, R., (1997), Prospects of wide-gap chalcopyrites for thin film photovoltaic modules, *Sol. Energ. Mater. Sol. C.*, 49, pp. 227–237.

Herron, J., Podewils, C. (2011), Im silbernen Käfig, *Photon*, September, pp. 110–116.

Hieslmair, H., Latchford, I., Mandrell, L., Chun, M., Adibi, B. (2012), Ion implantation for silicon solar cells, *Photovoltaics International*, 4th quarter, November, pp. 58–64.

Higuchi, M., Sugawa, S., Ikenaga, W., Ushio, J., Nohira, H., Maruizumi, T. (2007), Subnitride and valence band offset at Si3N4/Si interface formed using nitrogen-hydrogen radicals, *Appl. Phys. Lett.*, 90, pp. 123114-1–123114-3.

Hill, I. G., Rajagopal, A., Kahn, A. (1998), Molecular level alignment at organic semiconductor-metal interfaces, *Appl. Phys. Lett.*, 73, pp. 662–664.

Himpsel, F. J., McFeely, F. R., Taleb-Ibrahimi, A., Yarmoff, J. A., Hollinger, G. (1988), Microscopic structure of the SiO_2 / Si interface, *Phys. Rev. B*, 38, pp. 6084–6096.

Hoex, B., Schmidt, J., Pohl, P., van de Sanden, M. C. M., Kessles, W. M. M. (2008), Silicon surface passivation by atomic layer deposition Al_2O_3, *J. Appl. Phys.*, 104, pp. 0449003-1–044903-12.

Hotta, S., Rughooputh, S. D. D. V., Heeger, A. J. (1987), Conducting polymer composites of soluble polythiophenes in polystyrene, *Synt. Met.*, 22, pp. 79–87.

Hummelen, J. C., Knight, B. W., LePeq, F., Wudl, F. (1995), Preparation and characterization of fulleroid and methanofullerene derivatives, *J. Org. Chem.*, 60, pp. 532–538.

IEC (2008), Photovoltaic devices — part 3: measurement principles for terrestrial photovoltaic (PV) solar devices with reference spectral irradiance data, *IEC 60904-3*, ed. 2, pp. 1–62.

Istratov, A. A., Flink, C., Hieslmair, H., Weber, E. R., Heiser, T. (1998), Intrinsic diffusion coefficient of interstitial copper in silicon, *Phys. Rev. Lett.*, 81, pp. 1243–1246.

Jackson, W. B., Amer, N. M. (1982), Direct measurement of gap-state absorption in hydrogenated amorphous silicon by photothermal deflection spectroscopy, *Phys. Rev. B*, 25, pp. 5599–5562.

Jackson, P., Hariskos, D., Lotter, E., Paetel, S., Wuerz, R., Menner, R., Wischmann, W., Powalla, M. (2011), New world record efficiency for Cu(In,Ga)Se2 thin-film solar cells beyond 20%, *Prog. Photovolt.: Res. Appl.*, 19, pp. 894–897.

Jacoboni, C., Canali, C., Ottaviani, G., Alberigi Quarante, A. (1977), A review of some charge transport properties of silicon, *Solid-State Electron.*, 20, pp. 77–89.

Jaffe, J. E., Zunger, A. (1984), Theory of the band-gap anomaly in ABC_2 chalcopyrite semiconductors, *Phys. Rev. B*, 29, pp. 1882–1906.

Jasenek, A., Rau, U. (2001), Defect generation in Cu(In,Ga)Se2 heterojunction solar cells by high-energy electron and proton irradiation, *J. Appl. Phys.*, 90, pp. 650–658.

Jordan, D. C., Kurtz, S. R. (2013), Photovoltaic degradation rates — an analytical review, *Prog. Photovolt.: Res. Appl.*, 21, pp. 12–29.

Kahng, D. (1960), Electric field controlled semiconductor device, *US Patent*, 3,102,230.

Kayes, B. M., Nie, H., Twist, R., Sprytte, S. G., Reinhardt, F., Kizilyalli, I. C., Higashi, G. S. (2011), 27.6% conversion efficiency, a new record for single-junction solar cells under 1 sun illumination, *37th IEEE Photovoltaic Specialists Conference*, 4–8.

Keister, J. W., Rowe, J. E., Kolodziej, J. J., Niimi, H., Madey, T. E., Lucovski, G. (1999), Band offsets for ultrathin SiO_2 and Si_3N_4 films on Si(111) and Si(100) from photoemission spectroscopy, *J. Vac. Sci. Technol. B*, 17, pp. 1831–1835.

Keldysh, L. V. (1979), Coulomb interaction in thin films of semiconductors and semimetals, *Pisma Zh. Eksp. Teor. Fiz.*, 29, pp. 716–719.

Kepler, R. G., Hoestery, D. C. (1974), High-field mobility in anthracene crystals, *Phys. Rev. B*, 9, pp. 2743–2745.

Kern, W. (1984), Purification of Si and SiO_2 surfaces with hydrogen peroxide, *Semiconductor International*, 1984, pp. 94–99.

Kern, W., Puotinen, D. A. (1970), Cleaning solutions based on hydrogen peroxide for use in silicon semiconductor technology, *RCA Review*, 31, pp. 187–206.

Kieven, D., Dittrich, T., Belaidi, A., Tornow, J., Schwarzburg, K., Allsop, N., Lux-Steiner, M. (2008), *Appl. Phys. Lett.*, 92, pp. 153107-1–153107-3.

Kim, T.-J, Holloway, P. H. (1997), Ohmic contacts to GaAs epitaxial layers, *CRC Cr. Rev. Sol. State*, 22, pp. 239–273.

Kim, K. S., Lee, K. H., Cho, U.-I., Choi, J. S. (1986), A study of nonstoichiometric empirical formulas for semiconductive metal oxides, *Bull. Korean Chem. Soc.*, 7, pp. 29–35.

King, R. R., Haddad, M., Isshiki, T., Colter, P., Ermer, J., Yoon, H., Joslin, D. E., Karam, N. H. (2000), Metamorphic CaInP/GaInAs/Ge solar cells, *Proc. 28th IEEE Photovoltaic Specialists Conference*, pp. 982–985.

King, R. R., Law, D. C., Edmondson, K. M., Fetzer, C. M., Kinsey, G. S., Yoon, H., Sherif, R. A., Karam, N. H. (2007), 40% efficient metamorphic GaInP/GaInAs/Ge multijunction solar cell, *Appl. Phys. Lett.*, 90, pp. 183516-1–183516-3.

Klenk, R. (2001), Characterization and modeling of chalcopyrite solar cells, *Thin Solid Films*, 387, pp. 135–140.

Klenk, R., Klaer, J., Scheer, R., Lux-Steiner, M. Ch., Lick, I., Meyer, N., Rühle, U. (2005), Solar cells based on $CuInS_2$ — an overview, *Thin Solid Films*, 480–481, pp. 509–514.

Klenk, R., Steigert, A., Rissom, T., Greiner, D., Kaufmann, C. A., Unold, T., Lux-Steiner, M. C. (2014), Junction formation by Zn(O,S) sputtering yields CIGSe based cells with effciencies exceeding 18%, *Progr. Photovolt: Res. Appl.*, 22, pp. 161–165.

Knotter, D. M. (2000), Etching mechanism of vitreous silicon dioxide in HF based solutions, *J. Am. Chem. Soc.*, 122, pp. 4345–4351.

Kolodinski, S., Werner, J., Wittchen, T., Queisser, H. J. (1993), Quantum efficiencies exceeding unity due to impact ionization in silicon solar cells, *Appl. Phys. Lett.*, 63, pp. 2405–2507.

Konagai, M. (2011), Present status and future prospects of silicon thin-film solar cells, *Jpn. J. Appl. Phys.*, 50, pp. 039991-1–030001-12.

Kroto, H. W., Heath, J. R., O'Brien, S. C., Curl, R. F., Smalley, R. E. (1985), C_{60}: Buckminsterfullerene, *Nature*, 318, 162–163.

Kuech, T. F., Tischler, M. A., Wang, P.-J., Scilla, G. (1988), Controlled carbon doping of GaAs by metalorganic vapor phase epitaxy, *Appl. Phys. Lett.*, 53, pp. 1317–1319.

Kumar, N. S., Chyi, J. I., Peng, C. K., Morkoc, H. (1989), GaAs metal-semiconductor field-effect transistor with extremely low resistance non-alloyed ohmic contacts using InAs/GaAs superlattice, *Appl. Phys. Lett.*, 55, pp. 775–776.

Kupec, J., Stoop, R. L., Witzigmann, B. (2010), Light absorption and emission in nanowire array solar cells, *Optics Express*, 18, pp. 27589-1–27589-17.

Kurtin, S., McGill, T. C., Mead, C. A. (1970), Fundamental transition in the electronic nature of solids, *Phys. Rev. Lett.*, 22, pp. 1433–1436.

Lauinger, T., Schmidt, J., Aberle, A. G., Hezel, R. (1996), Record low surface recombination velocities on 1 Ωcm p-silicon using remote plasma silicon nitride passivation, *Appl. Phys. Lett.*, 68, pp. 1232–1234.

Law, M., Beard, M. C., Choi, S., Luther, J. M., Hanna, M. C., Nozik, A. J. (2008), Determining the internal quantum efficiency of PbSe nanocrystal solar cells with the aid of an optical model, *Nano Lett.*, 8, pp. 3904–3910.

Lebedev, E., Dittrich, T., Petrova-Koch, V., Karg, S., Brütting, W. (1997), Charge carrier mobility in poly-(p-phenylenevinylene) studied by time-of-flight technique, *Appl. Phys. Lett.*, 71, pp. 2686–2688.

Lee, C. H., Yu, G., Moses, D., Pakbaz, K., Zhang, C., Sariciftci, N. S., Heeger, A. J., Wudl, F. (1993), Sensitization of the photoconductivity of conducting polymers by C_{60}: photoinduced electron transfer, *Phys. Rev. B*, 48, pp. 15425–15433.

Leguijt, C., Lölgen, P., Eikelboom, J. A., Weber, A. W., Schuurmans, F. M., Sinke, W. C., Alkemade, P. F. E., Sarro, P. M., Merée, C. H. M., Verhoef, L. A. (1996), Low temperature surface passivation for silicon solar cells, *Sol. Energ. Mater. Sol. C.*, 40, pp. 297–345.

Liu, H., Zhitomirsky, D., Hoogland, S., Tang, J., Kramer, I. J., Ning, Z., Sargent, E. H. (2012), Systematic optimization of quantum junction colloidal quantum dot solar cells, *Appl. Phys. Lett.*, 101, 151112-1–151112-3.

Lu, S., Ji, L., He, W., Dai, P., Yang, H., Arimochi, M., Yoshida, H., Uchida, A., Ikeda, M. (2011), High-efficiency GaAs and GaInP solar cells grown by all solid-state molecular-beam-epitaxy, *Nanoscale Res. Lett.*, 6, pp. 576-1–576-4.

Luther, J. M., Law, M., Beard, M. C., Song, Q., Reese, M. O., Ellingson, R. J., Nozik, A. J. (2008a), Schottky solar cells based on colloidal nanocrystal films, *Nano Lett.*, 8, pp. 3488–3492.

Luther, J. M., Law, M., Song, Q., Perkins, C. L., Beard, M. C., Nozik, A. J. (2008b), Structural, optical, and electrical properties of self-assembled films of PbSe nanocrystals treated with 1,2-ethanedithiol, *ACS Nano*, 2, pp. 271–280.

Ma, W., Luther, J. M., Zheng, H., Wu, Y., Alivisatos, A. P. (2009), *Nano Lett.*, 9, pp. 1699–1703.

Macor, L., Gervaldo, M., Fungo, F., Otero, L., Dittrich, T., Lin, C.-Y, Chi, L.-C., Fang, F.C., Lii, S.-W., Wong, K.-T., Tsai, C.-H., Wu, C.-C. (2012), Photoinduced charge separation in donor–acceptor spiro compounds at metal and metal oxide surfaces: application to dye-sensitized solar cell, *RSC Adv.*, 2, pp. 4869–4878.

Mainz, R., Klenk, R. (2011), *In situ* analysis of elemental depth distributions in thin films by combined evaluation of synchrotron X-ray fluorescence and diffraction, *J. Appl. Phys.*, 109, pp. 123515-1–123515-13.

Markov, D. E., Amsterdam, A., Blom, P. W. M., Sieval, A. B., Hummelen, J. C. (2005), Accurate measurement of the exciton diffusion length in a conjugated polymer using a heterostructure with a side-chain cross-linked fullerene layer, *J. Phys. Chem. A*, 109, pp. 5266–5274.

Marshall, J. M., Street, R. A., Thompson, M. J. (1986), Electron drift mobility in amorphous Si:H, *Phil. Mag. B*, 54, 51–60.

Mårtensson, T., Carlberg, P., Borgström, M., Montelius, L., Seifert, W., Samuelson, L. (2004), Nanowire arrays defined by nanoimprint lithography, *Nano Lett.*, 4, pp. 699–702.

Massa, E., Mana, G., Kuetgens, U., Ferroglio, I. (2009), Measurement of the lattice parameter of a silicon crystal, *New J. Phys.*, 11, pp. 053013-1–053013-12.

McCandless, B., Birkmire, R. (1991), Analysis of post deposition processing for CdTe/CdS thin film solar cells, *Sol. Cells*, 31, pp. 527–535.

Mehdi, I., Reddy, U. K., Oh. J., East, J. R., Haddad, G. I. (1989), Nonalloyed and alloyed low-resistance ohmic contacts with good morphology for GaAs using a graded InGaAs cap layer, *J. Appl. Phys.*, 65, pp. 867–869.

Meier, J., Flückiger, R., Keppner, H., Shah, A. (1994), Complete microcrystalline p-i-n solar cell — crystalline or amorphous cell behavior?, *Appl. Phys. Lett.*, 65, pp. 860–862.

Merdes, S., Abou-Ras, D., Mainz, R., Klenk, R., Lux-Steiner, M. Ch., Meeder, A., Schock, H. W., Klaer, J. (2013), CdS/Cu(In,Ga)S$_2$ based solar cells with efficiencies reaching 12.9% prepared by a rapid thermal process, *Progr. Photovolt.: Res. Appl.*, 21, pp. 88–93.

Meyer, B. K., Polity, A., Farangis, B., He, Y., Hasselkjamp, D., Krämer, T., Wang, C. (2004), Structural properties and bandgap bowing of ZnO$_{1-x}$S$_x$ thin films deposited by reactive sputtering, *Appl. Phys. Lett.*, 85, pp. 4929–4931.

Miller, A., Abrahams, E. (1960), Impurity conduction at low concentrations, *Phys. Rev.*, 120, pp. 745–755.

Miller, O. D., Yablonovich, E., Kurtz, S. R. (2012), Strong internal and external luminescence as solar cells approach the Shockley–Queisser limit, *IEEE J. Photovolt.*, 2, pp. 303–311.

Mimura, T., Hiyamizu, S., Fuji, T., Nanbu, K. (1980), A new field-effect transistor with selectively doped GaAs/n-Al$_{1-x}$Ga$_x$As heterojunctions, *Jpn. J. Appl. Phys.*, 19, pp. L225–L227.

Miyazaki, S., Narasaki, M., Suyama, A., Yamoaka, M., Murakami, H. (2003), Electronic structure and energy band offsets for ultrathin silicon nitride on Si(100), *Appl. Surf. Sci.*, 216, pp. 252–257.

Moreels, I., Lambert, K., Smeets, D., de Muynck, D., Nollet, T., Martins, J. C., Vanhaecke, F., Vantomme, A., Delerue, C., Allan, G., Hnes, Z. (2009), Site-dependent optical properties of colloidal PbS quantum dots, *ACS Nano*, 3, pp. 3023–3030.

Mott, N. F. (1985), The pre-exponential factor in the conductivity of amorphous silicon, *Phil. Mag. B*, 51, 19–25.

Naghavi, N., Jehl, Z., Donsanti, F., Guilllemoles, J.-F., Gérard, I., Bouttemy, M., Etcheberry, A., Pelouard, J.-L., Collin, S., Colin, C., Péré-Leperne, N., Dahan, N., Greffet, J.-J., Morel, B., Djebbour, Z., Darga, A., Mencaraglia, D., Voorwinden, G., Dimmler, B., Powalla, M., Lincot, D. (2012), Toward high efficiency ultra-thin CIGSe based solar cells using light management techniques, *Proc. SPIE*, 8256, pp. 825617-1–825617-8.

Nahory, R. E., Pollack, M. A., Johnston, Jr., W. D., Barns, R. L. (1978), Band gap versus composition and demonstration of Vegards law for In$_{1-x}$Ga$_x$As$_y$P$_y$ lattice matched to InP, *Appl. Phys. Lett.*, 33, pp. 659–661.

Nakada, T., Mizutani, M. (2002), 18% efficiency Cd-free Cu(In,Ga)Se$_2$ thin-film solar cells fabricated using chemical bath deposition (CBD)-ZnS buffer layers, *Jpn. J. Appl. Phys.*, 41, L165–L167.

Nazeeruddin, M. K., Kay, A., Rodicio, I., Humphrey-Baker, R., Müller, E., Liska, P., Vlachopoulos, N., Grätzel, M. (1993), Conversion of light to electricity by cis-X$_2$Bis(2,2'-bipyridyl-4,4'-dicarboxylate)ruthenium(II) charge-transfer sensitizers (X = Cl$^-$, Br$^-$, I$^-$, CN$^-$, and SCN$^-$) on nanocrystalline TiO$_2$ electrodes, *J. Am. Chem. Soc.*, 115, pp. 6382–6390.

Nebel, C. E., Street, R. A., Johnson, N. M., Kocka, J. (1992), High electric field transport in a-Si:H. I. transient photoconductivity, *Phys. Rev. B*, 46, pp. 6789–6802.

Noufi, R., Axton, R., Herrington, C., Deb, S. K. (1984), Electronic properties versus composition of thin films of CuInSe$_2$, *Appl. Phys. Lett.*, 45, pp. 668–670.

Nusbaumer, H., Moser, J.-E., Zakeeruddin, S. M., Nazeeruddin, M. K., Grätzel, M. (2001), CoII(dbbip)22+ complex rivals tri-iodide/iodide redox mediator in dye-sensitized photovoltaic cells, *J. Phys. Chem. B*, 105, pp. 10461–10464.

Nusbaumer, H., Zakeeruddin, Moser, J.-E., Grätzel, M. (2003), An alternative efficient redox couple for the dye-sensitized solar cell system, *Chem. Eur. J.*, 9, pp. 3756–3763.

Ohnesorge, B., Weigand, R., Bacher, G., Forchel, A., Riedl, W., Karg, F. H. (1998), Minority-carrier lifetime and efficiency of $Cu(In,Ga)Se_2$ solar cells, *Appl. Phys. Lett.*, 73, pp. 1224–1226.

O'Regan, B., Grätzel, M. (1991), A low-cost, high-efficiency solar cell based on dye-sensitized colloidal TiO_2 films, *Nature*, 353, pp. 737–740.

Otero, P., Quelle, I., Fonrodona, M., Santos, S., Rodríguez, J. A., Grande, E., Mata, C., Vetter, M., Andreu, J. (2012), Improvement of very large area (5.7 m^2) a-Si:H PV module manufacturing by PECVD process control, *Proc. 27th European Photovoltaic Solar Energy Conference*, pp. 2100–2112.

Pacios, R., Nelson, J., Bradley, J., Bradley, D. C., Brabec, C. J. (2003), Composition dependence of electron and hole transport in polyfluorene: [6,6]-phenyl C61-butric acid methyl ester blend films, *Appl. Phys. Lett.*, 83, pp. 4764–4766.

Paire, M., Shams, A., Lombez, L., Péré-Leperne, N., Collin, S., Pelouard, J.-L., Guillemoles, J.-F., Lincot, D. (2012), Resistive and thermal scale effect for $Cu(In,Ga)Se_2$ polycrystalline thin film microcells under concentration, *Energy Environ. Sci*, 4, pp. 4972–4977.

Panasonic (2013), Panasonic HIT® solar cell achieves world's highest conversion efficiency of 24.7% at research level, *Press Release*, February 12, pp. 1–3.

Parker, I. D. (1994), Carrier tunneling and device characteristics in polymer light-emitting diodes, *J. Appl. Phys.*, 75, pp. 1656–1666.

Pattantyus-Abraham, A. G., Kramer, I. J., Barkhouse, A. R., Wang, X., Konstantatos, G., Debnath, R., Levina, L., Raabe, I., Nazeeruddin, M. K., Grätzel, M., Sargent, E. H. (2010), Depleted-heterojunction colloidal quantum dot solar cell, *ACS Nano*, 4, pp. 3374–3380.

Paulson, P. D., Birkmire, R. W., Shafarman, W. N. (2003), Optical characterization of $CuIn_{1-x}Ga_xSe_2$ alloys by spectroscopic ellipsometry, *J. Appl. Phys.*, 94, pp. 879–888.

Pawlak, B. J., Janssens, T., Singh, S., Kuzma-Filipek, I., Robbelein, J., Posthuma, N. E., Poortmans, J., Christiano, F., Bazizi, E. M. (2012), Studies of implanted boron emitters for solar cell applications, *Prog. Photovolt.: Res. Appl.*, 20, 106–110.

Perrin, J. (1995), Reactor design for a-Si:H deposition, in *Plasma Deposition of Amorphous Silicon-Based Materials*, eds. Capezzuto, P., Madan, A., Academic Press, Boston, pp. 177–203.

Persson, C. (2010), Electronic and optical properties of Cu_2ZnSnS_4 and $Cu_2ZnSnSe_4$, *J. Appl. Phys. Phys.*, 107, pp. 053710-1–053710-8.

Persson, C., Platzer-Björkmann, C., Malmström, J., Törndahl, T., Edoff, M. (2006), Strong valence band offset bowing of $ZnO_{1-x}S_x$ enhances p-type nitrogen doping of ZnO-like alloys, *Phys. Rev. Lett.*, 97, pp. 146403-1–146403-4.

Peter, L. (2009), "Sticky electrons" transport and interfacial transfer of electrons in the dye-sensitized solar cell, *Accounts Chem. Res.*, 42, pp. 1839–1847.

Pfann, W. G. (1957), Zone melting, *Metallurgical Reviews*, 2, pp. 29–76.

Pfeiffer, M. (2013), Private communication about I_{SC}, V_{OC} and FF concerning the organic record solar cell of Heliatek, to appear.

Philip, R. R., Pradeep, B., Shripathi, T. (2005), Photoconductivity in the ordered vacancy compound $CuIn_5Se_8$, *Appl. Surf. Sci.*, 250, pp. 216–222.

Picozzi, S., Asahi, R., Geller, C. B., Freeman, A. J. (2002), Accurate first-principles detailed-balance determination of Auger recombination and impact ionization rates in semiconductors, *Phys. Rev. Lett.*, 89, pp. 197601-1–197601-4.

Platzer-Björkman, C., Törndahl, T., Abou-Ras, D., Malmström, J., Kessler, J., Stolt, L. (2006), Zn(O,S) buffer layers by atomic layer deposition in $Cu(In,Ga)Se_2$ based thin film solar cells: band alignment and sulfur gradient, *J. Appl. Phys.*, 100, pp. 044506-1–044506-9.

Poindexter, E. H., Caplan, P. J., Deal, B. E., Razouk, R. R. (1981), Interface states and electron spin resonance centers in thermally oxidized (111) and (100) silicon wafers, *J. Appl. Phys.*, 52, pp. 879–884.

Polman, A., Atwater, H. A. (2012), Photonic design principles for ultrahigh-efficiency photovoltaics, *Nature Materials*, 11, pp. 174–177.

Ponomarev, E. A., Levy-Clement, C. (1998), Macropore formation on p-type Si in fluoride containing organic electrolytes, *Electrochem. Solid-State Lett.*, 1, pp. 42–45.

Prins, P., Grozema, F. C., Schins, J. M., Patil, S., Scherf, U., Siebbeles, L. D. A. (2006), High intrachain hole mobility on molecular wires of ladder-type poly(p-phenylenes), *Phys. Rev. Lett.*, 96, 1466011–146601-4.

Rao, R. A., Mathew, L., Sarkar, D., Smith, S., Saha, S., Garcia, R., Stout, R., Gurmu, A., Ainom, M., Onyegam, E., Xu, D., Jawarani, D., Fossum, J., Banerjee, S., Das, U., Upadhyaya, A., Rohatgi, A., Wang, Q. (2012), A low cost kerfless thin crystalline Si solar cell technology, *38th IEEE Photovoltaic Specialists Conference (PVSC2012)*, pp. 001837–001840.

Rauscher, S., Dittrich, T., Aggour, M., Rappich, J., Flietner, H., Lewerenz, H. J. (1995), Reduced interface state density after photocurrent oscillations and electrochemical hydrogenation of n-Si(111): a surface photovoltage investigation, *Appl. Phys. Lett.*, 66, pp. 3018–3020.

Ribeiro Jr., M., Fonseca, R. C., Ferreira, L. G. (2009), Accurate prediction of the Si/SiO_2 interface band offset using the self-consistent *ab initio* DFT/LDA-1/2 method, *Phys. Rev. B*, 79, pp. 241312-1–241312-4.

Robbins, H., Schwartz, B. (1959), Chemical etching of silicon: Part I, The system HF, HNO_3, and H_2O, *J. Electrochem. Soc.*, 106, pp. 505–508.

Robbins, H., Schwartz, B. (1960), Chemical etching of silicon: Part II, The system HF, HNO_3, H_2O and $HC_2H_3O_2$, *J. Electrochem. Soc.*, 107, pp. 108–111.

RReDc (2013), ASTM G173-03 reference spectra derived from SMARTS v. 2.9.2. Available at: http://rredc.nrel.gov/solar/spectra/am1.5/ASTMG173/ASTMG173.html.

Ryvkin, S. M. (1964), *Photoelectric Effects in Semiconductors*, Consulting Bureau, New York.

Sáez-Araoz, R., Krammer, J., Harndt, S., Koehler, T., Krueger, M., Pistor, P., Jasenek, A., Hergert, F., Lux-Steiner, M. Ch., Fischer, C.-H. (2012), *Prog. Photovolt.: Res. Appl.*, 20, pp. 855–861.

Samuelson, L., Junno, B., Paulsson, G., Fornell, J. O., Ledebo, L. (1992), CBE growth of (001) GaAs: RHEED and RD studies, *J. Cryst. Growth*, 124, pp. 23–29.

Sariciftci, S., Braun, D., Zhnag, C., Srdanov, V. I., Heeger, A. J., Stucky, G., Wudl, F. (1993), Semiconducting polymer-buckminsterfullerene heterojunctions: diodes, photodiodes, and photovoltaic cells, *Appl. Phys. Lett.*, 62, pp. 585–587.

Sariciftci, N. S., Smilowitz, L., Heeger, A. J., Wudl, F. (1992), Photoinduced electron transfer from a conducting polymer to buckminsterfullerene, *Science*, 258, pp. 1474–1476.

Sawada, T., Tereda, N., Tsuge, S., Baba, T., Takahama, T., Wakisaka, K., Tsuda, S., Nakano, S. (1994), High-efficiency a-Si/c-Si heterojunction solar cell, *Proc. 1st World Conf. on Photovoltaic Energy Conversion*, Hawaii, pp. 1219–1226.

Scanlon, W. W. (1958), Intrinsic optical absorption and the radiative recombination lifetime in PbS, *Phys. Rev.*, 109, pp. 47–50.

Schaller, R. D., Klimov, V. (2004), High efficiency carrier multiplication in PbSe nanocrystals: applications for solar energy conversion, *Phys. Rev. Lett.*, 92, pp. 186601-1–186601-4.

Scharber, M. C., Mühlbacher, D., Koppe, M., Denk, P., Waldauf, C., Heeger, A. J., Brabec, C. J. (2006), Design rules for donors in bulk-heterojunction solar cells — towards 10% energy-conversion efficiency, *Adv. Mater.*, 18, pp. 789–794.

Schechtman, B. H., Spicer, W. E. (1970), Near infrared to vacuum ultraviolet absorption spectra and the optical constants of phthalocyanine and porphyrin films, *J. Mol. Spectrosc.*, 33, pp. 28–48.

Scheer, R. (1999), Qualitative and quantitative analysis of thin film heterostructures by electron beam induced current, *Sol. St. Phen.*, 67–68, pp. 57–68.

Scheer, R., Knieper, C., Stolt, L. (1995), Depth dependent collection functions in thin film chalcopyrite solar cells, *Appl. Phys. Lett.*, 67, pp. 3007–3009.

Scheer, R., Messmann-Vera, L., Klenk, R., Schock, H.-W. (2011), On the role of non-doped ZnO in CIGSe solar cells, *Prog. Photovolt.: Res. Appl.*, 20, pp. 619–624.

Scheer, R., Walter, T., Schock, H. W., Fearheiley, M. L., Lewerenz, H. J. (1993), CuInS$_2$ based thin film solar cell with 10.2% efficiency, *Appl. Phys. Lett.*, 63, pp. 3294–3296.

Schmidt, M., Korte, L., Laades, A., Stangl, R., Schubert, Ch., Angermann, H., Conrad, E., Maydell, K. v. (2007), Physical aspects of c-Si:H/c-Si hetero-junction solar cells, *Thin Solid Films*, 515, pp. 7475–7480.

Schmidt, J., Merkle, A., Brendel, R., Hoex, B., van de Sanden, M. C. M., Kessels, W. M. M. (2008), Surface passivation of high-efficiency silicon solar cells by atomic-layer-deposited Al_2O_3, *Prog. Photovolt.: Res. Appl.*, 16, pp. 461–466.

Schneiderlöchner, E., Preu, R., Lüdemann, R., Glunz, S. W. (2002), Laser-fired rear contacts for crystalline silicon solar cells, *Prog. Photovolt.: Res. Appl.*, 10, pp. 29–34.

Scholes, G. D., Rumles, G. (2006), Excitons in nanoscale systems, *Nature Materials*, 5, pp. 683–696.

Schottky, W. (1938), Halbleitertheorie der Sperrschicht, *Naturwissenschaften*, 26, p. 843.

Schroeder, D. J., Hernandez, J. L., Berry, G. D., Rockett, A. A. (1998), Hole transport and doping in epitaxial $CuIn_{1-x}Ga_xSe_2$, *J. Appl. Phys.*, 83, pp. 1519–1526.

Schubert, G., Huster, F., Fath, P. (2006), Physical understanding of printed thick-film front contacts of crystalline Si solar cells–Review of existing models and recent developments, *Sol. Ener. Mat. Sol. C.*, 90, pp. 3399–3406.

Schumann, B., Neumann, H., Tempel, A., Kühn, G., Nowak, E. (1980), Structural and electrical properties of $CuIn_{1-x}Ga_xSe_2$ epitaxial layers on GaAs substrates, *Cryst. Res. Technol.*, 15, pp. 71–76.

Schüppel, R., Timmreck, R., Allinger, N., Mueller, T., Furno, M., Uhrich, C., Leo, K., Riede, M. (2010), Controlled current matching in small molecule organic tandem solar cells using doped spacer layers, *J. Appl. Phys.*, 107, pp. 044503-1–044503-6.

Schweickert, H., Reuschel, K., Gutsche, H. (1956), Vorrichtung zur Gewinnung reinsten Halbleitermaterials für elektrotechnische Zwecke, *Deutsches Patentamt*, Auslegeschrift 1 061 593, pp. 1–5.

Seeger, K. (2001), *Semiconductor Physics*, Springer Berlin, New York.

Shanks, H. R., Maycock, P. D., Sidler, P. H., Danielson, G. C. (1963), Thermal conductivity of silicon from 300 to 1400°K, *Phys. Rev.*, 130, pp. 1743–1748.

Sharp (2013), Sharp develops concentrator solar cells with world's highest conversion efficiency of 44.4%, *Press Release*, June 14 2013, pp. 1–2.

Shay, J. L., Tell, B., Kasper, H. M., Schiavone, L. M. (1972), p-d hybridization of the valence bands of I-III-VI compounds, *Phys. Rev. B*, 5, pp. 5003–5005.

Shay, J. L., Tell, B., Kasper, H. M., Schiavone, L. M. (1973), Electronic structure of $AgInSe_2$ and $CuInSe_2$, *Phys. Rev. B*, 7, pp. 4485–4490.

Shay, J. L., Wagner, S., Kasper, H. M. (1975), Efficient $CuInSe_2$/CdS solar cells, *Appl. Phys. Lett.*, 27, pp. 89–90.

Shenai-Khatkhate, D. V., Goyette, R. J., DiCarlo, R. L., Dripps, G. (2004), Environment, health and safety for sources used in MOVPE growth of compound semiconductors, *J. Cryst. Growth*, 272, pp. 816–821.

Shin, Y. J., Kim, S. K., Park, B. H., Jeong, T. S., Shin, H. K., Kim, T. S., Yu, P. Y. (1991), Photocurrent study of the splitting of the valence band for a CdS single-crystal platelet, *Phys. Rev. B*, 44, pp. 5522–5526.

Shockley, W., Queisser, H. J. (1961), Detailed balance limit of pn-junction solar cells, *J. Appl. Phys.*, 32, pp. 510–519.

Shockley, W., Read, W. T. (1952), Statistics of the recombination of holes and electrons, *Phys. Rev.*, 87, pp. 835–842.

Singh, R., Radu, I., Scholz, R., Himcinschi, C., Gösele, U., Christiansen, S. H. (2006), Low temperature In Player transfer onto Si by helium implantation and direct wafer bonding, *Semicond. Sci. Technol.*, 21, pp. 1311–1314.

Smilowitz, L., Sariciftci, N. S., Wu, R., Gettinger, C., Heeger, A. J., Wudl, F. (1993), Photoexcitation spectroscopy of conducting-polymer — C_0 composites: photoinduced electron transfer, *Phys. Rev. B*, 47, pp. 13835–13842.

Smith, R. L., Collin, S. D. (1992), Porous silicon formation mechanisms, *J. App. Phys.*, 71, pp. R1–R22.

Solar Junction (2012), Solar junction breaks its own world record — Silicon Valley based solar energy company achieves 44% cell efficiency, October 15. Available at: http://www.sj-solar.com.

Sollmann, D. (2009), Sechs Neuner sind das Ziel, *Photon*, April 2009, pp. 42–45.

Solmi, S., Parisini, A., Angelucci, R., Armigliato, A. (1996), Dopant and carrier concentration in Si in equilibrium with monoclinic SiP precipitates, *Phys. Rev. B*, 53, pp. 7836–7841.

Spear, W. E., Le Comber, P. G. (1975), Substitutional doping of amorphous silicon, *Solid State Commun.*, 17, pp. 1193–1196.

Spicer, W. E., Chye, P. W., Skeath, P. R., Su, C. Y., Lindau, I. (1979), New and unified model for Schottky barrier and III–V insulator interface states formation, *J. Vac. Sci. Technol.*, 16, pp. 1422–1433.

Spinelli, P., Verschuuren, M. A., Polman, A. (2012), Broadband omnidirectional antireflection coating based on subwavelength surface Mie resonators, *Nature Communications*, 3, pp. 692-1–692-5.

Staebler, D. L., Wronski, C. R. (1977), Reversible conductivity changes in discharge-produced amorphous silicon, *Appl. Phys. Lett.*, 31, pp. 292–294.

Stall, R. A., Wood, C. E. C., Board, K., Dandekar, N., Eastman, L. F., Devlin, J. (1981), A study of Ge/GaAs interfaces grown by molecular beam epitaxy, *J. Appl. Phys.*, 52, pp. 4062–4069.

Stathis, J. H. (1995), Dissociation kinetics of hydrogen-passivated (100) Si/SiO_2 interface defects, *J. Appl. Phys.*, 77, pp. 6205–6207.

Stephan, C., Schorr, S., Tovar, M., Schock, H.-W. (2011), Comprehensive insights into point defect and defect cluster formation in $CuInSe_2$, *Appl. Phys. Lett.*, 98, 091906-1–091906-3.

Stephens, A. W., Aberle, A. G., Green, M. A. (1994), Surface recombination velocity measurements at the silicon–silicon dioxide interface by microwave-detected photoconductance decay, *J. Appl. Phys.*, 76, pp. 363–370.

Strauss, U., Rühle, W. W., Köhler, K. (1993), Auger recombination in intrinsic GaAs, *Appl. Phys. Lett.*, 62, pp. 55–57.

Stutzmann, M., Biegelsen, D. K., Street, R. A. (1987), Detailed investigation of doping in hydrogenated amorphous silicon and germanium, *Phys. Rev. B*, 35, pp. 5666–5701.

Stutzmann, M., Jackson, W. B., Tsai, C. C. (1985), Light-induced metastable defects in hydrogenated amorphous silicon: a systematic study, *Phys. Rev. B*, 32, pp. 23–47.

Summitt, R., Marley, J. A., Borelli, N. F. (1964), The ultraviolet absorption edge of stannic oxide (SnO_2), *J. Phys. Chem. Solids*, 25, pp. 1465–1469.

Sun, X. W., Kwok, H. S. (1999), Optical properties of epitaxially grown zinc oxide films on sapphire by pulsed laser deposition, *J. Appl. Phys.*, 86, pp. 408–411.

Supasai, T., Rujisamphan, N., Ullrich, K., Chemseddine, A., Dittrich, T. (2013), Formation of a passivating $CH_3NH_3PbI_3/PbI_2$ interface during moderate heating of $CH_3NH_3PbI_3$ layers, *Appl. Phys. Lett.*, 103, pp. 183906–183906-3.

Szabo, N., Sagol, B. E., Seidel, U., Schwarzburg, K., Hannappel, T. (2008), InGaAsP/InGaAs tandem cells for a solar cell configuration with more than three junctions, *Phys. Stat. Sol. (RRL)*, 2, pp. 254–256.

Sze, S. M. (1981), *Physics of Semiconductor Devices*, 2nd ed., John Wiley & Sons, New York.

Sze, S. M., Irvine, J. C. (1968), Resistivity, mobility, and impurity levels in GaAs, Ge and Si at 300 K, *Solid-State Electron.*, 11, pp. 599–602.

Taguchi, M., Terakawa, A., Maruyana, E., Tanaka, M. (2005), Obtaining a higher V_{OC} in HIT solar cells, *Prog. Photovolt.: Res. Appl.*, 13, pp. 481–488.

Takahashi, K., Yamada, S., Unno, T. (2005), High-efficiency AlGaAs/GaAs tandem solar cells, *U.D.C.*, 621.383.51:523.9-7, pp. 1–6.

Talapin, D. V., Murray, C. B. (2005), PbSe nanocrystal solids for n- and p-channel thin film field-effect transistors, *Science*, 310, pp. 86–89.

Talapin, D. V., Rogach, A. L., Kornowski, A., Haase, M., Weller, H. (2001), Highly luminescent monodisperse CdSe and CdSe/ZnS nanocrystals synthesized in a hexadecylamine-trioctylphosphine oxide-trioctylphoshine mixture, *Nano Lett.*, 1, pp. 207–2011.

Tang, C. W. (1986), Two-layer organic photovoltaic cell, *Appl. Phys. Lett.*, 48, pp. 183–185.

Tang, K., Øvrelid, E. J., Tranell, G., Tangstad, M. (2009), Critical assessment of the impurity diffusivities in solid and liquid silicon, *J. Min. Met. Mat. Soc.*, 61, pp. 49–55.

Tang, H., Prasad, K., Sanijes, R., Schmid, P. E., Lévy, F. (1994), Electrical and optical properties of TiO_2 thin films, *J. Appl. Phys.*, 75, pp. 2042–2047.

Tell, B., Shay, J. L., Kasper, H. M. (1971), Electrical properties, optical properties, and band structure of $CuGaS_2$ and $CuInS_2$, *Phys. Rev. B*, 4, pp. 2463–2471.

Teng, C. W., Muth, J. F., Özgür, Ü., Bergmann, M. J., Everitt, H. O., Sharma, A. K., Jin, C., Narayan, J. (2000), Refractive indices and absorption of coefficients of $Mg_xZn_{1-x}O$ alloys, *Appl. Phys. Lett.*, 76, pp. 979–981.

Thomas, D. G. (1960), The exciton spectrum of zinc oxide, *J. Phys. Chem. Solids*, 15, pp. 86–96.

Thornton, J. A. (1974), Influence of apparatus geometry and deposition conditions on the structure and topography of thick sputtered coatings, *J. Vac. Sci. Technol.*, 11, pp. 666–670.

Thornton, J. A., Lomasson, T. C., Talieh, H., Tseng, B.-H. (1988), Reactive sputtered CuInSe$_2$, *Solar Cells*, 24, pp. 1–9.

Tiedje, T., Cebulka, J. M., Morel, D. L., Abeles, B. (1981), Evidence for exponential band tails in amorphous silicon hydride, *Phys. Rev. Lett.*, 46, pp. 1425–1428.

Tolcin, A. C. (2012), Indium, *U.S. Geological Survey, Mineral Commodity Summaries*, January, pp. 74–75.

Toor, F. (2012), Turning lemons into lemonade: opportunities in the turbulent PV equipment market, *Photovoltaics International*, 18th ed., 4th quarter, pp. 12–16.

Toyoda, T., Sano, T., Nakajima, J., Doi, S., Fukumoto, S., Ito, A., Tohyama, T., Yoshida, M., Kanagawa, T., Motohiro, T., Shiga, T., Higuchi, K., Tanaka, H., Takeda, Y., Fukano, T., Katoh, N., Takeichi, A., Takechi, K., Shiozawa, M. (2004), Outdoor performance of large scale DSC modules, *J. Photochem. Photobiol. A: Chemistry*, 164, pp. 203–207.

Triboulet, R., Marfaing, Y., Cornet, A., Siffert, P. (1974), Undoped high-resistivity cadmium telluride for nuclear radiation detectors, *J. Appl. Phys.*, 45, pp. 2759–2765.

Trumbore, F. A. (1960), Solid solubilities of impurity elements in germanium and silicon, *Bell. Syst. Tech. J.*, 39, pp. 205–233.

Tsunomura, Y., Yoshimine, Y., Taguchi, M., Baba, T., Kinoshita, T., Kanno, H., Sakata, H., Maruyama, E., Tanaka, M. (2009), Twenty-two percent efficiency HIT solar cell, *Sol. Energ. Mater. Sol. C.*, 93, pp. 670–673.

Turcu, M., Pakma, O., Rau, U. (2002), Interdependence of absorber composition and recombination mechanism in Cu(In,Ga)(Se,S)$_2$ heterojunction solar cells, *Appl. Phys. Lett.*, 80, pp. 2598–2600.

Turner, W. J., Reese, W. E., Pettit, G. D. (1964), Exciton absorption and emission in InP, *Phys. Rev.*, 136, pp. A1467–A1470.

Ulzhöfer, C., Hermann, S., Harder, N.-P., Altermatt, P. P., Brendel, R. (2008), The origin of reduced fill factors of emitter-wrap-through-solar cells, *Phys. Stat. Sol. (RRL)*, 2, pp. 251–253.

Ungelenk, J., Haug, V., Quintilla, A., Ahlswede, E. (2010), CuInSe$_2$ low-cost thin-film solar cells made from commercial elemental metallic nanoparticles, *Phys. Stat. Sol. (RRL)*, 4, pp. 58–60.

Van de Walle, C. G. (1994), Energies of various configurations of hydrogen in silicon, *Phys. Rev. B*, 49, pp. 4579–4585.

Van Roesbroeck, W., Shockley, W. (1954), Photon-radiative recombination of electrons and holes in germanium, *Phys. Rev.*, 94, pp. 1558–1560.

Van Sark, W. G. J. H. M. (2002), Methods of deposition of hydrogenated amorphous silicon for device applications, in *Handbook of Thin-Film Materials, Vol. 1, Deposition and processing of thin films*, ed. Nalwa, H. S., Academic Press, San Diego, pp. 1–102.

Varshni, Y. P. (1967a), Band-to-band radiative recombination in groups IV, VI and III-V semiconductors (II), *Phys. Stat. Sol.*, 20, pp. 9–36.

Varshni, Y. P. (1967b), Temperature dependence of the energy gap in semiconductors, *Physica*, 34, pp. 149–154.

Vautier, M., Guillard, C., Herrmann, J.-M. (2001), Photocatalytic degradation of dyes in water: case study of indigo and indigo carmine, *J. Catalysis*, 201, pp. 46–59.

Vegard, L. (1921), Die Konstitution der Mischkristalle und die Raumfüllung der Atome, *Z. Phys.*, 5, pp. 17–26.

Vetterl, O., Finger, F., Carius, R., Hapke, P., Houben, L., Kluth, O., Lambertz, A., Mück, A., Rech, B., Wagner, H. (2000), Intrinsic microcrystalline silicon: a new material for photovoltaics, *Sol. Energ. Mater. Sol. C.*, 62, pp. 97–108.

Vogel, H. (1995), *Gerthsen Physik*, Springer-Verlag, Berlin Heidelberg.

Voncken, M. M. A. J., Schermer, J. J., Bauhuis, G. J., Mulder, P., Larsen, P. K. (2004), Multiple release layer study of the intrinsic lateral etch rate of the epitaxial lift-off process, *Appl. Phys. A*, pp. 1801–1807.

Vurgaftman, I., Meyer, J. R., Ram-Mohan, L. R. (2001), Band parameters for III-V compound semiconductors and their alloys, *J. Appl. Phys. — Appl. Physics Rev.*, 89, pp. 5815–5873.

Wadia, C., Alivisatos, A. P., Kammen, D. M. (2009), Materials availability expands the opportunity for large-scale photovoltaics deployment, *Environ. Sci. Technol.*, 43, 2072–2077.

Wagner, P., Helbig, R. (1974), Halleffekt und Anisotropie der Beweglichkeit der Elektronen in ZnO, *J. Phys. Chem. Solids*, 35, pp. 327–335.

Wagner, S., Shay, J. L., Migliorato, P., Kasper, H. M. (1974), $CuInSe_2/CdS$ heterojunction photovoltaic detectors, *Appl. Phys. Lett.*, 25, pp. 434–435.

Wallentin, J., Anttu, N., Asoli, D., Huffmann, M., Åberg, I., Magnusson, M. H., Siefer, G., Fuss-Kailuweit, P., Dimroth, F., Witzigmann, B., Xu, H. Q., Samuelson, L., Deppert, K., Borgström, M. T. (2013), InP nanowire array solar cells achieving 13.8% efficiency by exceeding the ray optics limit, *Science*, 339, pp. 1057–1060.

Walter, M. G., Warren, E. L., McKone, J. R., Boettcher, S. W., Mi, Q., Santori, E. A., Lewis, N. S. (2010), Solar water splitting cells, *Chem. Rev.*, 110, pp. 6446–6473.

Walukiewicz, W. (1988), Mechanism of Fermi-level stabilization in semiconductors, *Phys. Rev. B*, 37, pp. 4760–4763.

Wang, P., Klein, C., Humphrey-Baker, R., Zakeeruddin, S. M., Grätzel, M. (2005), Stable $\geq$ 8% efficient nanocrystalline dye-sensitized solar cell based on an electrolyte of low volatility, *Appl. Phys. Lett.*, 86, pp. 123508-1–123508-3.

Wang, C.-L., Lan, C.-M., Hong, S.H., Wang, Y.-F., Pan, T.-Y., Chang, C.-W., Kuo, H.-H., Kuo, M.-Y., Diau, E. W.-G., Lin, C.-Y. (2012), Enveloping porphyrins for efficient dye-sensitized solar cells, *Energy Environ. Sci.*, pp. 6933–6940.

Wang, M., Liu, J., Cevey-Ha, N.-L., Moon, S.-J., Liska, P., Humphrey-Baker, R., Moser, J. E., Grätzel, C., Wang, P., Zakeeruddin, S. M., Grätzel, M. (2010), High efficiency solid-state sensitized heterojunction photovoltaic device, *Nano Today*, 5, 169–174.

Wang, S., Macdonald, D. (2012), Temperature dependence of Auger recombination in highly injected crystalline silicon, *J. Appl. Phys.*, 112, pp. 113708-1– 113708-4.

Wang, E., White, T., Catchpole, K. (2013), to be submitted.

Wang, A., Zhao, J., Green, M. A. (1990), 24% efficient silicon solar cells, *Appl. Phys. Lett.*, 57, pp. 602–604.

Wei, H.S., Zunger, A. (1993), Band offsets at the CdS/CuInSe$_2$ heterojunction, *Appl. Phys. Lett.*, 63, 2549–2551.

Wei, S.-H., Zunger, A. (1998), Calculated natural band offsets of all II-VI and III-V semiconductors: chemical trends and the role of cation *d* orbitals, *Appl. Phys. Lett.*, 72, pp. 2011–2013.

Weiher, R. L. (1966), Optical properties of free electrons in ZnO, *Phys. Rev.*, 152, pp. 736–739.

Weinberg, Z. A., Rubloff, G. W., Bassous, E. (1979), Transmission, photoconductivity, and the experimental band gap of thermally grown SiO$_2$ films, *Phys. Rev. B*, 19, pp. 3107–3116.

Wenham, S. R., Honsberg, C. B., Green, M. A. (1994), Buried contact silicon solar cells, *Sol. Energ. Mater. Sol. C.*, 34, pp. 101–110.

Werner, F., Cosceev, A., Schmidt, J. (2012), Silicon surface passivation by Al$_2$O$_3$: recombination parameters and inversion layer solar cells, *Energy Procedia*, 27, pp. 319–324.

Werner, J., Kolodinski, S., Rau, U., Arch, J. K., Bauser, E. (1993), Silicon solar cell of 16.8 μ thickness and 14.7% efficiency, *Appl. Phys. Lett.*, 62, pp. 2998–3000.

Wild-Scholten, M., de, Sturm, M., Butturi, M. A., Noack, M., Narec, K. H., Timò, G. (2010), Environmental sustainability of concentrator PV systems: preliminary LCA results of the Apollon project, *Proc. 25th European Photovoltaic Solar Energy Conference*, 3908–3911.

Wilke, A., Endres, J., Hörmann, U., Niederhaüsen, J., Schlesinger, R., Frisch, J., Amsalem, P., Wagner, J., Gruber, M., Opitz, A., Vollmer, A., Brütting, W., Kahn, A., Koch, N. (2012), Correlation between interface energetics and open circuit voltage in inorganic photovoltaic cells, *Appl. Phys. Lett.*, 101, pp. 233301-1– 2333301-4.

Wu, X. (2004), High-efficiency polycrystalline CdTe thin-film solar cells, *Sol. Energy*, 77, pp. 803–814.

Würfel, P. (2005), *Physics of Solar Cells. From Principles to New Concepts*, Wiley-VCH, Weinheim.

Yablonovitch, E. (1982), Statistical ray optics, *J. Opt. Soc. Am.*, 72, pp. 899–907.

Yablonovitch, E., Allara, D. L., Chang, C. C., Gmitter, T, Bright, T. B. (1986b), Unusually low surface recombination velocity on silicon and germanium surfaces, *Phys. Rev. Lett.*, 57, pp. 249–252.

Yablonovitch, E., Gmitter, T. (1986a), Auger recombination in silicon at low carrier densities, *Appl. Phys. Lett.*, 49, pp. 587–589.

Yablonovitch, E., Gmitter, T., Harbison, J. P., Bhat, R. (1987), Extreme selectivity in the lift-off of epitaxial GaAs films, *Appl. Phys. Lett.*, 51, pp. 2222–2224.

Yamada, S. (1960), On the electrical and optical properties of p-type cadmium telluride crystals, *J. Phys. Soc. Jpn.*, 15, pp. 1940–1944.

Yatsurugi, Y., Akiyama, N., Endo, Y., Nozaki, T. (1973), Concentration, solubility, and equilibrium distribution coefficient of nitrogen and oxygen in semiconductor silicon, *J. Electrochem. Soc.*, 120, pp. 975–979.

Yella, A., Lee, H.-W., Tsao, H. N., Yi, C., Chandiran, A. K., Nezeeruddin, M. K., Diau, E. W.-G., Yeh, C. Y., Zakeeruddin, S. M., Grätzel, M. (2011), Porphyrin-sensitized solar cells with cobalt (II/III)-based redox electrolyte exceed 12 percent efficiency, *Science*, 334, pp. 629–634.

Yoon, J., Jo, S., Chun, I. S., Jung, I., Kim, H.-S., Meitl, M., Menard, E., Li, X., Coleman, J. J., Paik, U., Rogers, J. A. (2010), GaAs photovoltaics and optoelectronics using releasable multilayer epitaxial assemblies, *Nature Letters*, 465, pp. 329–334.

Yu, A. Y. (1970), Electron tunneling and contact resistance of metal-silicon contact barriers, *Solid-State Electron.*, 13, pp. 239–247.

Yu, G., Gao, J., Hummelen, J. C., Wudl, F., Heeger, A. J. (1995), Polymer photovoltaic cells: enhanced efficiency via a network of internal donor-acceptor heterojunctions, *Science*, 270, pp. 1789–1791.

Zahler, J. M., Tanabe, K., Ladous, C., Pinnington, T., Newman, F. D., Atwater, H. A. (2007), High efficiency InGaAs solar cells on Si by InP layer transfer, *Appl. Phys. Lett.*, 91, pp. 012108-1–012108-3.

Zhang, B., Tian, Y., Zhang, J. X., Cai, W. (2011), The studies of the role of fluorine in SnO_2:F films prepared by spray pyrolysis with $SnCl_4$, *J. Optoel. Adv. Mats.*, 13, pp. 89–93.

Zhang, S. B., Wei, S.-H., Zunger, A., Katayama-Yoshida, H. (1998), Defect physics of the $CuInSe_2$ chalcopyrite semiconductor, *Phys. Rev. B*, 57, 9642–9656.

Zhao, J., Wang, A., Altermatt, P., Green, M. A., (1995), Twenty-four percent efficient silicon solar cells with double layer antireflection coatings and reduced resistance loss, *Appl. Phys. Lett.*, 66, pp. 3636–3638.

Zhao, J., Wang, A., Green, M. A., Ferreza, F. (1998), Novel 19.8% efficient "honeycomb" textured multicrystalline and 24.4% monocrystalline silicon solar cells, *Appl. Phys. Lett.*, 73, pp. 1991–1993.

Zide, J. M. O., Kleiman-Schwarzstein, A., Strandwitz, N. C., Zimmerman, J. D., Steenblock-Smith, T., Gossard, A. C., Forman, A., Ivanovskaya, A., Stucky, G. D. (2006), Increased efficiency in multijunction solar cells through the incorporation of semimetallic ErAs nanoparticles into the tunnel junction, *Appl. Phys. Lett.*, 88, pp. 162103-1–162103-3.

Zillner, E., Fengler, S., Niyamakom, P., Rauscher, F., Köhler, K., Dittrich, T. (2012), Role of ligand exchange at CdSe quantum dot layers for charge separation, *J. Phys. Chem. C*, 116, pp. 16747–16754.

Zulehner, W. (2000), Historical overview of silicon crystal pulling development, *Mater. Sci. Eng. B*, 73, pp. 7–15.

Index

Printed in the United States
By Bookmasters